Mathematics for Computer Graphics Applications

M.E. Mortenson

SECOND EDITION

Library of Congress Cataloging-in-Publication Data

Mortenson, Michael E.,
    Mathematics for computer graphics applications : an introduction
to the mathematics and geometry of CAD/CAM, geometric modeling,
scientific visualization, and other CG applications/Michael
Mortenson.—2nd ed.
        p.   cm.
    Rev. ed. of Computer graphics. c1989.
    ISBN 0-8311-3111-X
    1. Computer graphics—Mathematics.   I. Mortenson, Michael E.,
Computer graphics. II. Title.
T385.M668   1999
006.6'01'51—dc21   99-10096
                                                      CIP

The first edition was published under the title
Computer Graphics: An Introduction
To the Mathematics and Geometry

Industrial Press, Inc.
200 Madison Avenue
New York, NY 10016-4078

Second Edition, 1999

Sponsoring Editor: John Carleo
Project Editor: Sheryl A. Levart
Interior Text and Cover Designer: Janet Romano

Printed in the United States of America

10 9 8 7 6 5 4 3 2

To

JAM

PREFACE

Computer graphics is more than an isolated discipline at an intersection of computer science, mathematics, and engineering. It is the very visible leading edge of a revolution in how we use computers and automation to enhance our lives. It has changed forever the worlds of engineering and science, medicine, and the media, too. The wonderful special effects that entertain and inform us at the movies, on TV, or while surfing cyberspace are possible only through powerful computer graphics applications. From the creatures of *Jurassic Park*, and the imaginary worlds of *Myst* and *Star Wars*, to the dance of virtual molecules on a chemist's computer display screen, these effects applied in the arts, entertainment, and science are possible, in turn, only because an equally wonderful world of mathematics is at work behind the scenes. Some of the mathematics is traditional, but with a new job to do, and some is brand new.

Mathematics for Computer Graphics Applications introduces the mathematics that is the foundation of many of today's most advanced computer graphics applications. These include computer-aided design and manufacturing, geometric modeling, scientific visualization, robotics, computer vision, virtual reality, ... and, oh yes, special effects in cinematography. It is a textbook for the college student majoring in computer science, engineering, or applied mathematics and science, whose special interests are in computer graphics, CAD/CAM systems, geometric modeling, visualization, or related subjects. Some exposure to elementary calculus and a familiarity with introductory linear algebra as usually presented in standard college algebra courses is helpful but not strictly necessary. (Careful attention to the material of Chapter 5, which discusses limit and continuity, may preempt the calculus prerequisite.) The textbook is also useful for on-the-job training of employees whose skills can be profitably expanded into this area, and as a tutorial for independent study. The professional working in these fields will find it to be a comprehensive reference.

This second edition includes thoroughly updated subject matter, some major organizational changes, and several new topics: chapters on symmetry, limit and continuity, constructive solid geometry, and the Bézier curve have been added. Many new figures and exercises are also presented. This new edition brings to the fore the basic mathematical tools of computer graphics, including vectors, matrices, and transformations. The text then uses these tools to present the geometry, from the elementary building blocks of points, lines, and planes, to complex three-dimensional constructions and the Bézier curve. Here is an outline of what you will find in each chapter.

Chapter 1 Vectors There are several ways to teach and to learn about vectors. Today most of these methods are within the context of some other area of study. While this approach usually works, it too often unnecessarily delays a student's experience with vectors until more specialized courses are taken. Experience shows, however, that students are capable of learning both the abstract and the practical natures of vectors much earlier. Vectors, it turns out, have a strong intuitive appeal because of their innate geometric character. This chapter builds upon this intuition.

Chapter 2 Matrix Methods As soon as a student learns how to solve simultaneous linear equations, he or she is also ready to learn simple matrix algebra. The usefulness of the matrix formulation is immediately obvious, and allows the student to see that there are ways to solve even very large sets of equations. There is ample material here for both student and teacher to apply to the understanding of concepts in later chapters. Subjects covered include definitions, matrix equivalence, matrix arithmetic, partitioning, determinants, inversion, and eigenvectors.

Chapter 3 Transformations explores geometric transformations and their invariant properties. It introduces systems of equations that produce linear transformations and develops the algebra and geometry of translations, rotations, reflection, and scaling. Vectors and matrix algebra are put to good use here, in ways that clearly demonstrate their effectiveness. The mathematics of transformations is indispensable to animation and special effects, as well as many areas of physics and engineering.

Chapter 4 Symmetry and Groups This chapter uses geometry and intuition to define symmetry and then introduces easy-to-understand but powerful analytic methods to further explore this subject. It defines a *group* and the more specialized *symmetry group*. The so-called *crystallographic restriction* shows how symmetry depends on the nature of space itself. This chapter concludes with a look at the rotational symmetry of polyhedra.

Chapter 5 Limit and Continuity Limit processes are at work in computer graphics applications to ensure the display of smooth, continuous-looking curves and surfaces. Continuity is now an important consideration in the construction and rendering of shapes for CAD/CAM and similar applications. These call forth the most fundamental and elementary concepts of the calculus, because the big difference between calculus and the subjects that precede it is that calculus uses limiting processes and requires a definition of continuity. The ancient Greek *method of exhaustion* differs from integral calculus mainly because the former lacks the concept of the limit of a function. This chapter takes a look at some old and some new ways to learn about limit and continuity, including the method of exhaustion, sequences and series, functions, limit theorems, continuity and continuous functions.

Chapter 6 Topology What makes a sphere different from a torus? Why are left and right not reliable directions on a Möbius strip? This chapter answers questions like these. It shows that topology is the study of continuity and connectivity and how these characteristics are preserved when geometric figures are deformed. Although historically a relative newcomer, topology is an important part of many disciplines, including geometric modeling and mathematical physics. In fact, for the latter, the Möbius strip and its analog in a higher dimension, the Klein bottle, both appear in superstring theory as aspects of Feynman diagrams representing the interactions of elementary particles.

Topological equivalence, topology of a closed path, piecewise flat surfaces, closed curved surfaces, orientation, and curvature are among the many topics covered here.

Chapter 7 Halfspaces Halfspaces are always fun to play with because we can combine very simple halfspaces to create complex shapes that are otherwise extremely difficult, if not impossible, to represent mathematically. This chapter explores both two- and three-dimensional halfspaces. It first reviews elementary set theory and how to interpret it geometrically in Venn diagrams and then explains the Boolean operators *union, intersection* and *difference* and how to use them to create more complex shapes.

Chapter 8 Points, Chapter 9 Lines, and *Chapter 10 Planes* define and discuss these three simple geometric objects and show how they are the basic building blocks for other geometric objects. Arrays of points, absolute and relative points, line intersections, and the relationship between a point and a plane are some of the topics explored.

Chapter 11 Polygons Polygons and polyhedra were among the first forms studied in geometry. Their regularity and symmetry made them the center of attention of mathematicians, philosophers, artists, architects, and scientists for thousands of years. Polygons are still important today. For example, their use in computer graphics is widespread, because it is easy to subdivide and approximate the surfaces of solids with planes bounded by them. Topics covered here include definitions of the various types of polygons, their geometric properties, convex hulls, construction of the regular polygons, and symmetry.

Chapter 12 Polyhedra This chapter defines convex, concave, and stellar polyhedra, with particular attention to the five regular polyhedra, or Platonic solids. It defines *Euler's Formula* and shows how to use it to prove that only five regular polyhedra are possible in a space of three dimensions. Other topics include definitions of the various families of polyhedra, nets, the convex hull of a polyhedron, the connectivity matrix, halfspace representations of polyhedra, model data structures, and maps.

Chapter 13 Constructive Solid Geometry Traditional geometry, plane and solid— or analytic—does not tell us how to create even the simplest shapes we see all around us. Constructive solid geometry is a way to describe these shapes as combinations of even simpler shapes. The chapter revisits and applies some elementary set theory, halfspaces, and Boolean algebra. Binary trees are introduced, and many of their more interesting properties are explored.

Chapter 14 Curves explores the mathematical definition of a curve as a set of parametric equations, a form that is eminently computationally useful to CAD/CAM, geometric modeling, and other computer graphic applications. Parametric equations are the basis for Bézier, NURBS, and Hermite curves. Both plane and space curves are introduced, followed by discussions of the tangent vector, blending functions, conic curves, reparameterization, and continuity.

Chapter 15 The Bézier Curve This curve is not only an important part of almost every computer-graphics illustration program, it is a standard tool of animation techniques. The chapter begins by describing a surprisingly simple geometric construction of a Bézier curve, followed by a derivation of its algebraic definition, basis functions, control points, and how to join two curves end-to-end to form a single, more complex curve.

Chapter 16 Surfaces develops the parametric equations of surfaces, a natural extension of the mathematics of curves discussed in Chapter 14. Topics include the surface patch, plane and cylindrical surfaces, the bicubic surface, the Bézier surface, and the surface normal.

Chapter 17 Computer Graphics Display Geometry introduces some of the basic geometry and mathematics of computer graphics, including display coordinate systems, windows, line and polygon clipping, polyhedra edge visibility, and silhouettes.

Chapter 18 Display and Scene Transformations discusses orthographic and perspective transformations, and explores some scene transformations: orbit, pan, and aim. Chapters 17 and 18 draw heavily on concepts introduced earlier in the text and lay the foundation for more advanced studies in computer graphics applications.

Most chapters include many exercises, with answers to selected exercises provided following the last chapter. A separate *Solutions Manual*, which presents hints and solutions for all the exercises, is available for instructors. Here is a suggestion that was offered in the first edition and must be repeated here: Read all of the exercises (even those not assigned). They are a great help to the reader who chooses to use this textbook as a tutorial, and for the student in the classroom setting who is serious about mastering all the concepts. An annotated bibliography is also included. Use it to find more advanced coverage of these subjects or to browse works both contemporary and classic that put these same subjects into a broader context, both mathematical and cultural.

Mathematics for Computer Graphics Applications is a textbook with a purpose, and that is to provide a strong and comprehensive base for later more advanced studies that the student will encounter in math, computer science, physics, and engineering, including specialized subjects such as algebraic and computational geometry, geometric modeling, and CAD/CAM. At the same time, the material is developed enough to be immediately put to work by the novice or experienced professional. Because the textbook's content is relevant to contemporary development and use of computer graphics applications, and because it introduces each subject from an elementary standpoint, *Mathematics for Computer Graphics Applications* is suitable for industry or government on-the-job training programs aimed at increasing the skills and versatility of employees who are working in related but less mathematically demanding areas. It can be used as a primary textbook, as a supplementary teaching resource, as a tutorial for the individual who wants to master this material on his or her own, or as a professional reference.

The creation of this second edition was made possible by the generous help and cooperation of many people. First, thanks to all the readers of the first edition who took the time to suggest improvements and corrections. Thanks to the necessarily anonymous reviewers whose insightful comments also contributed to this new edition. Thanks to John Carleo, Sheryl Levart, and Janet Romano of Industrial Press, Inc. whose many editorial and artistic talents transformed my manuscript-on-a-disk into an attractive book. Finally, thanks to my wife Janet, who read and reread the many drafts of this edition, and who was a relentless advocate for simplicity and clarity.

TABLE OF CONTENTS

CHAPTER 1

VECTORS

Perhaps the single most important mathematical device used in computer graphics and many engineering and physics applications is the *vector*. A vector is a geometric object of a sort, because, as we will soon see, it fits our notion of a displacement or motion. (We can think of displacement as a change in position. If we move a book from a shelf to a table, we have displaced it a specific distance and direction.) Vector methods offer a distinct advantage over traditional analytic geometry by minimizing our computational dependence on a specific coordinate system until the later stages of solving a problem. Vectors are direct descendants of complex numbers and are generalizations of hyper-complex numbers. Interpreting these numbers as directed line segments makes it easier for us to understand their properties and to apply them to practical problems.

Length and direction are the most important vector properties. Scalar multiplication of a vector (that is, multiplication by a constant), vector addition, and scalar and vector products of two vectors reveal more geometric subtleties. Representing straight lines and planes using vector equations adds to our understanding of these elements and gives us powerful tools for solving many geometry problems. Linear vector spaces and basis vectors provide rigor and a deeper insight into the subject of vectors. The history of vectors and vector geometry tells us much about how mathematics develops, as well as how gifted mathematicians established a new discipline.

1.1 Introduction

In the 19th century, mathematicians developed a new mathematical object— a new kind of number. They were motivated in part by an important observation in physics: Physicists had long known that while some phenomena can be described by a single number—the temperature of a beaker of water (6°C), the mass of a sample of iron (17.5g), or the length of a rod (31.736 cm)—other phenomena require something more.

A ball strikes the side rail of a billiard table at a certain speed and angle (Figure 1.1); we cannot describe its rebound by a single number. A pilot steers north with an air speed of 800 kph in a cross wind of 120 kph from the west (Figure 1.2); we cannot describe the airplane's true motion relative to the ground by a single number. Two spherical bodies collide; if we know the momentum (mass × velocity) of each body before impact, then we can determine their speed and direction after impact (Figure 1.3). The

1

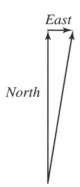

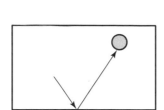

Figure 1.1 *Billiard ball rebound.*

Figure 1.2 *Effect of a cross wind on an airplane's course.*

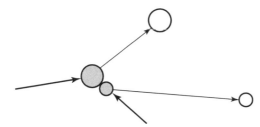

Figure 1.3 *Collision of two spherical bodies.*

description of the momentum of each body seems to require more information than a single number can convey.

Billiard balls, airplanes, and colliding bodies need a number that describes both the speed and the direction of their motion. Mathematicians found that they could do this by using a *super number* made up of two or more normal numbers, called *components*. When the component member numbers of this super number are combined according to certain rules, the results define magnitude and direction.

For the example of the airplane flying in a crosswind, its magnitude is given in kilometers per hour, and its direction is given by a compass reading. For any specified elapsed time t and velocity v, the problem immediately reduces to one of directed distance or displacement d because we know that $d = vt$.

When a super number is associated with a distance and direction, it is called a *vector*, and we can use vectors directly to solve problems of the kind just described. Vectors are derived from a special class of numbers called *hypercomplex numbers*. We will use a simple version of a hyper-complex number and call it a *hypernumber*. In the next section we look at hypernumbers in a way that lets us interpret them geometrically.

1.2 Hypernumbers

Hypernumbers are a generalization of complex numbers. Recall that a complex number such as $a + bi$ consists of a real part a and an imaginary part bi, $i = \sqrt{-1}$. If we eliminate the coefficient i of the imaginary part and retain a and b as an ordered pair of real numbers (a, b), we create a two-dimensional hypernumber. Ordered triples, quadruples, or n-tuples, for that matter, of real numbers comprise and define higher-dimension hypernumbers.

Order is very important, as we shall soon see, and we will often need to use subscripts to indicate that order. For example, suppose we have the ordered triple of real numbers (a, b, c). None of the letters a, b, or c alone tells us about its position in the sequence. However, if we make the substitutions $a_1 = a$, $a_2 = b$, and $a_3 = c$, the hypernumber becomes (a_1, a_2, a_3). This notation not only lets us know the position of each number in the sequence, but it also frees b and c for other duties.

It is tedious to write out the sequence of numbers for a hypernumber. Therefore, whenever we can, we will use an italicized uppercase letter to represent a hypernumber. Thus, we let $A = (a_1, a_2, a_3)$, $B = (b_1, b_2, b_3)$, and so on. We generalize this to any number of dimensions so that, for example, A might represent the n-dimensional hypernumber $(a_1, a_2, \ldots, a_n)$, where $a_1, a_2, \ldots, a_n$ are the *components* of A.

Hypernumbers have their own special arithmetic and algebra. One of the most basic questions we can ask about two hypernumbers is: Are they equal? Two hypernumbers A and B are equal, that is $A = B$, if and only if their corresponding components are equal. This means that if $A = B$, then $a_1 = b_1$, $a_2 = b_2, \ldots a_n = b_n$. This also means, of course, that A and B must have the same number of components. We say that they must have the same *dimension*. Is the hypernumber $A = (5, 2)$ equal to the hypernumber $B = (2, 5)$? No, because $a_1 = 5$, $b_1 = 2$, and $a_1 \neq b_1$. What about $C = (-1, 6, 2)$ and $D = (-1, 6, 2)$? Clearly, here $C = D$.

We can add or subtract two hypernumbers of the same dimension by adding (or subtracting) their corresponding components. Thus,

$$A + B = (a_1, a_2, \ldots, a_n) + (b_1, b_2, \ldots, b_n) \tag{1.1}$$

or

$$A + B = (a_1 + b_1, a_2 + b_2, \ldots, a_n + b_n) \tag{1.2}$$

We find that the sum of two hypernumbers is another hypernumber; for example, $A + B = C$, where $c_1 = a_1 + b_1$, and $c_2 = a_2 + b_2, \ldots, c_n = a_n + b_n$. For the two-dimensional hypernumbers, $A = (14, -5)$, $B = (0, 8)$, and $A + B = (14, 3)$.

The simplest kind of multiplication involving a hypernumber is scalar multiplication, where we multiply each component by a common factor k. We write this as follows:

$$kA = k(a_1, a_2, a_3) = (ka_1, ka_2, ka_3) \tag{1.3}$$

We find that k has the effect of scaling each component equally. In fact, k is called a *scalar* to distinguish it from the hypernumber. The two hypernumbers $(3, 5)$ and $(9, 15)$ differ by a scale factor of $k = 3$. And, as it turns out, division is not defined for hypernumbers.

When we multiply two hypernumbers, we expect to find a product such as $(a_1, a_2, a_3) \times (b_1, b_2, b_3)$. In Section 1.7 we will see that there are actually two kinds of multiplication, one of which is not commutative (i.e., $AB \neq BA$).

1.3 Geometric Interpretation

To see the full power of hypernumbers, we must give them a geometric interpretation. Distance and direction are certainly important geometric properties, and we will now see how to derive them from hypernumbers.

Imagine that an ordered pair of numbers, a hypernumber, is really just a set of instructions for moving about on a flat two-dimensional surface. For the moment, let's agree that the first number of the pair represents a displacement (how we are to move) east if plus (+), or west if minus (−), and that the second number represents a displacement north (+) or south (−). Then we can interpret the two-dimensional hypernumber, (16.3, −10.2) for example, as a displacement of 16.3 units of length (feet, meters, light-years, or whatever) to the east, followed by a displacement of 10.2 units of length (same as the east-west units) to the south (Figure 1.4). Note something interesting here: By applying the Pythagorean theorem we find that this is equivalent to a total (resultant) displacement of $\sqrt{(16.3)^2 + (-10.2)^2} = 19.23$ units of length in a southeasterly direction. We can, of course, be more precise about the direction. The compass heading from our initial position would be

$$90° + \arctan \frac{10.2}{16.3} = 122° \text{ SE} \tag{1.4}$$

We can apply this interpretation to any ordered pair of numbers (a_1, a_2) so that each pair produces both a magnitude (the total displacement) and a direction. Think of this interpretation as an algorithm, or mathematical machine, that uses as input an ordered pair of real numbers and produces as output two real numbers that we interpret as magnitude and direction. In fact, this interpretation is so different from that of an ordinary hypernumber that we are justified in giving the set $A = (a_1, a_2)$ a new name and its own notation. W. R. Hamilton (1805–1865) was the first to use the term *vector* (from the Latin word *vectus*, to carry over) to describe this new mathematical object, and that is what we will do. We will use boldface lowercase letters to represent vectors, and list the components inside brackets without separating commas, so that $\mathbf{a} = [a_1 \quad a_2]$, for example.

We treat ordered triples of numbers (a_1, a_2, a_3) in the same way, by creating a third dimension to complement the compass headings of the two-dimensional plane. This is easy enough: Adding up (+) and down (−) is all that we need to do. We have,

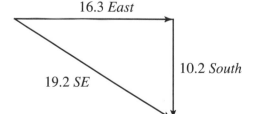

16.3 *East*

19.2 *SE*

10.2 *South*

Figure 1.4 *Displacement as a vector.*

then, for an ordered triple, an agreement that the first number represents a displacement east $(+)$ or west $(-)$, the second number is a displacement north $(+)$ or south $(-)$, and the third number is a displacement up $(+)$ or down $(-)$ (Figure 1.5). Later we will see how an ordered triple yields a distance and direction.

Before doing this, let's see how all of this relates to adding hypernumbers. We will consider only ordered pairs, with the understanding that the treatment easily extends to ordered triples in three dimensions and n-tuples in n dimensions. Given two ordered pairs A and B, we saw that their sum, $A + B$, is simply $(a_1 + b_1, a_2 + b_2)$, or $C = A + B$ where $c_1 = a_1 + b_1$ and $c_2 = a_2 + b_2$. Nothing we discussed earlier prevents us from interpreting $c_1 = a_1 + b_1$ as the total east-west displacement and $c_2 = a_2 + b_2$ as the total north-south displacement. This means that the resulting grand total displacement is $\sqrt{(a_1 + b_1)^2 + (a_2 + b_2)^2}$ or $\sqrt{c_1^2 + c_2^2}$, with the compass direction given by $90° - \arctan\left[(a_2 + b_2)/(a_1 + b_1)\right]$. Thus, the distance-and-direction interpretation holds true for addition, as well.

The distance-and-direction interpretation suggests a powerful way for us to visualize a vector, and that is as a directed line segment or arrow (Figure 1.6). The length of the arrow (at some predetermined scale) represents the magnitude of the vector, and the orientation of the segment and placement of the arrowhead (at one end of the segment or the other) represent its direction. The figure shows several examples lying in the plane of the paper. Two vectors are equal if they have the same length and direction, so that $\mathbf{a} = \mathbf{b}$. Although $\mathbf{c}$ is the same length as $\mathbf{a}$, it is in the opposite direction, so $\mathbf{a} \neq \mathbf{c}$. Clearly, neither $\mathbf{d}$ nor $\mathbf{e}$ is the equivalent of $\mathbf{a}$. Another way of stating this is that if any vector $\mathbf{a}$ can be transported, remaining parallel to its initial orientation, into coincidence with another vector $\mathbf{b}$, then $\mathbf{a} = \mathbf{b}$.

We can use this idea of parallel transport of directed line segments (arrows) to add two vectors $\mathbf{a}$ and $\mathbf{b}$ as follows: Transport $\mathbf{b}$ until its tail is coincident with the head

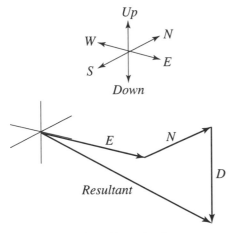

Figure 1.5 *A three-dimensional displacement.*

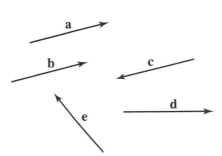

Figure 1.6 *Vectors as directed line segments.*

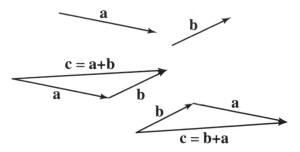

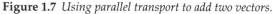

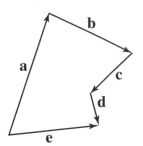

Figure 1.7 *Using parallel transport to add two vectors.*

Figure 1.8 *Head-to-tail chain of vectors.*

of **a** (Figure 1.7). Then their sum, **a** + **b**, is the directed line segment beginning at the tail of **a** and ending at the head of **b**. Note that transporting **a** so that its tail coincides with the head of **b** produces the same result. Thus, **a** + **b** = **b** + **a**. We extend this method of adding two vectors to adding many vectors by simply transporting each vector so as to form a head-to-tail chain (Figure 1.8). Connecting the tail of the first vector to the head of the last vector in the chain produces the resultant vector. In the figure we see that this construction yields **e** = **a** + **b** + **c** + **d**.

This process leads directly to the *parallelogram law* for adding two vectors, where we transport **a** and/or **b** so that their tails are coincident and then complete the construction of the suggested parallelogram. The diagonal **c** represents their sum. (Figure 1.9). The parallelogram law of addition suggests a way to find the components of a vector **a** along any two directions L and M (Figure 1.10). We construct lines L and M through the tail of **a** and their parallel images L' and M' through the head of **a**. This construction produces a parallelogram whose adjacent sides $\mathbf{a}_L$ and $\mathbf{a}_M$ are the components of **a** along L and M, respectively. We see, of course, that $\mathbf{a} = \mathbf{a}_L + \mathbf{a}_M$ are not unique! We could just as easily construct other lines, say PQ and $P'Q'$, to find $\mathbf{a}_P$ and $\mathbf{a}_Q$. Obviously, $\mathbf{a} = \mathbf{a}_P + \mathbf{a}_Q$ and, in general, $\mathbf{a}_P \neq \mathbf{a}_L, \mathbf{a}_M$ or $\mathbf{a}_Q \neq \mathbf{a}_L, \mathbf{a}_M$. This property of the non-uniqueness of the vector's components is a very powerful feature, which we will see often in the sections to follow.

So far we have not constrained vectors to any particular location, so we call them *free vectors*. We have moved them around and preserved their properties of length and orientation. This is possible only if we always move them parallel to themselves. It

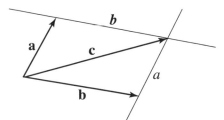

Figure 1.9 *The parallelogram law of vector addition.*

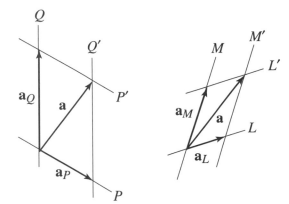

Figure 1.10 *Using parallelogram construction and vector components.*

is also true in three dimensions (Figure 1.11). The components must not be coplanar, and in the figure we use them to construct a rectangular parallelepiped.

When we use vectors simply to represent pure displacements, then we use unrestricted parallel transport, and the solution is in terms of free vectors. Many problems, however, require a *bound* or *fixed vector*, which is often the case in physics, computer graphics, and geometric modeling.

Fixed vectors always begin at a common point, usually (but not necessarily) the origin of a coordinate system. If we use the origin of a rectangular Cartesian coordinate system, then the components of a fixed vector lie along the principal axes and correspond to the coordinates of the point at the tip of the arrowhead (Figure 1.12). Thus, the vector $\mathbf{p}$ in the figure has vector components $\mathbf{p}_x$ and $\mathbf{p}_y$, also known as the x and y component. This means that $\mathbf{p} = \mathbf{p}_x + \mathbf{p}_y$, and in three dimensions $\mathbf{p} = \mathbf{p}_x + \mathbf{p}_y + \mathbf{p}_z$ (Figure 1.13).

The distinction between free and fixed vectors is often blurred by the nature of the problem and because most arithmetic and algebraic operations are identical for both kinds of vectors. The distinction may be as important for visualization and intuition as for any other reason.

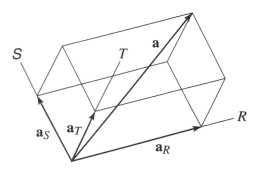

Figure 1.11 *Free vectors and parallel transport in three dimensions.*

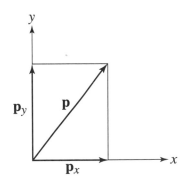

Figure 1.12 *Fixed vector in two dimensions.*

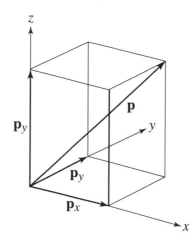

Figure 1.13 *Fixed vector in three dimensions.*

1.4 Vector Properties

It is time to introduce some special vectors, denoted **i**, **j**, **k**, each of which has a length equal to one. The vector **i** lies along the x axis, **j** lies along the y axis, and **k** lies along the z axis (Figure 1.14), where

$$
\begin{aligned}
\mathbf{i} &= [1 \quad 0 \quad 0] \\
\mathbf{j} &= [0 \quad 1 \quad 0] \\
\mathbf{k} &= [0 \quad 0 \quad 1]
\end{aligned}
\tag{1.5}
$$

Because we can multiply a vector by some constant that changes its magnitude but not its direction, we can express any fixed vector **a** as

$$
\mathbf{a} = a_x\mathbf{i} + a_y\mathbf{j} + a_z\mathbf{k}
\tag{1.6}
$$

This follows from $\mathbf{a} = \mathbf{a}_x + \mathbf{a}_y + \mathbf{a}_z$, where $\mathbf{a}_x = a_x\mathbf{i}$, $\mathbf{a}_y = a_y\mathbf{j}$, and $\mathbf{a}_z = a_z\mathbf{k}$ (Figure 1.15).

We can describe the vector **a** more simply as an ordered triple and enclose it in brackets. Thus, $\mathbf{a} = [a_x \quad a_y \quad a_z]$, where a_x, a_y, and a_z are the components of **a**. The

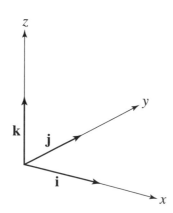

Figure 1.14 *Unit vectors along the coordinate axes.*

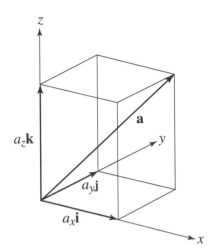

Figure 1.15 *Fixed-vectors components along the coordinate axes.*

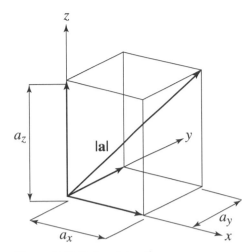

Figure 1.16 *Determining the magnitude of a vector.*

components may be negative, depending on the direction of the vector. We reverse the direction of any vector by multiplying each of its components by −1. Thus, the reverse of $\mathbf{a} = [3 \quad 2 \quad -7]$ is $-\mathbf{a}$, or $-\mathbf{a} = [-3 \quad -2 \quad 7]$.

Magnitude (length) and direction are the most important properties of a vector. Because a vector's magnitude is best described as a length, it is always positive. The length of $\mathbf{a}$ is a scalar, denoted as $|\mathbf{a}|$ and given by

$$|\mathbf{a}| = \sqrt{a_x^2 + a_y^2 + a_z^2} \tag{1.7}$$

which is a simple application of the Pythagorean theorem for finding the length of the diagonal of a rectangular solid (Figure 1.16).

We define a unit vector as any vector whose length or magnitude is equal to one, independent of its direction, of course. As we saw, $\mathbf{i}$, $\mathbf{j}$, and $\mathbf{k}$ are special cases, with specific directions assigned to them. A *unit vector* in the direction of $\mathbf{a}$ is denoted as $\hat{\mathbf{a}}$, where

$$\hat{\mathbf{a}} = \frac{\mathbf{a}}{|\mathbf{a}|} \tag{1.8}$$

and its components are

$$\hat{\mathbf{a}} = \begin{bmatrix} \dfrac{a_x}{|\mathbf{a}|} & \dfrac{a_y}{|\mathbf{a}|} & \dfrac{a_z}{|\mathbf{a}|} \end{bmatrix} \tag{1.9}$$

We can make this more concise with the following substitutions:

$$\hat{\mathbf{a}}_x = \frac{a_x}{|\mathbf{a}|}, \quad \hat{\mathbf{a}}_y = \frac{a_y}{|\mathbf{a}|}, \quad \hat{\mathbf{a}}_z = \frac{a_z}{|\mathbf{a}|} \tag{1.10}$$

so that

$$\hat{\mathbf{a}} = [\hat{a}_x \quad \hat{a}_y \quad \hat{a}_z] \tag{1.11}$$

Note that if α, β, and γ are the angles between $\mathbf{a}$ and the x, y, and z axes, respectively, then

$$\hat{a}_x = \frac{a_x}{|\mathbf{a}|} = \cos\alpha, \quad \hat{a}_y = \frac{a_y}{|\mathbf{a}|} = \cos\beta, \quad \hat{a}_z = \frac{a_z}{|\mathbf{a}|} = \cos\gamma \tag{1.12}$$

This means that $\hat{a}_x$, $\hat{a}_y$, and $\hat{a}_z$ are also the *direction cosines* of $\mathbf{a}$.

1.5 Scalar Multiplication

Multiplying any vector $\mathbf{a}$ by a scalar k produces a vector $k\mathbf{a}$ or, in component form,

$$k\mathbf{a} = [ka_x \quad ka_y \quad ka_z] \tag{1.13}$$

If k is positive, then $\mathbf{a}$ and $k\mathbf{a}$ are in the same direction. If k is negative, then $\mathbf{a}$ and $k\mathbf{a}$ are in opposite directions. The magnitude (length) of $k\mathbf{a}$ is

$$|k\mathbf{a}| = \sqrt{k^2 a_x^2 + k^2 a_y^2 + k^2 a_z^2} \tag{1.14}$$

so that

$$|k\mathbf{a}| = k|\mathbf{a}| \tag{1.15}$$

We can see that scalar multiplication is well named, because it changes the scale of the vector. Here are the possible effects of a scalar multiplier k:

$k > 1$	Increases length
$k = 1$	No change
$0 < k < 1$	Decreases length
$k = 0$	Null vector (0 length, direction undefined)
$-1 < k < 0$	Decreases length and reverses direction
$k = -1$	Reverses direction only
$k < -1$	Reverses direction and increases length

1.6 Vector Addition

Vector addition (or subtraction) in terms of components is perhaps the simplest of all vector operations (except for multiplication of a vector by a scalar). Given $\mathbf{a} = [a_x \quad a_y \quad a_z]$ and $\mathbf{b} = [b_x \quad b_y \quad b_z]$, then

$$\mathbf{a} + \mathbf{b} = [a_x + b_x \quad a_y + b_y \quad a_z + b_z] \tag{1.16}$$

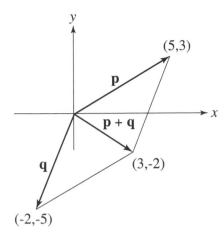

Figure 1.17 *Confirming the parallelogram law of vector addition.*

For example, given the two-dimensional vectors $\mathbf{p} = [5 \quad 3]$ and $\mathbf{q} = [-2 \quad -5]$, we readily obtain their sum

$$\mathbf{p} + \mathbf{q} = [5 + (-2) \quad 3 + (-5)] = [3 \quad -2] \tag{1.17}$$

In Figure 1.17 we see that the parallelogram law of addition is satisfied.

Given vectors $\mathbf{a}$, $\mathbf{b}$, and $\mathbf{c}$ and scalars k and l, then vector addition and scalar multiplication have the following properties:

1. $\mathbf{a} + \mathbf{b} = \mathbf{b} + \mathbf{a}$
2. $\mathbf{a} + (\mathbf{b} + \mathbf{c}) = (\mathbf{a} + \mathbf{b}) + \mathbf{c}$
3. $k(l\mathbf{a}) = kl\mathbf{a}$
4. $(k + l)\mathbf{a} = k\mathbf{a} + l\mathbf{a}$
5. $k(\mathbf{a} + \mathbf{b}) = k\mathbf{a} + k\mathbf{b}$

1.7 Scalar and Vector Products

We multiply two vectors $\mathbf{a}$ and $\mathbf{b}$ in two very different ways. One way produces a single real number, or scalar, identified as the *scalar product*. The other way produces a vector, identified as the *vector product*. Both kinds of multiplication require that $\mathbf{a}$ and $\mathbf{b}$ have the same dimension. To avoid problems beyond the scope of this textbook, we will work only with three-dimensional vectors.

The scalar product of two vectors $\mathbf{a}$ and $\mathbf{b}$ is the sum of the products of their corresponding components:

$$\mathbf{a} \cdot \mathbf{b} = a_x b_x + a_y b_y + a_z b_z \tag{1.18}$$

which is a scalar, not another vector. Occasionally you will see or hear it referred to as the *dot product*, particularly in older texts. It is easy to show that the scalar product is commutative, that is, $\mathbf{a} \cdot \mathbf{b} = \mathbf{b} \cdot \mathbf{a}$.

It is interesting to note that the scalar product of a vector with itself produces the square of the vector's length $\mathbf{a} \cdot \mathbf{a} = a_x^2 + a_y^2 + a_z^2$, so that

$$\mathbf{a} \cdot \mathbf{a} = |\mathbf{a}|^2 \tag{1.19}$$

Using the law of cosines we can demonstrate that the angle θ between two vectors $\mathbf{a}$ and $\mathbf{b}$ satisfies the equation

$$\mathbf{a} \cdot \mathbf{b} = |\mathbf{a}| \, |\mathbf{b}| \cos \theta \tag{1.20}$$

Solving this equation for θ yields

$$\theta = \cos^{-1} \frac{\mathbf{a} \cdot \mathbf{b}}{|\mathbf{a}| \, |\mathbf{b}|} \tag{1.21}$$

This means that if $\mathbf{a} \cdot \mathbf{b} = 0$, then $\mathbf{a}$ and $\mathbf{b}$ are perpendicular. If $\theta = 0$, then they are parallel. The scalar product has the following properties:

1. $\mathbf{a} \cdot \mathbf{b} = |\mathbf{a}| \, |\mathbf{b}| \cos \theta$, where θ is the angle between $\mathbf{a}$ and $\mathbf{b}$
2. $\mathbf{a} \cdot \mathbf{a} = |\mathbf{a}|^2$
3. $\mathbf{a} \cdot \mathbf{b} = \mathbf{b} \cdot \mathbf{a}$, commutative property
4. $\mathbf{a} \cdot (\mathbf{b} + \mathbf{c}) = \mathbf{a} \cdot \mathbf{b} + \mathbf{a} \cdot \mathbf{c}$, distributive property
5. $(k\mathbf{a}) \cdot \mathbf{b} = \mathbf{a} \cdot (k\mathbf{b}) = k(\mathbf{a} \cdot \mathbf{b})$, associative property
6. If $\mathbf{a}$ is perpendicular to $\mathbf{b}$, then $\mathbf{a} \cdot \mathbf{b} = 0$

The vector product of two vectors $\mathbf{a}$ and $\mathbf{b}$ is

$$\mathbf{a} \times \mathbf{b} = (a_y b_z - a_z b_y)\mathbf{i} - (a_x b_z - a_z b_x)\mathbf{j} + (a_x b_y - a_y b_x)\mathbf{k} \tag{1.22}$$

In component form this becomes

$$\mathbf{a} \times \mathbf{b} = [(a_y b_z - a_z b_y) \quad - (a_x b_z - a_z b_x) \quad (a_x b_y - a_y b_x)] \tag{1.23}$$

You might wonder how such a collection of terms arises. Although the detailed derivation of this expression is too long and complex to include here (W. R. Hamilton spent years investigating the problem of multiplying hyper-complex numbers), we can find meaningful patterns. For example, each possible permutation of the product $a_i b_j$ (where $i, j = x, y, z$) appears only once. There are no a_x or b_x terms in the first, or x, component of the product, no a_y or b_y terms in the y component, and no a_z or b_z terms in the z component. One way to remember this multi-term expression is as the expansion of the following determinant:

$$\mathbf{a} \times \mathbf{b} = \begin{vmatrix} \mathbf{i} & \mathbf{j} & \mathbf{k} \\ a_x & a_y & a_z \\ b_x & b_y & b_z \end{vmatrix} \tag{1.24}$$

If $\mathbf{c} = \mathbf{a} \times \mathbf{b}$, then $\mathbf{c}$ is perpendicular to both $\mathbf{a}$ and $\mathbf{b}$ and, thus, it is also perpendicular to the plane defined by $\mathbf{a}$ and $\mathbf{b}$. We can prove this assertion by computing

$\mathbf{a} \cdot \mathbf{c}$ and $\mathbf{b} \cdot \mathbf{c}$:

$$\mathbf{a} \cdot \mathbf{c} = a_x(a_y b_z - a_z b_y) - a_y(a_x b_z - a_z b_x) + a_z(a_x b_y - a_y b_x)$$
$$= a_x a_y b_z - a_x a_z b_y - a_x a_y b_z + a_y a_z b_x + a_x a_z b_y - a_y a_z b_x \qquad (1.25)$$
$$= 0$$

Because $\mathbf{a} \cdot \mathbf{c} = 0$, we know that $\mathbf{a}$ and $\mathbf{c}$ are perpendicular. We can also show that $\mathbf{b} \cdot \mathbf{c} = 0$.

If two vectors $\mathbf{a}$ and $\mathbf{b}$ are parallel, then $\mathbf{a} \times \mathbf{b} = 0$. To prove this, we let $\mathbf{b} = k\mathbf{a}$; this guarantees that $\mathbf{a}$ and $\mathbf{b}$ are parallel. Then we compute $\mathbf{a} \times k\mathbf{a}$:

$$\mathbf{a} \times k\mathbf{a} = [(ka_y a_z - ka_z a_y) \quad -(ka_x a_z - ka_z a_x) \quad (ka_x a_y - ka_y a_x)] \qquad (1.26)$$

This reduces to

$$\mathbf{a} \times k\mathbf{a} = [0 \quad 0 \quad 0] \qquad (1.27)$$

and $[0 \quad 0 \quad 0]$ is the so-called *null vector*, or $\mathbf{0}$. Obviously, this means that $\mathbf{a} \times \mathbf{a} = 0$, because we have put no restrictions on k.

The vector product is not commutative; in fact, $\mathbf{b} \times \mathbf{a} = -(\mathbf{a} \times \mathbf{b})$. Thus, reversing the order of the two vectors reverses the direction of their vector product. This is easy to verify.

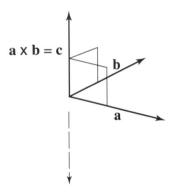

Figure 1.18 *Vector product.*

To say that if $\mathbf{c} = \mathbf{a} \times \mathbf{b}$, then $\mathbf{c}$ is perpendicular to both $\mathbf{a}$ and $\mathbf{b}$ only gives the line of action of the vector $\mathbf{c}$. In which of two directions does $\mathbf{c}$ point? (Figure 1.18) The direction is, of course, inherent in the components of the vector product. However, an intuitive rule applies: Imagine rotating $\mathbf{a}$ into $\mathbf{b}$ through the smallest of the angles formed by their lines of action, curling the fingers of your right hand in this angular sense. The extended thumb of your right hand points in the direction of $\mathbf{c}$, where $\mathbf{c} = \mathbf{a} \times \mathbf{b}$. For $\mathbf{d} = \mathbf{b} \times \mathbf{a}$, $\mathbf{d}$ points in the opposite direction, that is, $\mathbf{d} = -\mathbf{c}$.

The vector product has the following properties:

1. $\mathbf{a} \times \mathbf{b} = \mathbf{c}$, where $\mathbf{c}$ is perpendicular to both $\mathbf{a}$ and $\mathbf{b}$

2. $\mathbf{a} \times \mathbf{b} = \begin{vmatrix} \mathbf{i} & \mathbf{j} & \mathbf{k} \\ a_x & a_y & a_z \\ b_x & b_y & b_z \end{vmatrix}$, the expansion of the determinant

3. $\mathbf{a} \times \mathbf{b} = |\mathbf{a}|\,|\mathbf{b}|\,\hat{\mathbf{n}}\,\sin\theta$, where $\hat{\mathbf{n}}$ is the unit vector perpendicular to the plane of $\mathbf{a}$ and $\mathbf{b}$ and θ is the angle between them
4. $\mathbf{a} \times \mathbf{b} = -\mathbf{b} \times \mathbf{a}$
5. $\mathbf{a} \times (\mathbf{b} + \mathbf{c}) = \mathbf{a} \times \mathbf{b} + \mathbf{a} \times \mathbf{c}$
6. $(k\mathbf{a}) \times \mathbf{b} = \mathbf{a} \times (k\mathbf{b}) = k(\mathbf{a} \times \mathbf{b})$
7. $\mathbf{i} \times \mathbf{j} = \mathbf{k},\ \mathbf{j} \times \mathbf{k} = \mathbf{i},\ \mathbf{k} \times \mathbf{i} = \mathbf{j}$
8. If $\mathbf{a}$ is parallel to $\mathbf{b}$, then $\mathbf{a} \times \mathbf{b} = 0$
9. $\mathbf{a} \times \mathbf{a} = \mathbf{0}$

1.8 Elements of Vector Geometry

We can use vector equations to describe many geometric objects, from points, lines, and curves to very complex surfaces. We do this by writing a vector equation in terms of one or more variables. We will be exploring only straight lines and planes here.

Lines

The vector equation of a line through some point $\mathbf{p}_0$ and parallel to another vector $\mathbf{t}$ is

$$\mathbf{p}(u) = \mathbf{p}_0 + u\mathbf{t} \qquad (1.28)$$

where u is a scalar variable multiplying $\mathbf{t}$ (Figure 1.19). We see that as u takes on successive numerical values, the equation generates points on a straight line. The components of $\mathbf{p}$ are the coordinates of a point on this line. In other words, because $\mathbf{p}_0$ and $\mathbf{t}$ are constant for any specific line, any real value of u generates a point on that line.

We can expand this equation by writing it in its component form. This time we will list the components in a vertical or column array instead of the horizontal or row array we have used. (The row and column forms are mathematically equivalent, demanding only that we do not mix the two and that we use some simple bookkeeping

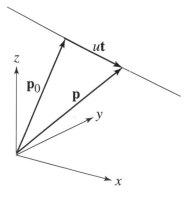

Figure 1.19 *Vector equation of a line.*

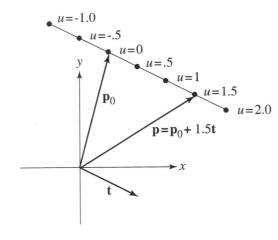

Figure 1.20 *Example of a vector equation of a straight line.*

techniques when doing algebra on them.) This produces

$$\begin{bmatrix} x \\ y \\ z \end{bmatrix} = \begin{bmatrix} x_0 \\ y_0 \\ z_0 \end{bmatrix} + u \begin{bmatrix} t_x \\ t_y \\ t_z \end{bmatrix} \tag{1.29}$$

In ordinary algebraic form we have

$$\begin{aligned} x &= x_0 + ut_x \\ y &= y_0 + ut_y \\ z &= z_0 + ut_z \end{aligned} \tag{1.30}$$

where u is the independent variable; x, y, and z are dependent variables; and x_0, y_0, z_0, t_x, t_y, and t_z are constants. Mathematicians call this set of equations the *parametric equations* of a straight line: (x, y, z) are the coordinates of any point on the line, (x_0, y_0, z_0) are the coordinates of a given point on the line, and (t_x, t_y, t_z) are the components of a vector (in point form) parallel to the line.

Here is a simple example in two dimensions (Figure 1.20): Let's find the vector equation for a straight line that passes through the point described by the vector $\mathbf{p}_0 = [1 \quad 4]$ and parallel to the vector $\mathbf{t} = [2 \quad -1]$. In vector component form, we have

$$\mathbf{p} = \begin{bmatrix} x \\ y \end{bmatrix} = \begin{bmatrix} 1 \\ 4 \end{bmatrix} + u \begin{bmatrix} 2 \\ -1 \end{bmatrix} \tag{1.31}$$

or, in algebraic form,

$$\begin{aligned} x &= 1 + 2u \\ y &= 4 - u \end{aligned} \tag{1.32}$$

Now we can compute the coordinates of points on this line for a series of values of u and tabulate the results (Table 1.1).

Of course, we can easily expand this tabulation in a variety of ways. We can, for example, use values of u closer together or farther apart, as well as values beyond (in

Table 1.1 Points on a line

u	x	y
−1.0	−1.0	5.0
−0.5	0	4.5
0	1.0	4.0
0.5	2.0	3.5
1.0	3.0	3.0
1.5	4.0	2.5
2.0	5.0	2.0

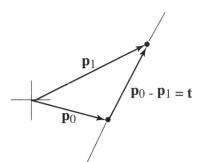

Figure 1.21 *Vector equation of a line through two points.*

either direction) those tabulated. Using equations like this as part of a computer-graphics program, a computer scientist can construct and display straight lines.

An interesting variation of this vector geometry of a straight line will help us find the vector equation of a line through two given points, say, $\mathbf{p}_0$ and $\mathbf{p}_1$ (Figure 1.21). In this figure we see that we can define $\mathbf{t}$ as $\mathbf{p}_1 - \mathbf{p}_0$. Making the appropriate substitution into our original vector equation for a straight line, we obtain

$$\mathbf{p} = \mathbf{p}_0 + u(\mathbf{p}_1 - \mathbf{p}_0) \tag{1.33}$$

If we limit the allowable value of u to the interval $0 \le u \le 1$, then this equation defines a line segment extending from $\mathbf{p}_0$ to $\mathbf{p}_1$. In algebraic form, this vector equation expands to

$$\begin{aligned}
x &= x_0 + u(x_1 - x_0) \\
y &= y_0 + u(y_1 - y_0) \\
z &= z_0 + u(z_1 - z_0)
\end{aligned} \tag{1.34}$$

(Note that we have arbitrarily used both fixed and free vector interpretations. Compare Figures 1.20 and 1.21.)

Planes

There are four ways to define a plane in three dimensions using vector equations. One way is by the vector equation of a plane through $\mathbf{p}_0$ and parallel to two independent vectors $\mathbf{s}$ and $\mathbf{t}$:

$$\mathbf{p} = \mathbf{p}_0 + u\mathbf{s} + w\mathbf{t} \tag{1.35}$$

where $\mathbf{s} \neq k\mathbf{t}$ and where u and w are scalar independent variables multiplying $\mathbf{s}$ and $\mathbf{t}$, respectively (Figure 1.22). The vector $\mathbf{p}$ represents the set of points defining a plane as the parameters u and w vary independently. In terms of the vector's components, the dependent variables x, y, and z, we have the three equations

$$
\begin{aligned}
x &= x_0 + us_x + wt_x \\
y &= y_0 + us_y + wt_y \\
z &= z_0 + us_z + wt_z
\end{aligned}
\tag{1.36}
$$

We can also write Equation 1.36 as the matrix equation

$$
\begin{bmatrix} x \\ y \\ z \end{bmatrix} = \begin{bmatrix} x_0 \\ y_0 \\ z_0 \end{bmatrix} + u \begin{bmatrix} s_x \\ s_y \\ s_z \end{bmatrix} + w \begin{bmatrix} t_x \\ t_y \\ t_z \end{bmatrix}
\tag{1.37}
$$

A second way: three points $\mathbf{p}_0$, $\mathbf{p}_1$, and $\mathbf{p}_2$ are sufficient to define a plane in space if they are not collinear (Figure 1.23). We can rewrite Equation 1.35 in terms of these points:

$$
\mathbf{p} = \mathbf{p}_0 + u(\mathbf{p}_1 - \mathbf{p}_0) + w(\mathbf{p}_2 - \mathbf{p}_1)
\tag{1.38}
$$

Any vector perpendicular to a plane is called a *normal vector* to that plane. We usually denote it as $\mathbf{n}$. Thus far we have two ways to compute it:

$$
\mathbf{n} = \mathbf{s} \times \mathbf{t}
\tag{1.39}
$$

or

$$
\mathbf{n} = (\mathbf{p}_1 - \mathbf{p}_0) \times (\mathbf{p}_2 - \mathbf{p}_1)
\tag{1.40}
$$

Note that we can construct a normal at any point on the plane and that, of course, all normals to the plane are parallel to one another. If the magnitude of $\mathbf{n}$ is not of interest,

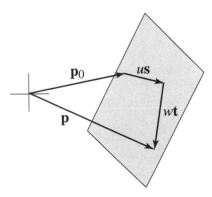

Figure 1.22 *Vector equation of a plane.*

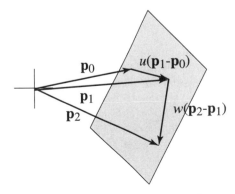

Figure 1.23 *Three points defining a plane.*

we can work with the unit normal $\hat{\mathbf{n}}$, where

$$\hat{\mathbf{n}} = \frac{\mathbf{n}}{|\mathbf{n}|} \qquad (1.41)$$

A third way to define a plane is by using a point it passes through and the normal vector to the plane. This means that any point $\mathbf{p}$ lies on the plane if and only if $\mathbf{p} - \mathbf{p}_0$ is perpendicular to $\hat{\mathbf{n}}$, because $\hat{\mathbf{n}}$ is perpendicular to all lines in the plane. In terms of a vector equation, this statement becomes

$$(\mathbf{p} - \mathbf{p}_0) \cdot \hat{\mathbf{n}} = 0 \qquad (1.42)$$

(Remember: The scalar product of two mutually perpendicular vectors is zero.) Performing the indicated scalar product yields

$$(x - x_0)\,\hat{n}_x + (y - y_0)\,\hat{n}_y + (z - z_0)\,\hat{n}_z = 0 \qquad (1.43)$$

where $\hat{n}_x$, $\hat{n}_y$, and $\hat{n}_z$ are the components of $\hat{\mathbf{n}}$.

The fourth way we can define a plane is a variation of the third way: Given the vector $\mathbf{d}$ to a point on the plane and where $\mathbf{d}$ is itself perpendicular to the plane, then any point $\mathbf{p}$ on the plane must satisfy

$$(\mathbf{p} - \mathbf{d}) \cdot \mathbf{d} = 0 \qquad (1.44)$$

Point of Intersection between a Plane and a Straight Line

We can use vector equations to solve many kinds of problems in geometry. For example, given the plane $\mathbf{p}_P = \mathbf{a} + u\mathbf{b} + w\mathbf{c}$ and the straight line $\mathbf{p}_L = \mathbf{d} + t\mathbf{e}$, at their point of intersection it must be true that $\mathbf{p}_P = \mathbf{p}_L$, so that

$$\mathbf{a} + u\mathbf{b} + w\mathbf{c} = \mathbf{d} + t\mathbf{e} \qquad (1.45)$$

This represents, of course, a system of three linear equations in three unknowns, u, w, and t. In algebraic form this equation becomes

$$\begin{aligned}
a_x + ub_x + wc_x &= d_x + te_x \\
a_y + ub_y + wc_y &= d_y + te_y \\
a_z + ub_z + wc_z &= d_z + te_z
\end{aligned} \qquad (1.46)$$

However, we can use the properties of vectors to solve Equation 1.46 directly. We solve for u, w, and t by isolating each in turn. To solve for t, we take the scalar product of both sides of the equation with the vector product $(\mathbf{b} \times \mathbf{c})$ as follows:

$$(\mathbf{b} \times \mathbf{c}) \cdot (\mathbf{a} + u\mathbf{b} + w\mathbf{c}) = (\mathbf{b} \times \mathbf{c}) \cdot (\mathbf{d} + t\mathbf{e}) \qquad (1.47)$$

Because $(\mathbf{b} \times \mathbf{c})$ is perpendicular to both $\mathbf{b}$ and $\mathbf{c}$, Equation 1.47 reduces to

$$(\mathbf{b} \times \mathbf{c}) \cdot \mathbf{a} = (\mathbf{b} \times \mathbf{c}) \cdot \mathbf{d} + t(\mathbf{b} \times \mathbf{c}) \cdot \mathbf{e} \qquad (1.48)$$

Solving Equation 1.48 for t yields

$$t = \frac{(\mathbf{b} \times \mathbf{c}) \cdot \mathbf{a} - (\mathbf{b} \times \mathbf{c}) \cdot \mathbf{d}}{(\mathbf{b} \times \mathbf{c}) \cdot \mathbf{e}} \tag{1.49}$$

We continue in the same way to solve for u and w:

$$u = \frac{(\mathbf{c} \times \mathbf{e}) \cdot \mathbf{d} - (\mathbf{c} \times \mathbf{e}) \cdot \mathbf{a}}{(\mathbf{c} \times \mathbf{e}) \cdot \mathbf{b}} \tag{1.50}$$

$$w = \frac{(\mathbf{b} \times \mathbf{e}) \cdot \mathbf{d} - (\mathbf{b} \times \mathbf{e}) \cdot \mathbf{a}}{(\mathbf{b} \times \mathbf{e}) \cdot \mathbf{c}} \tag{1.51}$$

1.9 Linear Vector Spaces

If we algebraically treat vectors as if they originated at a common point, then we work in a *linear vector space*. Not only do all vectors have a common origin, but any vector combines with any other vector according to the parallelogram law of addition. Vectors must be subjected to the following two operations to qualify as members of a linear vector space:

1. Addition of any two vectors must produce a third vector, identified as their sum: $\mathbf{a} + \mathbf{b} = \mathbf{c}$.
2. Multiplication of a vector $\mathbf{a}$ by a scalar k must produce another vector $k\mathbf{a}$ as the product.

The set of all vectors is *closed* with respect to these two operations, which means that both the sum of two vectors and the product of a vector and a scalar are themselves vectors. These two operations have the following properties, some of which we have seen before:

1. Commutativity: $\mathbf{a} + \mathbf{b} = \mathbf{b} + \mathbf{a}$
2. Associativity: $(\mathbf{a} + \mathbf{b}) + \mathbf{c} = \mathbf{a} + (\mathbf{b} + \mathbf{c})$
3. Identity element: $\mathbf{a} + \mathbf{0} = \mathbf{a}$
4. Inverse: $\mathbf{a} - \mathbf{a} = \mathbf{0}$
5. Identity under scalar multiplication: $k\mathbf{a} = \mathbf{a}$, when $k = 1$
6. $c(d\mathbf{a}) = (cd)\mathbf{a}$
7. $(c + d)\mathbf{a} = c\mathbf{a} + d\mathbf{a}$
8. $k(\mathbf{a} + \mathbf{b}) = k\mathbf{a} + k\mathbf{b}$

A set of vectors that can be subjected to the two operations with these eight properties forms a linear vector space. The other vector operations we have discussed, such as the scalar and vector products, are not pertinent to this definition of a linear vector space. This brings us back to hypernumbers. It is easy to show that the set of all vectors of the form $\mathbf{r} = [r_1 \quad r_2 \quad \cdots \quad r_n]$, where $r_1, r_2, \ldots, r_n$ are real numbers, constitutes a linear vector space.

The set of all linear combinations of a given set of vectors (none of which is a scalar multiple of any other in the set) forms a vector space. For example, if we let $x_1, x_2, \ldots, x_n$ be any n vectors, then $a_1x_1 + a_2x_2 + \cdots + a_nx_n$ (where $a_1, a_2, \ldots, a_n$ are scalars) is a linear combination of the vectors $x_1, x_2, \ldots, x_n$. But that is not all we can do. If we let

$$\begin{aligned} s &= a_1x_1 + a_2x_2 + \cdots + a_nx_n \\ t &= b_1x_1 + b_2x_2 + \cdots + b_nx_n \end{aligned} \tag{1.52}$$

so that the vectors s and t are linear combinations of the vectors $x_1, x_2, \ldots, x_n$, then

$$s + t = (a_1 + b_1)x_1 + (a_2 + b_2)x_2 + \cdots + (a_n + b_n)x_n \tag{1.53}$$

We also have

$$ks = (ka_1)x_1 + (ka_2)x_2 + \cdots + (ka_n)x_n \tag{1.54}$$

which is also a linear combination of $x_1, x_2, \ldots, x_n$. Mathematicians point out that the space of all linear combinations of a given set of vectors is the space generated by that set.

Given a single vector x, the space generated by all scalar multiples of x is a straight line collinear with x (for $x \neq 0$). Given two vectors s and t, where t is not a scalar multiple of s, then the space generated by their linear combinations is the plane containing s and t. For example, if we let $r = as + bt$, then from the parallelogram law of addition we know that vectors r, s, and t are coplanar. Of course, we could continue this process, generating spaces of three and more dimensions simply by increasing the number of vectors in the generating set. To do this, we must impose certain conditions, as in the previous example where we did not allow s and t to be scalar multiples of each other. This leads us to the concepts of linear independence and dependence.

1.10 Linear Independence and Dependence

Vectors $x_1, x_2, \ldots, x_n$ are *linearly dependent* if and only if there are real numbers $a_1, a_2, \ldots, a_n$ not all equal to zero, such that

$$a_1x_1 + a_2x_2 + \cdots + a_nx_n = 0 \tag{1.55}$$

If this equation is true only if $a_1, a_2, \ldots, a_n$ are all zero, then $x_1, x_2, \ldots, x_n$ are *linearly independent*.

If $x_1, x_2, \ldots, x_n$ are linearly dependent, then we can express any one of them as a linear combination of the others. On the other hand, if one of the vectors $x_1, x_2, \ldots, x_n$ is a linear combination of the others, then the vectors are linearly dependent. Another way of saying this is that vectors $x_1, x_2, \ldots, x_n$ are linearly dependent if and only if one of them belongs to the space generated by the remaining $n - 1$ vectors. We can now define the dimension of a linear space as equal to the maximum number of linearly independent vectors that it can contain. This fact underlies our study of basis vectors.

We observe that any two vectors are dependent if and only if they are parallel (or collinear); three vectors are dependent if and only if they are coplanar; four vectors are dependent in a space of three dimensions; n vectors are dependent in a space of $n-1$ dimensions. Finally, in a space of three dimensions, a set of three vectors $\mathbf{r}$, $\mathbf{s}$, and $\mathbf{t}$ is linearly dependent if and only if the following determinant is equal to zero:

$$\begin{vmatrix} r_x & r_y & r_z \\ s_x & s_y & s_z \\ t_x & t_y & t_z \end{vmatrix} = 0 \tag{1.56}$$

If the vectors $\mathbf{x}_1, \mathbf{x}_2, \ldots, \mathbf{x}_n$ are linearly independent, it is impossible to represent any one of them as a linear combination of the other $n-1$ vectors.

1.11 Basis Vectors and Coordinate Systems

Fixed vectors always relate to some frame of reference. All vectors have components, but only the components of fixed vectors are also coordinates. In our discussion of basis vectors we will use Cartesian systems. The familiar rectangular Cartesian coordinates are a special case of the more general Cartesian systems. Associated with each dimension, or principal direction, of a general Cartesian system is a family of parallel straight lines, with a uniform scale or metric. All principal directions may have the same scale, or each may be different. The principal directions need not be mutually orthogonal, but one point must serve as the origin (Figure 1.24).

We create a set of linearly independent vectors $\mathbf{e}_1, \mathbf{e}_2, \ldots, \mathbf{e}_n$ to form the *basis* of a Cartesian space S_n of n dimensions, expressing any point vector $\mathbf{r}$ in S_n as a linear combination of these basis vectors. In a space of three dimensions, the set of basis vectors $\mathbf{e}_1, \mathbf{e}_2$, and $\mathbf{e}_3$ originating at a common point O defines three families of parallel lines and forms a Cartesian system. The three lines X_1, X_2, and X_3 concurrent at O and collinear with $\mathbf{e}_1, \mathbf{e}_2$, and $\mathbf{e}_3$, respectively, define the principal coordinate axes (Figure 1.25). This system is a right-handed one, but it could just as well be left-handed. The basis vectors are analogous to the set of $\mathbf{i}, \mathbf{j}, \mathbf{k}$ that we introduced previously for a rectangular Cartesian coordinate system, although $\mathbf{e}_1, \mathbf{e}_2$, and $\mathbf{e}_3$ are not necessarily unit vectors or mutually perpendicular.

We find the components (coordinates) of any point $\mathbf{r}$ by constructing a parallelepiped, with the origin, O, and one vertex $\mathbf{r}$ as a body diagonal, and concurrent edges

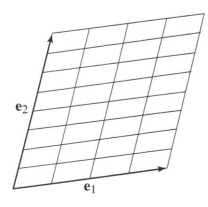

Figure 1.24 *Principal directions of basis vectors.*

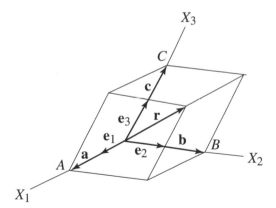

Figure 1.25 *Basis vectors and principal coordinate axes.*

collinear with the basis vectors. The three directed line segments corresponding to edges *OA*, *OB*, and *OC* define the vector components of **r** in this basis system. We denote them as **a**, **b**, and **c**. Note that this construction technique guarantees that the parallelogram law of vector addition applies, so that

$$\mathbf{r} = \mathbf{a} + \mathbf{b} + \mathbf{c} \tag{1.57}$$

The coordinates of **r** with respect to this basis are:

$$r_1 = \frac{|\mathbf{a}|}{|\mathbf{e}_1|}, \quad r_2 = \frac{|\mathbf{b}|}{|\mathbf{e}_2|}, \quad r_3 = \frac{|\mathbf{c}|}{|\mathbf{e}_3|} \tag{1.58}$$

Using these coordinates, we rewrite Equation 1.57 to obtain

$$\mathbf{r} = r_1\mathbf{e}_1 + r_2\mathbf{e}_2 + r_3\mathbf{e}_3 \tag{1.59}$$

where r_1, r_2, and r_3 are the coordinates or components of **r** with respect to the coordinate system defined by $\mathbf{e}_1$, $\mathbf{e}_2$, and $\mathbf{e}_3$. Mathematicians call these terms the parallel coordinates of the point **r**. These coordinates coincide with the coordinates of an affine three-dimensional space if the basis vectors are unit vectors.

1.12 A Short History

The complete history of vectors and vector geometry is far too complex to cover completely here. Many textbook-length works do an excellent job of it. The development of vectors could serve as a model for the development of the most important subdisciplines of mathematics. The subject of vectors has an ill-defined beginning and, as an important branch of applied mathematics, it is still sprouting new surprises and directions of inquiry.

Vectors are taught in courses on linear algebra, structural mechanics, introductory physics, and many other courses in engineering and the physical sciences. They are an indispensable tool in computer graphics, geometric modeling, and computer-aided design and manufacturing.

The story of vectors goes back at least to the sixteenth century, when Rafael Bombelli (1526?–1572?), an Italian mathematician, first treated $\sqrt{-1}$ as a number, albeit an *imaginary number*, and defined arithmetic operations on imaginary numbers. He combined real and imaginary numbers to form *complex numbers* and defined arithmetic operations on them.

Caspar Wessel (1745–1818), a Norwegian surveyor, gave complex numbers a geometric interpretation. Wessel's work, and in 1806 the work of Jean Argand (1768–1822)—the Swiss-French mathematician—led to the association of complex numbers with points on a plane. A complex number $a + ib$ is associated with the real number pair (a, b), which is then interpreted as the coordinates of a point in the plane (Figure 1.26). We can represent this complex number as a directed line segment, with its tail, or initial point, at the origin and its terminal point (or arrowhead) at the point (a, b). An example of the "geometry" of complex numbers is that each successive multiplication of a complex number by i rotates its equivalent directed line segment by $90°$ counterclockwise, so that $i(a + ib) = -b + ia$ (Figure 1.27).

William Rowan Hamilton (1805–1865), the great Irish mathematician and scientist, was a child prodigy. He had mastered over a dozen languages before becoming a teenager and was Ireland's preeminent mathematician before he was 20 years old. Hamilton treated complex numbers as merely an ordered pair of real numbers and generalized this concept to create hyper-complex numbers. For 15 years he struggled to develop an arithmetic for hyper-complex numbers, looking for operations on them that would correspond to operations on ordinary numbers and that would obey the customary laws of arithmetic (closure, commutativity, associativity, distributivity). This, he finally found, could not be done. So Hamilton created a new, logically consistent arithmetic, dropping the commutative law of multiplication. He focused on quadruples of real numbers which, with their special properties, became known as *quaternions*. These mathematical objects contain what we now recognize to be a scalar part and a vector part.

Hermann Grassman (1809–1877), a German geometer, soon formulated a more general algebra of hyper-complex numbers. The British mathematician Arthur Cayley (1821–1895) also generalized some of Hamilton's ideas and in the process developed

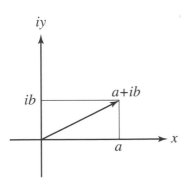

Figure 1.26 *The complex plane.*

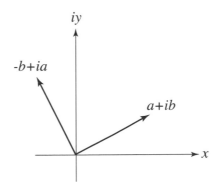

Figure 1.27 *Rotation in the complex plane.*

matrix algebra. However, it was Josiah Willard Gibbs (1839–1903), a Yale professor of mathematical physics, who refined vectors and vector analysis into the discipline as we know it today.

Exercises

1.1 Perform the indicated operations on the following complex numbers:
    a. $(3 + 2i) + (1 - 4i)$ d. $(6 - 5i) \times i$
    b. $(7 - i) + (1 + 3i)$ e. $(a + bi)(a - bi)$
    c. $4i + (-3 + 2i)$

1.2 Perform the indicated addition or subtraction on the following hypernumbers:
    a. $(6, 4, 1) + (1, -3, 0)$ d. $(5, 9, 1, 12) + (5, -3, 6, -1)$
    b. $(9, -8) - (2, -2)$ e. $(0, 0, 8) - (4, 1, 8)$
    c. $(a, b) + (c, d)$

1.3 Given the five vectors shown in Figure 1.28, write them in component form.

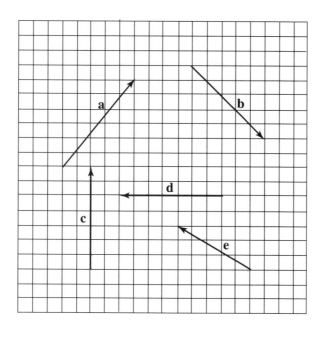

Figure 1.28 *Exercise 1.3.*

1.4 Compute the magnitudes of the vectors given in Exercise 1.3.

1.5 Compute the direction cosines of the vectors given in Exercise 1.3.

1.6 Given **a**, **b**, and **c** are three-component vectors, express the compact vector equation
    $\mathbf{a}x + \mathbf{b}y = \mathbf{c}$
    a. in expanded vector form
    b. in ordinary algebraic form

1.7 Compute the direction cosines of the following vectors:
   a. $\mathbf{a} = \begin{bmatrix} 5 & 6 \end{bmatrix}$ d. $\mathbf{d} = \begin{bmatrix} -7 & 0 \end{bmatrix}$
   b. $\mathbf{b} = \begin{bmatrix} 5 & -5 \end{bmatrix}$ e. $\mathbf{e} = \begin{bmatrix} -5 & 3 \end{bmatrix}$
   c. $\mathbf{c} = \begin{bmatrix} 0 & 7 \end{bmatrix}$

1.8 Given $\mathbf{a} = \begin{bmatrix} -2 & 0 & 7 \end{bmatrix}$ and $\mathbf{b} = \begin{bmatrix} 4 & 1 & 3 \end{bmatrix}$, compute
   a. $\hat{\mathbf{a}}$ d. $\mathbf{c} = 3\mathbf{a}$
   b. $\hat{\mathbf{b}}$ e. $\mathbf{c} = \mathbf{a} + \mathbf{b}$
   c. $\mathbf{c} = \mathbf{a} - 2\mathbf{b}$

1.9 Given $\mathbf{a} = \begin{bmatrix} 6 & 2 & -5 \end{bmatrix}$ and $\mathbf{b} = 2\mathbf{a}$, compare the unit vectors $\hat{\mathbf{a}}$ and $\hat{\mathbf{b}}$.

1.10 Given $\mathbf{a} = \begin{bmatrix} 2 & 3 & 5 \end{bmatrix}$ and $\mathbf{b} = \begin{bmatrix} 6 & -1 & 3 \end{bmatrix}$, compute
   a. $|\mathbf{a}|$ d. $\mathbf{c} = \mathbf{a} - \mathbf{b}$
   b. $|\mathbf{b}|$ e. $\mathbf{c} = 2\mathbf{a} + 3\mathbf{b}$
   c. $\mathbf{c} = \mathbf{a} + \mathbf{b}$

1.11 Compute the following scalar products:
   a. $\mathbf{i} \cdot \mathbf{i}$ d. $\mathbf{j} \cdot \mathbf{j}$
   b. $\mathbf{i} \cdot \mathbf{j}$ e. $\mathbf{j} \cdot \mathbf{k}$
   c. $\mathbf{i} \cdot \mathbf{k}$ f. $\mathbf{k} \cdot \mathbf{k}$

1.12 Given $\mathbf{a} = \begin{bmatrix} 2 & 0 & 5 \end{bmatrix}$, $\mathbf{b} = \begin{bmatrix} -1 & 3 & 1 \end{bmatrix}$, and $\mathbf{c} = \begin{bmatrix} 6 & -2 & -4 \end{bmatrix}$, compute
   a. $\mathbf{a} \cdot \mathbf{a}$ f. $\mathbf{c} \cdot \mathbf{c}$
   b. $\mathbf{a} \cdot \mathbf{b}$ g. $\mathbf{b} \cdot \mathbf{a}$
   c. $\mathbf{a} \cdot \mathbf{c}$ h. $\mathbf{c} \cdot \mathbf{b}$
   d. $\mathbf{b} \cdot \mathbf{b}$ i. $\hat{\mathbf{a}} \cdot \hat{\mathbf{a}}$
   e. $\mathbf{b} \cdot \mathbf{c}$ j. $\hat{\mathbf{a}} \cdot \hat{\mathbf{b}}$

1.13 Compute the magnitude and direction cosines for each of the following vectors:
   a. $\mathbf{a} = \begin{bmatrix} 3 & 4 \end{bmatrix}$ d. $\mathbf{d} = \begin{bmatrix} 1 & 4 & -3 \end{bmatrix}$
   b. $\mathbf{b} = \begin{bmatrix} 0 & -2 \end{bmatrix}$ e. $\mathbf{e} = \begin{bmatrix} x & y & z \end{bmatrix}$
   c. $\mathbf{c} = \begin{bmatrix} -3 & -5 & 0 \end{bmatrix}$

1.14 Compute the scalar product of the following pairs of vectors:
   a. $\mathbf{a} = \begin{bmatrix} 0 & -2 \end{bmatrix}$, $\mathbf{b} = \begin{bmatrix} 1 & 3 \end{bmatrix}$ d. $\mathbf{a} = \begin{bmatrix} 3 & 0 & -2 \end{bmatrix}$, $\mathbf{b} = \begin{bmatrix} 0 & -1 & -3 \end{bmatrix}$
   b. $\mathbf{a} = \begin{bmatrix} 4 & -1 \end{bmatrix}$, $\mathbf{b} = \begin{bmatrix} 2 & 1 \end{bmatrix}$ e. $\mathbf{a} = \begin{bmatrix} 5 & 1 & 7 \end{bmatrix}$, $\mathbf{b} = \begin{bmatrix} -2 & 4 & 1 \end{bmatrix}$
   c. $\mathbf{a} = \begin{bmatrix} 1 & 0 \end{bmatrix}$, $\mathbf{b} = \begin{bmatrix} 0 & 4 \end{bmatrix}$

1.15 Compute the following vector products:
   a. $\mathbf{i} \times \mathbf{i}$ f. $\mathbf{k} \times \mathbf{i}$
   b. $\mathbf{j} \times \mathbf{j}$ g. $\mathbf{j} \times \mathbf{i}$
   c. $\mathbf{k} \times \mathbf{k}$ h. $\mathbf{k} \times \mathbf{j}$
   d. $\mathbf{i} \times \mathbf{j}$ i. $\mathbf{i} \times \mathbf{k}$
   e. $\mathbf{j} \times \mathbf{k}$

1.16 Show that $\mathbf{b} \times \mathbf{a} = -(\mathbf{a} \times \mathbf{b})$.

1.17 Given $a = [1 \quad 0 \quad -2]$, $b = [3 \quad 1 \quad 4]$, and $c = [-1 \quad 6 \quad 2]$, compute
a. $a \times a$ d. $b \times c$
b. $a \times b$ e. $c \times a$
c. $b \times a$

1.18 Given $a = [4 \quad -1]$, $b = [2 \quad 8]$, $c = [-4 \quad 1]$, and $d = [3 \quad 2]$, compute the angle between
a. a and b d. a and d
b. a and c e. c and d
c. c and b

1.19 Compute the angle between the pairs of vectors given in Exercise 1.14.

1.20 Compute the vector product for each of the following pairs of vectors:
a. $a = [3 \quad -1 \quad 2]$, $b = [2 \quad 0 \quad 2]$
b. $a = [4 \quad 1 \quad -5]$, $b = [3 \quad 6 \quad 2]$
c. $a = [2 \quad -1 \quad 3]$, $b = [-4 \quad 2 \quad -6]$
d. $a = [0 \quad 1 \quad 0]$, $b = [1 \quad 0 \quad 0]$
e. $a = [0 \quad 0 \quad 1]$, $b = [1 \quad 0 \quad 0]$

1.21 Show that the vectors $a = \left[-\frac{1}{3} \quad \frac{2}{3} \quad \frac{2}{3}\right]$, $b = \left[\frac{2}{3} \quad -\frac{1}{3} \quad \frac{2}{3}\right]$, and $c = \left[-\frac{2}{3} \quad -\frac{2}{3} \quad \frac{1}{3}\right]$ are mutually perpendicular.

1.22 Show that the line joining the midpoints of two sides of a triangle is parallel to the third side and has one half its magnitude.

1.23 Find the midpoint of the line segment between $p_0 = [3 \quad 5 \quad 1]$ and $p_1 = [-2 \quad 6 \quad 4]$.

1.24 Given $a = [6 \quad -1 \quad -2]$, $b = [3 \quad 2 \quad 4]$, and $c = [7 \quad 0 \quad 2]$, write the vector equation of a line
a. through a and parallel to b
b. through b and parallel to c
c. through c and parallel to a
d. through a and parallel to a
e. through b and parallel to the a axis

1.25 Find the equations of the x, y, and z vector components for the line segments given by the following pairs of endpoints:
a. $p_0 = [0 \quad 0 \quad 0]$, $p_1 = [1 \quad 1 \quad 1]$
b. $p_0 = [-3 \quad 1 \quad 6]$, $p_1 = [2 \quad 0 \quad 7]$
c. $p_0 = [1 \quad 1 \quad -4]$, $p_1 = [5 \quad -3 \quad 9]$
d. $p_0 = [6 \quad 8 \quad 8]$, $p_1 = [-10 \quad 0 \quad -3]$
e. $p_0 = [0 \quad 0 \quad 1]$, $p_1 = [0 \quad 0 \quad -1]$

1.26 Given $x = 3 + 2u$, $y = -6 + u$, and $z = 4$, find p_0 and p_1.

1.27 What is the difference between the following two line segments: for line 1, $p_0 = [2 \quad 1 \quad -2]$ and $p_1 = [3 \quad -3 \quad 1]$; for line 2, $p_0 = [3 \quad -3 \quad 1]$ and $p_1 = [2 \quad 1 \quad -2]$.

1.28 Write the vector equation of the plane passing through a and parallel to b and c.

1.29 Write the vector equation of a plane that passes through the origin and is perpendicular to the y axis.

1.30 Is the set of all vectors lying in the first quadrant of the x, y plane a linear space? Why?

1.31 Determine nontrivial linear relations for the following sets of vectors:
   a. $\mathbf{p} = [1 \quad 0 \quad -2]$, $\mathbf{q} = [3 \quad -1 \quad 3]$, $\mathbf{r} = [5 \quad -2 \quad 8]$
   b. $\mathbf{p} = [2 \quad 0 \quad 1]$, $\mathbf{q} = [0 \quad 5 \quad 1]$, $\mathbf{r} = [6 \quad -5 \quad 4]$
   c. $\mathbf{p} = [3 \quad 0]$, $\mathbf{q} = [1 \quad 4]$, $\mathbf{r} = [2 \quad -1]$

1.32 Are the vectors $\mathbf{r} = [1 \quad -1 \quad 0]$, $\mathbf{s} = [0 \quad 2 \quad -1]$, and $\mathbf{t} = [2 \quad 0 \quad -1]$ linearly dependent? Why?

CHAPTER 2

MATRIX METHODS

A *matrix* is a set of numbers or other mathematical elements arranged in a rectangular array of rows and columns. The structure of a matrix makes it easy to assemble and work with certain kinds of mathematical data. The coefficients of a set of simultaneous linear equations or the coordinates of a point can be written as a matrix, for example. However, a matrix is not just a tool to organize data. One matrix can operate on another matrix to change the data in some way. For example, we can use matrices to solve complex systems of equations, to represent geometric objects in computer data bases, and to perform geometric transformations, such as translation, rotation, and scaling. The rules of matrix algebra define allowable operations in these areas. But before we can do any of these things, we must study some simple properties of matrices and some matrix arithmetic and algebra. Along the way, we will review determinants, which look much like matrices but really aren't.

2.1 Definition of a Matrix

A matrix is a rectangular array of numbers, or their algebraic equivalents, arranged in m rows and n columns. We denote a matrix with a boldface uppercase letter, such as $\mathbf{A}$, $\mathbf{B}$, $\mathbf{C}$, $\mathbf{P}$, ..., $\mathbf{T}$, and use brackets to enclose the array. For example,

$$\mathbf{A} = \begin{bmatrix} a_{11} & a_{12} & a_{13} \\ a_{21} & a_{22} & a_{23} \\ a_{31} & a_{32} & a_{33} \\ a_{41} & a_{42} & a_{43} \end{bmatrix} \tag{2.1}$$

The lowercase subscripted letters, such as a_{32}, a_{41}, and a_{43}, are the elements of the matrix. The double subscript gives the position of each element in the array by row and column number. For example, a_{32} is in the third row and second column, and a_{ij} is in row i and column j. We use a comma between subscript indices or numbers, when necessary.

We can represent the coefficients of a system of simultaneous linear equations by a matrix. For example, given

$$\begin{aligned} x - 3y + z &= 5 \\ 4x + y - 2z &= -2 \\ -2x + 3y &= 1 \end{aligned} \tag{2.2}$$

we construct a matrix **A** whose elements are the coefficients:

$$\mathbf{A} = \begin{bmatrix} 1 & -3 & 1 \\ 4 & 1 & -2 \\ -2 & 3 & 0 \end{bmatrix} \tag{2.3}$$

We also represent the set of constants on the right sides of the equations using another matrix **B**:

$$\mathbf{B} = \begin{bmatrix} 5 \\ -2 \\ 1 \end{bmatrix} \tag{2.4}$$

What about the variables x, y, and z? We do the same for them:

$$\mathbf{X} = \begin{bmatrix} x \\ y \\ z \end{bmatrix} \tag{2.5}$$

These are all very different looking matrices. Even so, we assemble them into an equation that looks something like Equation 2.2, as follows:

$$\begin{bmatrix} 1 & -3 & 1 \\ 4 & 1 & -2 \\ -2 & 3 & 0 \end{bmatrix} \begin{bmatrix} x \\ y \\ z \end{bmatrix} = \begin{bmatrix} 5 \\ -2 \\ 1 \end{bmatrix} \tag{2.6}$$

or more simply as

$$\mathbf{AX} = \mathbf{B} \tag{2.7}$$

We will see why this is so later, as well as how we actually use matrix algebra to solve this system of equations for x, y, and z.

The number of rows and columns of a matrix determines its *order*. Matrix **A** in Equation 2.1 is of order 4×3, read as "4 by 3," and matrix **X** above is of order 3×1. Note that we state the order by giving the number of rows first, followed by the number of columns. In general, a matrix with m rows and n columns is of order $m \times n$.

2.2 Special Matrices

For any given matrix, if the number of rows equals the number of columns, $m = n$, then it is a *square matrix*. For example,

$$\mathbf{C} = \begin{bmatrix} c_{11} & c_{12} \\ c_{21} & c_{22} \end{bmatrix} \tag{2.8}$$

A *row matrix* has a single row of elements, and a *column matrix* has a single column of elements. In the example below, **A** is a row matrix and **B** is a column matrix.

$$\mathbf{A} = \begin{bmatrix} a_{11} & a_{12} & a_{13} \end{bmatrix}, \quad \mathbf{B} = \begin{bmatrix} b_{11} \\ b_{21} \\ b_{31} \end{bmatrix} \tag{2.9}$$

Row and column matrices are sometimes called *vectors*.

A square matrix that has zero elements everywhere except on the main diagonal is a *diagonal matrix*. (The *main diagonal* runs from the upper-left-corner element to the lower-right-corner element.)

$$\mathbf{A} = \begin{bmatrix} a_{11} & 0 & 0 & \cdots & 0 \\ 0 & a_{22} & 0 & \cdots & 0 \\ 0 & 0 & a_{33} & \cdots & 0 \\ \vdots & \vdots & \vdots & \ddots & \vdots \\ 0 & 0 & 0 & \cdots & a_{nn} \end{bmatrix} \tag{2.10}$$

This means that $a_{ij} = 0$ if $i \neq j$.

If all the a_{ii} are equal, then the diagonal matrix is a *scalar matrix*. For example,

$$\mathbf{A} = \begin{bmatrix} 2 & 0 & 0 \\ 0 & 2 & 0 \\ 0 & 0 & 2 \end{bmatrix} \tag{2.11}$$

A very important matrix in matrix algebra is the *identity matrix*. This is a special diagonal matrix that has unit elements on the main diagonal. It is denoted by the symbol **I**. The 3×3 identity matrix is

$$\mathbf{I} = \begin{bmatrix} 1 & 0 & 0 \\ 0 & 1 & 0 \\ 0 & 0 & 1 \end{bmatrix} \tag{2.12}$$

Elements of **I** are denoted by δ_{ij}, where

$$\begin{aligned} \delta_{ij} &= 0 \quad \text{if} \quad i \neq j \\ &= 1 \quad \text{if} \quad i = j \end{aligned} \tag{2.13}$$

δ_{ij} is called the *Kronecker delta*, named after Leopold Kronecker (1823–1891), a German mathematician whose great program was the arithmetization of mathematics.

A *null matrix* is one whose elements are all zero. It is usually denoted by Ø.

A matrix whose elements are symmetric about the main diagonal is a *symmetric matrix*. If **A** is a symmetric matrix, then $a_{ij} = a_{ji}$; for example,

$$\mathbf{A} = \begin{bmatrix} 3 & 0 & -4 & 1 \\ 0 & 6 & 2 & 3 \\ -4 & 2 & 5 & 7 \\ 1 & 3 & 7 & -2 \end{bmatrix} \tag{2.14}$$

A matrix is *antisymmetric* (or *skew symmetric*) if $a_{ij} = -a_{ji}$. For example,

$$\mathbf{A} = \begin{bmatrix} 3 & 0 & -4 & 1 \\ 0 & 6 & 2 & 3 \\ 4 & -2 & 5 & 7 \\ -1 & -3 & -7 & -2 \end{bmatrix} \tag{2.15}$$

2.3 Matrix Equivalence

Two matrices are equal if all of their corresponding elements are equal. This means that for two matrices to be equal they must be of the same order. No equality occurs between matrices of different orders. Given the three matrices **A**, **B**, and **C**, where

$$\mathbf{A} = \begin{bmatrix} a_{11} & a_{12} \\ a_{21} & a_{22} \\ a_{31} & a_{32} \end{bmatrix}, \quad \mathbf{B} = \begin{bmatrix} b_{11} & b_{12} \\ b_{21} & b_{22} \\ b_{31} & b_{32} \end{bmatrix}, \quad \mathbf{C} = \begin{bmatrix} c_{11} & c_{12} & c_{13} \\ c_{21} & c_{22} & c_{23} \\ c_{31} & c_{32} & c_{33} \end{bmatrix} \tag{2.16}$$

Matrices **A** and **B** are the same order, so **A** = **B** if and only if $a_{ij} = b_{ij}$ for $i = 1, 2, 3$ and $j = 1, 2$. This means that **A** = **B** if and only if $a_{11} = b_{11}, a_{12} = b_{12}, a_{21} = b_{21}, a_{22} = b_{22}, a_{31} = b_{31}$, and $a_{32} = b_{32}$. Obviously, **A** ≠ **C** and **B** ≠ **C** because their elements cannot be equated one to one. An $m \times n$ matrix cannot equal an $n \times m$ matrix if $m \neq n$. Even though they have the same number of elements, they are not the same order.

2.4 Matrix Arithmetic

Adding two matrices **A** and **B** produces a third matrix **C**, whose elements are equal to the sum of the corresponding elements of **A** and **B**. Thus,

$$\mathbf{A} + \mathbf{B} = \mathbf{C} \tag{2.17}$$

or

$$a_{ij} + b_{ij} = c_{ij} \tag{2.18}$$

Of course, we can add or subtract two matrices if and only if they are the same order. Similarly, the difference of two matrices is another matrix, so that

$$\mathbf{A} - \mathbf{B} = \mathbf{D} \tag{2.19}$$

or

$$a_{ij} - b_{ij} = d_{ij} \qquad (2.20)$$

We can add or subtract a sequence of more than two matrices as long as they are the same order:

$$\mathbf{A} + \mathbf{B} - \mathbf{C} = \mathbf{D} \qquad (2.21)$$

or

$$a_{ij} + b_{ij} - c_{ij} = d_{ij} \qquad (2.22)$$

Given two 3×2 matrices $\mathbf{A}$ and $\mathbf{B}$, for example, where

$$\mathbf{A} = \begin{bmatrix} 2 & -3 \\ 1 & 4 \\ 0 & -1 \end{bmatrix} \quad \text{and} \quad \mathbf{B} = \begin{bmatrix} 5 & -1 \\ 2 & 9 \\ 2 & 7 \end{bmatrix} \qquad (2.23)$$

then

$$\mathbf{A} + \mathbf{B} = \mathbf{C} = \begin{bmatrix} 2+5 & -3-1 \\ 1+2 & 4+9 \\ 0+2 & -1+7 \end{bmatrix} = \begin{bmatrix} 7 & -4 \\ 3 & 13 \\ 2 & 6 \end{bmatrix} \qquad (2.24)$$

and

$$\mathbf{A} - \mathbf{B} = \mathbf{C} = \begin{bmatrix} 2-5 & -3+1 \\ 1-2 & 4-9 \\ 0-2 & -1-7 \end{bmatrix} = \begin{bmatrix} -3 & -2 \\ -1 & -5 \\ -2 & -8 \end{bmatrix} \qquad (2.25)$$

Matrix addition is commutative, thus $\mathbf{A} + \mathbf{B} = \mathbf{B} + \mathbf{A}$.

Multiplying a matrix $\mathbf{A}$ by a scalar k produces a new matrix $\mathbf{B}$ of the same order. Each element of $\mathbf{B}$ is obtained by multiplying the corresponding element of $\mathbf{A}$ by the scalar k:

$$k\mathbf{A} = \mathbf{B} \qquad (2.26)$$

or

$$ka_{ij} = b_{ij} \qquad (2.27)$$

The product $\mathbf{AB}$ of two matrices is another matrix $\mathbf{C}$. This operation is possible if and only if the number of columns of the first matrix is equal to the number of rows of the second matrix. If $\mathbf{A}$ is $m \times n$ and $\mathbf{B}$ is $n \times p$, then $\mathbf{C}$ is $m \times p$. When this condition is satisfied, we say that the matrices $\mathbf{A}$ and $\mathbf{B}$ are *conformable* for multiplication. We write this as

$$\mathbf{AB} = \mathbf{C} \qquad (2.28)$$

For this example, we say that $\mathbf{A}$ premultiplies $\mathbf{B}$, or $\mathbf{B}$ postmultiplies $\mathbf{A}$.

In general, the product of two matrices is not commutative unless one of them is the identity matrix. Thus,

$$\mathbf{AB} \neq \mathbf{BA} \tag{2.29}$$

but

$$\mathbf{AI} = \mathbf{IA} \tag{2.30}$$

The product of two matrices in terms of their elements is

$$c_{ij} = a_{i1}b_{1j} + a_{i2}b_{2j} + \cdots + a_{in}b_{nj} \tag{2.31}$$

This unusual expression intuitively makes more sense with an example: Given the 3×3 matrix $\mathbf{A}$ and the 3×2 matrix $\mathbf{B}$, they are conformable, and their product is the 3×2 matrix $\mathbf{C}$. We arrange the matrices as follows:

$$
\begin{bmatrix} b_{11} & b_{12} \\ b_{21} & b_{22} \\ b_{31} & b_{32} \end{bmatrix}
$$
$$
\begin{bmatrix} a_{11} & a_{12} & a_{13} \\ a_{21} & a_{22} & a_{23} \\ a_{31} & a_{32} & a_{33} \end{bmatrix}
\begin{bmatrix} c_{11} & c_{12} \\ c_{21} & c_{22} \\ c_{31} & c_{32} \end{bmatrix}
\tag{2.32}
$$

Then we determine each element of c_{ij} by summing the pairwise products of elements in row i of $\mathbf{A}$ with corresponding elements in column j of $\mathbf{B}$. For example,

$$c_{21} = a_{21}b_{11} + a_{22}b_{21} + a_{23}b_{31} \tag{2.33}$$

This also illustrates why $\mathbf{A}$ and $\mathbf{B}$ must be conformable and how their orders determine the order of $\mathbf{C}$.

Here is another example: Let

$$\mathbf{A} = \begin{bmatrix} 2 & -1 \\ 3 & 4 \end{bmatrix} \quad \text{and} \quad \mathbf{B} = \begin{bmatrix} 1 & 0 & 3 \\ 5 & 2 & 4 \end{bmatrix} \tag{2.34}$$

then for $AB = C$

$$
\begin{aligned}
\mathbf{C} &= \begin{bmatrix} 2 \times 1 - 1 \times 5 & 2 \times 0 - 1 \times 2 & 2 \times 3 - 1 \times 4 \\ 3 \times 1 + 4 \times 5 & 3 \times 0 + 4 \times 2 & 3 \times 3 + 4 \times 4 \end{bmatrix} \\
&= \begin{bmatrix} -3 & -2 & 2 \\ 23 & 8 & 25 \end{bmatrix}
\end{aligned}
\tag{2.35}
$$

Multiplying any matrix $\mathbf{A}$ of order $m \times n$ by the identity matrix reproduces the original matrix. Thus,

$$\mathbf{AI} = \mathbf{A} \tag{2.36}$$

or

$$\begin{bmatrix} a_{11} & a_{12} & \cdots & a_{1n} \\ a_{21} & a_{22} & \cdots & a_{2n} \\ \vdots & \vdots & \ddots & \vdots \\ a_{m1} & a_{m2} & \cdots & a_{mn} \end{bmatrix} \begin{bmatrix} 1 & 0 & \cdots & 0 \\ 0 & 1 & \cdots & 0 \\ \vdots & \vdots & \ddots & \vdots \\ 0 & 0 & \cdots & 1 \end{bmatrix} = \begin{bmatrix} a_{11} & a_{12} & \cdots & a_{1n} \\ a_{21} & a_{22} & \cdots & a_{2n} \\ \vdots & \vdots & \ddots & \vdots \\ a_{m1} & a_{m2} & \cdots & a_{mn} \end{bmatrix} \tag{2.37}$$

where $\mathbf{I}$ must be order $n \times n$ if it postmultiplies $\mathbf{A}$, and of order $m \times m$ if it premultiplies $\mathbf{A}$.

By interchanging the rows and columns of a matrix $\mathbf{A}$ we obtain its transpose, $\mathbf{A}^T$, so that $a_{ij}^T = a_{ji}$, where the a_{ij}^T are the elements of the transpose of $\mathbf{A}$. For example, if

$$\mathbf{A} = \begin{bmatrix} a & c & e \\ b & d & f \end{bmatrix}, \quad \text{then} \quad \mathbf{A}^T = \begin{bmatrix} a & b \\ c & d \\ e & f \end{bmatrix} \tag{2.38}$$

Following is a summary of matrix properties.
For matrix addition and scalar multiplication:

1. $\mathbf{A} + \mathbf{B} = \mathbf{B} + \mathbf{A}$
2. $\mathbf{A} + (\mathbf{B} + \mathbf{C}) = (\mathbf{A} + \mathbf{B}) + \mathbf{C}$
3. $b(\mathbf{A} + \mathbf{B}) = b\mathbf{A} + b\mathbf{B}$
4. $(b + d)\mathbf{A} = b\mathbf{A} + d\mathbf{A}$
5. $b(d\mathbf{A}) = (bd)\mathbf{A} = d(b\mathbf{A})$

For matrix multiplication:

1. $(\mathbf{AB})\mathbf{C} = \mathbf{A}(\mathbf{BC})$
2. $\mathbf{A}(\mathbf{B} + \mathbf{C}) = \mathbf{AB} + \mathbf{AC}$
3. $(\mathbf{A} + \mathbf{B})\mathbf{C} = \mathbf{AC} + \mathbf{BC}$
4. $\mathbf{A}(k\mathbf{B}) = k(\mathbf{AB}) = (k\mathbf{A})\mathbf{B}$

For matrix transpose:

1. $(\mathbf{A} + \mathbf{B})^T = \mathbf{A}^T + \mathbf{B}^T$
2. $(k\mathbf{A})^T = k\mathbf{A}^T$
3. $(\mathbf{AB})^T = \mathbf{B}^T\mathbf{A}^T$
4. If $\mathbf{AA}^T = \mathbf{I}$, then $\mathbf{A}$ is an *orthogonal matrix*.

2.5 Partitioning a Matrix

We may find it computationally efficient to partition a matrix into submatrices and treat it as a matrix whose elements are these submatrices. For example, given

$$T = \begin{bmatrix} t_{11} & t_{12} & \vdots & t_{13} \\ t_{21} & t_{22} & \vdots & t_{23} \\ t_{31} & t_{32} & \vdots & t_{33} \\ \text{---} & \text{---} & \text{---} \\ t_{41} & t_{42} & \vdots & t_{43} \\ t_{51} & t_{52} & \vdots & t_{53} \end{bmatrix} = \begin{bmatrix} \mathbf{T}_{11} & \mathbf{T}_{12} \\ \mathbf{T}_{21} & \mathbf{T}_{22} \end{bmatrix} \tag{2.39}$$

we can partition $\mathbf{T}$ into the four submatrices $\mathbf{T}_{11}$, $\mathbf{T}_{12}$, $\mathbf{T}_{21}$, and $\mathbf{T}_{22}$, where

$$\mathbf{T}_{11} = \begin{bmatrix} t_{11} & t_{12} \\ t_{21} & t_{22} \\ t_{31} & t_{32} \end{bmatrix} \quad \mathbf{T}_{12} = \begin{bmatrix} t_{13} \\ t_{23} \\ t_{33} \end{bmatrix}$$

$$\mathbf{T}_{21} = \begin{bmatrix} t_{41} & t_{42} \\ t_{51} & t_{52} \end{bmatrix} \quad \mathbf{T}_{22} = \begin{bmatrix} t_{43} \\ t_{53} \end{bmatrix} \tag{2.40}$$

Note that $\mathbf{T}_{11}$ and $\mathbf{T}_{12}$ must necessarily have the same number of rows, similarly for $\mathbf{T}_{21}$ and $\mathbf{T}_{22}$. Furthermore, $\mathbf{T}_{11}$ and $\mathbf{T}_{21}$ must have the same number of columns, similarly for $\mathbf{T}_{12}$ and $\mathbf{T}_{22}$.

Here is an example of adding partitioned matrices: Let

$$\mathbf{A} = \begin{bmatrix} \mathbf{A}_{11} & \mathbf{A}_{12} & \mathbf{A}_{13} \\ \mathbf{A}_{21} & \mathbf{A}_{22} & \mathbf{A}_{23} \end{bmatrix} \quad \text{and} \quad \mathbf{B} = \begin{bmatrix} \mathbf{B}_{11} & \mathbf{B}_{12} & \mathbf{B}_{13} \\ \mathbf{B}_{21} & \mathbf{B}_{22} & \mathbf{B}_{23} \end{bmatrix} \tag{2.41}$$

then

$$\mathbf{A} + \mathbf{B} = \begin{bmatrix} \mathbf{A}_{11} + \mathbf{B}_{11} & \mathbf{A}_{12} + \mathbf{B}_{12} & \mathbf{A}_{13} + \mathbf{B}_{13} \\ \mathbf{A}_{21} + \mathbf{B}_{21} & \mathbf{A}_{22} + \mathbf{B}_{22} & \mathbf{A}_{23} + \mathbf{B}_{23} \end{bmatrix} \tag{2.42}$$

where the $\mathbf{A}_{ij}$ and $\mathbf{B}_{ij}$ are conformable.

Finally, here is an example of the product of two partitioned matrices. The matrices must be conformable for multiplication before and after partitioning. If

$$\mathbf{A} = \begin{bmatrix} a_{11} & a_{12} & a_{13} & a_{14} & a_{15} \\ a_{21} & a_{22} & a_{23} & a_{24} & a_{25} \\ a_{31} & a_{32} & a_{33} & a_{34} & a_{35} \\ a_{41} & a_{42} & a_{43} & a_{44} & a_{45} \end{bmatrix} = \begin{bmatrix} \mathbf{A}_{11} & \mathbf{A}_{12} & \mathbf{A}_{13} \\ \mathbf{A}_{21} & \mathbf{A}_{22} & \mathbf{A}_{23} \end{bmatrix} \tag{2.43}$$

and

$$\mathbf{B} = \begin{bmatrix} b_{11} & b_{12} & b_{13} \\ b_{21} & b_{22} & b_{23} \\ b_{31} & b_{32} & b_{33} \\ b_{41} & b_{42} & b_{43} \\ b_{51} & b_{52} & b_{53} \end{bmatrix} = \begin{bmatrix} \mathbf{B}_{11} & \mathbf{B}_{12} \\ \mathbf{B}_{21} & \mathbf{B}_{22} \\ \mathbf{B}_{31} & \mathbf{B}_{32} \end{bmatrix} \tag{2.44}$$

then

$$\mathbf{AB} = \begin{bmatrix} \mathbf{A}_{11}\mathbf{B}_{11} + \mathbf{A}_{12}\mathbf{B}_{21} + \mathbf{A}_{13}\mathbf{B}_{31} & \mathbf{A}_{11}\mathbf{B}_{12} + \mathbf{A}_{12}\mathbf{B}_{22} + \mathbf{A}_{13}\mathbf{B}_{32} \\ \mathbf{A}_{21}\mathbf{B}_{11} + \mathbf{A}_{22}\mathbf{B}_{21} + \mathbf{A}_{23}\mathbf{B}_{31} & \mathbf{A}_{21}\mathbf{B}_{12} + \mathbf{A}_{22}\mathbf{B}_{22} + \mathbf{A}_{23}\mathbf{B}_{32} \end{bmatrix} \tag{2.45}$$

2.6 Determinants

A *determinant* is an operator in the form of a square array of numbers that produces a single value. Determinants are an important part of many vector and matrix operations, and sometimes we evaluate a square matrix as a determinant. The determinant of a 2 × 2 matrix **A** is |**A**|, where

$$\mathbf{A} = \begin{bmatrix} a_{11} & a_{12} \\ a_{21} & a_{22} \end{bmatrix} \tag{2.46}$$

and

$$|\mathbf{A}| = a_{11}a_{22} - a_{12}a_{21} \tag{2.47}$$

For a 3 × 3 matrix,

$$\mathbf{A} = \begin{bmatrix} a_{11} & a_{12} & a_{13} \\ a_{21} & a_{22} & a_{23} \\ a_{31} & a_{32} & a_{33} \end{bmatrix} \tag{2.48}$$

and

$$|\mathbf{A}| = a_{11} \begin{vmatrix} a_{22} & a_{23} \\ a_{32} & a_{33} \end{vmatrix} - a_{12} \begin{vmatrix} a_{21} & a_{23} \\ a_{31} & a_{33} \end{vmatrix} + a_{13} \begin{vmatrix} a_{21} & a_{22} \\ a_{31} & a_{32} \end{vmatrix} \tag{2.49}$$

The *minor* of an element a_{ij} of a determinant |**A**| is another determinant $|\mathbf{A}'_{ij}|$ obtained by deleting elements of the ith row and jth column of |**A**|. If |**A**| is $n \times n$, then all $|\mathbf{A}'_{ij}|$ are $n - 1 \times n - 1$.

The *cofactor* of element a_{ij} of a determinant |**A**| is obtained by the product of the minor of the element with a plus or minus sign determined according to

$$c_{ij} = (-1)^{i+j} \left| \mathbf{A}'_{ij} \right| \tag{2.50}$$

We can arrange these elements into a matrix of cofactors, **C**, of **A**, where $|\mathbf{A}'_{ij}|$ denotes the minor of |**A**|. The value of a determinant is simply equal to the sum of the products of each element of any row (or column) and its cofactor. By successively applying this process we reduce the determinant to a computable expression consisting of the multiplication and summation of elements of the array.

Here are important properties of determinants:

1. The determinant of a square matrix is equal to the determinant of its transpose: $|\mathbf{A}| = |\mathbf{A}^T|$.
2. Interchanging any two rows (or any two columns) of **A** changes the sign of |**A**|.
3. If we obtain **B** by multiplying one row (or column) of **A** by a constant, k, then $|\mathbf{B}| = k|\mathbf{A}|$.
4. If two rows (or columns) of **A** are identical, then $|\mathbf{A}| = 0$.
5. If we derive **B** from **A** by adding a multiple of one row (or column) of **A** to another row (or column) of **A**, then $|\mathbf{B}| = |\mathbf{A}|$.

6. If **A** and **B** are both $n \times n$ matrices, then the determinant of their product is $|\mathbf{AB}| = |\mathbf{A}||\mathbf{B}|$.
7. If every element of a row (or column) is zero, then the value of the determinant is zero.
8. If the determinant of a square matrix **A** is equal to one, $|\mathbf{A}| = +1$, then it is orthogonal and proper. If $|\mathbf{A}| = -1$, it is orthogonal and improper. The orthogonal property is important in geometric transformations theory and application.

2.7 Matrix Inversion

Matrix arithmetic does not define a division operation, but it does include a process for finding the *inverse* of a matrix. Dividing any number by itself equals one: $\frac{x}{x} = 1$. The inverse of a square matrix **A** is $\mathbf{A}^{-1}$, which satisfies the condition

$$\mathbf{AA}^{-1} = \mathbf{A}^{-1}\mathbf{A} = \mathbf{I} \tag{2.51}$$

The elements of $\mathbf{A}^{-1}$ are a_{ij}^{-1}, where

$$a_{ij}^{-1} = \frac{(-1)^{i+j}|\mathbf{A}'_{ji}|}{|\mathbf{A}|} \tag{2.52}$$

or, more compactly,

$$\mathbf{A}^{-1} = \frac{\mathbf{C}^T}{|\mathbf{A}|} \tag{2.53}$$

The numerator is simply the determinant of the cofactors of $|\mathbf{A}|$ transposed (note the subscript order on $\mathbf{A}_{ji}^{-1}$).

For example, find $\mathbf{A}^{-1}$ when

$$\mathbf{A} = \begin{bmatrix} 2 & -1 & 3 \\ 1 & 6 & -4 \\ 5 & 0 & 8 \end{bmatrix} \tag{2.54}$$

First, we compute $|\mathbf{A}| = 34$, then we replace each element of **A** with its cofactor and transpose the result to obtain

$$\mathbf{C}^T = \begin{bmatrix} 48 & 8 & -14 \\ -28 & 1 & 11 \\ -30 & -5 & 13 \end{bmatrix} \tag{2.55}$$

Then, using $\mathbf{A}^{-1} = \mathbf{C}^T/|\mathbf{A}|$, we have, for the example,

$$\mathbf{A}^{-1} = \frac{1}{34} \begin{bmatrix} 48 & 8 & -14 \\ -28 & 1 & 11 \\ -30 & -5 & 13 \end{bmatrix} \tag{2.56}$$

We note that for $\mathbf{A}^{-1}$ to exist at all, $|\mathbf{A}| \neq 0$. It is easy to check the correctness of $\mathbf{A}^{-1}$ because, again. $\mathbf{A}\mathbf{A}^{-1} = \mathbf{A}^{-1}\mathbf{A} = \mathbf{I}$.

We can compare matrix inversion to taking the reciprocal of an expression in the course of solving an algebraic equation. For the expression

$$3x = 8 \tag{2.57}$$

we compute the value of x by operating on 8 with the reciprocal of 3, so that

$$\left(\frac{1}{3}\right) 3x = \left(\frac{1}{3}\right) 8$$

$$x = \left(\frac{8}{3}\right) \tag{2.58}$$

A similar procedure applies to matrix algebra, using matrix inversion. For example, given a set of n simultaneous linear equations with n unknowns:

$$a_{11}x_1 + a_{12}x_2 + \cdots + a_{1n}x_n = b_1$$

$$a_{21}x_1 + a_{22}x_2 + \cdots + a_{2n}x_n = b_2$$

$$\vdots \qquad\qquad\qquad \vdots \tag{2.59}$$

$$a_{n1}x_1 + a_{n2}x_2 + \cdots + a_{nn}x_n = b_n$$

We easily rewrite this as a matrix equation:

$$\mathbf{AX} = \mathbf{B} \tag{2.60}$$

We premultiply both sides by $\mathbf{A}^{-1}$ to obtain

$$\mathbf{A}^{-1}\mathbf{AX} = \mathbf{A}^{-1}\mathbf{B} \tag{2.61}$$

Because $\mathbf{A}^{-1}\mathbf{A} = \mathbf{I}$ and $\mathbf{IX} = \mathbf{X}$, we simplify Equation 2.60 to

$$\mathbf{X} = \mathbf{A}^{-1}\mathbf{B} \tag{2.62}$$

Now let's use matrix inversion to solve the set of equations in Equation 2.2. From Equation 2.6 we compute

$$\begin{vmatrix} 1 & -3 & 1 \\ 4 & 1 & -2 \\ -2 & 3 & 0 \end{vmatrix} = 8 \tag{2.63}$$

and

$$\begin{bmatrix} 1 & -3 & 1 \\ 4 & 1 & -2 \\ -2 & 3 & 0 \end{bmatrix}^{-1} = \frac{1}{8}\begin{bmatrix} 6 & 3 & 5 \\ 4 & 2 & 6 \\ 14 & 3 & 13 \end{bmatrix} \tag{2.64}$$

To find $\mathbf{X}$, we compute $\mathbf{X} = \mathbf{A}^{-1}\mathbf{B}$:

$$\frac{1}{8}\begin{bmatrix} 6 & 3 & 5 \\ 4 & 2 & 6 \\ 14 & 3 & 13 \end{bmatrix}\begin{bmatrix} 5 \\ -2 \\ 1 \end{bmatrix} = \begin{bmatrix} 29/8 \\ 22/8 \\ 77/8 \end{bmatrix} \tag{2.65}$$

or

$$\begin{bmatrix} x \\ y \\ z \end{bmatrix} = \begin{bmatrix} 29/8 \\ 22/8 \\ 77/8 \end{bmatrix} \qquad (2.66)$$

This result is easy to verify by substituting $x = 29/8$, $y = 22/8$, and $z = 77/8$ into Equations 2.2.

Some conditions would prohibit this approach. For example, it would not work if $|\mathbf{A}| = 0$. Difficulties also arise when, depending on the limits of precision, we must compute the difference of two nearly equal large numbers in the course of evaluating the determinate. Many good textbooks on numerical analysis address these and other problems of matrix inversion.

2.8 Scalar and Vector Products

We can use matrices to represent vectors, and we can use matrix multiplication to generate their scalar and vector products. Let matrices $\mathbf{A}$ and $\mathbf{B}$ represent the vectors $\mathbf{a}$ and $\mathbf{b}$, where $\mathbf{A} = [a_1 \quad a_2 \quad a_3]$ and $\mathbf{B} = [b_1 \quad b_2 \quad b_3]$. Then the scalar product $\mathbf{a} \cdot \mathbf{b}$ is

$$\mathbf{a} \cdot \mathbf{b} = \mathbf{A}\mathbf{B}^T \qquad (2.67)$$

or

$$\mathbf{a} \cdot \mathbf{b} = [a_1 \quad a_2 \quad a_3] \begin{bmatrix} b_1 \\ b_2 \\ b_3 \end{bmatrix} = a_1 b_1 + a_2 b_2 + a_3 b_3 \qquad (2.68)$$

If we use the components of $\mathbf{a}$ to form the antisymmetric matrix

$$\begin{bmatrix} 0 & -a_3 & a_2 \\ a_3 & 0 & -a_1 \\ -a_2 & a_1 & 0 \end{bmatrix} \qquad (2.69)$$

then

$$\mathbf{a} \times \mathbf{b} = \begin{bmatrix} 0 & -a_3 & a_2 \\ a_3 & 0 & -a_1 \\ -a_2 & a_1 & 0 \end{bmatrix} \begin{bmatrix} b_1 \\ b_2 \\ b_3 \end{bmatrix} \qquad (2.70)$$

2.9 Eigenvalues and Eigenvectors

We can use matrix multiplication to change the magnitude or direction of a vector. For example, if $\mathbf{P}$ is an $n \times 1$ column matrix representing a vector, we transform (make restricted changes to) $\mathbf{P}$ by multiplying it by an $n \times n$ matrix $\mathbf{A}$ so that

$$\mathbf{P}' = \mathbf{A}\mathbf{P} \qquad (2.71)$$

where $\mathbf{P}'$ is the new, or transformed, vector. It is possible to find a scalar constant λ (lowercase Greek lambda) such that $\mathbf{P}' = \lambda\mathbf{P}$ or

$$\mathbf{AP} = \lambda\mathbf{P} \tag{2.72}$$

Every vector for which this is true for a given $\mathbf{A}$ is an *eigenvector* of $\mathbf{A}$, and λ is the *eigenvalue* of $\mathbf{A}$ corresponding to the vector $\mathbf{P}$. Matrix $\mathbf{A}$ transforms its eigenvector(s) into a collinear vector. The corresponding eigenvalue(s) is equal to the ratio of the magnitudes of the two collinear vectors. (*Eigenvalue* is from the German *eigenwerte*, meaning proper value(s).)

Equation 2.72 is equivalent to

$$(\mathbf{A} - \lambda\mathbf{I})\mathbf{P} = 0 \tag{2.73}$$

which has nontrivial solutions if $\mathbf{P} \neq 0$, so that

$$|\mathbf{A} - \lambda\mathbf{I}| = 0 \tag{2.74}$$

This is what mathematicians call the *characteristic equation*, and its solutions are eigenvalues of $\mathbf{A}$. Then the solutions of $(\mathbf{A} - \lambda_i\mathbf{I})\mathbf{P} = 0$ are the eigenvectors corresponding to λ_i.

Here is an example for a 2×2 matrix. We will let

$$\mathbf{A} = \begin{bmatrix} 5 & -1 \\ 3 & 1 \end{bmatrix} \tag{2.75}$$

Setting $|\mathbf{A} - \lambda\mathbf{I}| = 0$, we write

$$\begin{vmatrix} 5 - \lambda & -1 \\ 3 & 1 - \lambda \end{vmatrix} = 0 \tag{2.76}$$

When we solve this determinant, we obtain the characteristic equation

$$\lambda^2 - 6\lambda + 8 = 0 \tag{2.77}$$

whose solution yields the eigenvalues $\lambda_1 = 2$ and $\lambda_2 = 4$. From Equation 2.72 we have

$$\begin{bmatrix} 5 & -1 \\ 3 & 1 \end{bmatrix} \begin{bmatrix} x \\ y \end{bmatrix} = \lambda \begin{bmatrix} x \\ y \end{bmatrix} \tag{2.78}$$

and for the eigenvalue $\lambda_1 = 2$ we have

$$\text{a.} \quad 5x - y = 2x \qquad \text{b.} \quad 3x + y = 2y \tag{2.79}$$

or

$$\text{a.} \quad 3x = y \qquad \text{b.} \quad 3x = y \tag{2.80}$$

This tells us that a family of vectors is associated with the eigenvalue $\lambda_1 = 2$ that has the form

$$\begin{bmatrix} k \\ 3k \end{bmatrix} \tag{2.81}$$

where $k \neq 0$. In other words, any vector whose y component is exactly three times as great as its x component is an eigenvector of **A**. For the other eigenvalue, $\lambda_2 = 4$, we have

$$\text{a.} \quad 5x - y = 4x \qquad \text{b.} \quad 3x + y = 4y \tag{2.82}$$

or

$$\text{a.} \quad x = y \qquad \text{b.} \quad x = y \tag{2.83}$$

So that λ_2 has eigenvectors of the form

$$\begin{bmatrix} k \\ k \end{bmatrix} \tag{2.84}$$

In this example, any vector whose x and y components are equal is an eigenvector of **A**. For the general case of a 2×2 matrix, if

$$\mathbf{A} = \begin{bmatrix} a & b \\ c & d \end{bmatrix} \tag{2.85}$$

then we compute the eigenvalues from

$$\begin{vmatrix} a - \lambda & b \\ c & d - \lambda \end{vmatrix} = 0 \tag{2.86}$$

or

$$\lambda^2 - (a + d)\lambda + (ad - bc) = 0 \tag{2.87}$$

There are, of course, two roots to Equation 2.87, two eigenvalues. These may be real and distinct, equal or complex. Using the eigenvalues we can compute values of the corresponding eigenvectors. By doing this we find two possible forms for the ratio of the vector components, depending on the value of λ; thus,

$$\frac{\mathbf{P}_x}{\mathbf{P}_y} = -\frac{b}{a - \lambda_i} \quad \text{or} \quad -\frac{d - \lambda_i}{c} \tag{2.88}$$

where $i = 1, 2$. We can generalize this process to apply to $n \times n$ matrices, but this is beyond the scope of this book.

2.10 Similarity Transformation

If a matrix **A** is premultiplied and postmultiplied by another matrix and its inverse, respectively, then the resulting matrix **B** is a *similarity transformation* of **A**; thus,

$$\mathbf{B} = \mathbf{TAT}^{-1} \tag{2.89}$$

and **A** and **B** are *similar* matrices. Similar matrices have equal determinants, the same characteristic equation, and the same eigenvalues, although not necessarily the same eigenvectors. Similarity transformations preserve eigenvalues.

If **A** is similar to a diagonal matrix **D** then

$$\mathbf{D} = \begin{bmatrix} \lambda_1 & 0 & \cdots & 0 \\ 0 & \lambda_2 & \cdots & 0 \\ \vdots & \vdots & \ddots & \vdots \\ 0 & 0 & \cdots & \lambda_n \end{bmatrix} \tag{2.90}$$

where $\lambda_1, \lambda_2, \ldots, \lambda_n$ are the eigenvalues of **A**. However, for this to be true, there must be a nonsingular matrix **S** such that $\mathbf{SAS}^{-1} = \mathbf{D}$. Obviously, the elements on the main diagonal of a diagonal matrix **D** are its eigenvalues. Therefore, the eigenvalues of **D** are the eigenvalues of **A**. Finally, if the eigenvalues of **A** are distinct, then it is similar to a diagonal matrix.

2.11 Symmetric Transformations

As we have already seen, by definition a real symmetric matrix satisfies the condition $a_{ij} = a_{ji}$, or $\mathbf{A}^T = \mathbf{A}$. If **A** and **B** are symmetric so that $[\mathbf{AB}]^T = \mathbf{B}^T\mathbf{A}^T = \mathbf{BA}$, then the product of two symmetric matrices is symmetric if and only if the two matrices commute. The eigenvalues of a real symmetric matrix are real. If **A** is a real symmetric matrix, then there is an orthogonal matrix $\mathbf{R}(\mathbf{RR}^T) = \mathbf{IR}$ such that $\mathbf{R}^{-1}\mathbf{AR}$ is a diagonal matrix. Furthermore, if **A** is a real symmetric matrix, then the eigenvectors of **A** associated with the distinct eigenvalues are mutually orthogonal vectors.

2.12 Diagonalization of a Matrix

As we have seen previously, we associate an eigenvector $\mathbf{P}_i$ with each eigenvalue λ_i, which we write as

$$\mathbf{AP}_i = \lambda_i \mathbf{P}_i \tag{2.91}$$

Now we can readily construct a square matrix **E** of order n whose columns are the eigenvectors $\mathbf{P}_i$ of **A**, and we rewrite Equation 2.91:

$$\mathbf{AE} = \mathbf{E}\Lambda \tag{2.92}$$

where Λ is a diagonal matrix whose elements are the eigenvalues of **A**:

$$\Lambda = \begin{bmatrix} \lambda_1 & 0 & \cdots & 0 \\ 0 & \lambda_2 & \cdots & 0 \\ \vdots & \vdots & \ddots & \vdots \\ 0 & 0 & \cdots & \lambda_n \end{bmatrix} \tag{2.93}$$

If the eigenvalues in Equation 2.93 are distinct, then the matrix $\mathbf{E}$ is nonsingular, and we can premultiply both sides of the equation by $\mathbf{E}^{-1}$ to yield

$$\mathbf{E}^{-1}\mathbf{A}\mathbf{E} = \Lambda \tag{2.94}$$

We see that by using the matrix of eigenvectors and its inverse we can transform any matrix $\mathbf{A}$ with distinct eigenvalues into a diagonal matrix whose elements are the eigenvalues of $\mathbf{A}$. Mathematicians call this process the *diagonalization* of the matrix $\mathbf{A}$.

Exercises

2.1 Given $\mathbf{A} = \begin{bmatrix} 7 & 3 & -1 \\ 2 & -5 & 6 \end{bmatrix}$ and $\mathbf{B} = \begin{bmatrix} 1 & 5 & 6 \\ -4 & -2 & 3 \end{bmatrix}$, find

   a. $\mathbf{A} + \mathbf{B}$
   b. $\mathbf{A} - \mathbf{B}$

2.2 Given $\mathbf{A} = \begin{bmatrix} 7 & 4 & 4 \\ 9 & 1 & 3 \\ 0 & 2 & 5 \end{bmatrix}$ and $\mathbf{B} = \begin{bmatrix} 6 & 5 \\ 8 & 1 \\ 3 & 9 \end{bmatrix}$, find

   a. a_{23}
   b. a_{12}
   c. a_{31}
   d. b_{11}
   e. b_{32}
   f. What is the order of $\mathbf{A}$?
   g. What is the order of $\mathbf{B}$?
   h. Which matrix, if any, is a square matrix?
   i. List the elements, in order, on the main diagonal of $\mathbf{A}$.
   j. Change $a_{12}, a_{13},$ and a_{23} so that $\mathbf{A}$ is a symmetric matrix.

2.3 Find the values of the following δ_{ij}:
   a. $\delta_{3,2}$ d. $\delta_{7,10}$
   b. $\delta_{1,4}$ e. $\delta_{1,1}$
   c. $\delta_{3,3}$

2.4 Given $\mathbf{A} = \begin{bmatrix} 5 & 4 & 9 \\ 2 & 1 & 0 \\ 6 & 7 & 1 \end{bmatrix}$,

   a. Change $a_{12}, a_{13},$ and a_{23} so that $\mathbf{A}$ becomes antisymmetric.
   b. What other changes, if any, are necessary?

2.5 Write out the 2×2 null matrix.

2.6 Given $A = \begin{bmatrix} 1 & 0 & 0 \\ 0 & 1 & 0 \\ 0 & 0 & 1 \end{bmatrix}$, $B = \begin{bmatrix} 7 \\ 4 \\ 9 \\ 5 \end{bmatrix}$, and $C = \begin{bmatrix} 1 & -2 & 4 & 6 \end{bmatrix}$, find

a. a_{23} f. What is the order of B?

b. a_{32} g. What is the order of C?

c. b_{31} h. Which is the column matrix?

d. c_{14} i. Which is the row matrix?

e. What is the order of A?   j. Which is the identity matrix?

2.7 Given $A = \begin{bmatrix} 3 & 7 & -2 \end{bmatrix}$, find $-A$.

2.8 Given $A = \begin{bmatrix} 1 & 5 & 2 \\ 0 & -1 & 4 \end{bmatrix}$, $B = \begin{bmatrix} 6 & 1 & 3 \\ 0 & 9 & 2 \end{bmatrix}$ and $C = \begin{bmatrix} 4 & 1 & 1 \\ 5 & 8 & 3 \end{bmatrix}$, find

a. $A + 2A$ d. $A - B + C$

b. $B + B$ e. $A - 2B - C$

c. $2A + B$

2.9 Given $A = \begin{bmatrix} 1 & -4 \\ 3 & 0 \end{bmatrix}$, find $1.5A$.

2.10 Find I^T.

2.11 Given $A = \begin{bmatrix} 5 & 3 & 8 \\ -1 & 4 & 7 \\ 0 & 1 & 1 \end{bmatrix}$ and $B = \begin{bmatrix} 6 & 7 \\ 10 & 9 \\ 2 & -3 \end{bmatrix}$ find AB.

2.12 Given $A = \begin{bmatrix} 2 & 1 \\ 3 & 4 \end{bmatrix}$ and $B = \begin{bmatrix} 6 \\ 3 \end{bmatrix}$, find AB.

2.13 Given $A = \begin{bmatrix} 4 & 0 & 7 \\ 5 & 1 & 2 \end{bmatrix}$, find A^T.

2.14 Given $A = \begin{bmatrix} 4 & -2 \\ 1 & 0 \\ 6 & 7 \end{bmatrix}$ and $B = \begin{bmatrix} 1 & 1 \\ 5 & 2 \\ 2 & 4 \end{bmatrix}$, find

a. $(A^T)^T$ c. $A^T + B^T$

b. $(A + B)^T$ d. $B^T + A^T$

2.15 Find the product $\begin{bmatrix} t^2 & t & 1 \end{bmatrix} \begin{bmatrix} a_x & a_y \\ b_x & b_y \\ c_x & c_y \end{bmatrix}$.

2.16 Are the following matrices orthogonal, proper, or improper?

a. $\begin{bmatrix} \dfrac{\sqrt{2}}{2} & \dfrac{\sqrt{2}}{2} \\ -\dfrac{\sqrt{2}}{2} & \dfrac{\sqrt{2}}{2} \end{bmatrix}$

b. $\begin{bmatrix} 1 & \dfrac{1}{2} \\ 2 & 0 \end{bmatrix}$

c. $\begin{bmatrix} 3 & 1 \\ 5 & 2 \end{bmatrix}$

d. $\begin{bmatrix} -\dfrac{3}{2} & \dfrac{1}{2} \\ \dfrac{1}{2} & \dfrac{3}{2} \end{bmatrix}$

e. $\begin{bmatrix} 0 & 0 & 1 \\ 1 & 0 & 0 \\ 0 & -1 & 0 \end{bmatrix}$

f. $\begin{bmatrix} \dfrac{1}{2} & -\dfrac{\sqrt{3}}{2} & 0 \\ \dfrac{\sqrt{3}}{2} & \dfrac{1}{2} & 0 \\ 0 & 0 & 2 \end{bmatrix}$

g. $\begin{bmatrix} \dfrac{\sqrt{3}}{2} & \dfrac{\sqrt{3}}{4} & 1 \\ \dfrac{1}{2} & -\dfrac{3}{4} & -\dfrac{\sqrt{3}}{4} \\ 0 & \dfrac{1}{2} & -\dfrac{\sqrt{3}}{2} \end{bmatrix}$

2.17 Find $\begin{bmatrix} 0 & 0 & 1 \\ 1 & 0 & 0 \\ 0 & -1 & 0 \end{bmatrix}^{-1}$

2.18 If $\mathbf{P} = \mathbf{ABC}$ and the order of $\mathbf{A}$ is 1×4, the order of $\mathbf{B}$ is 4×4, and the order of $\mathbf{C}$ is 4×3, then what is the order of $\mathbf{P}$?

2.19 Show that the inverse of the orthogonal matrix $\mathbf{A}$ is an orthogonal matrix, where

$$A = \begin{bmatrix} \dfrac{2}{3} & -\dfrac{2}{3} & \dfrac{1}{3} \\ \dfrac{1}{3} & \dfrac{2}{3} & \dfrac{2}{3} \\ \dfrac{2}{3} & \dfrac{1}{3} & -\dfrac{2}{3} \end{bmatrix}$$

2.20 Compute the following determinants:

a. $\begin{vmatrix} 2 & 0 \\ -3 & 2 \end{vmatrix}$

d. $\begin{vmatrix} 1 & 2 \\ 2 & 4 \end{vmatrix}$

b. $\begin{vmatrix} 1 & 2 \\ 4 & -5 \end{vmatrix}$

e. $\begin{vmatrix} 2 & 5 \\ -3 & 1 \end{vmatrix}$

c. $\begin{vmatrix} 0 & 3 \\ 0 & 0 \end{vmatrix}$

2.21 Given $|\mathbf{A}| = \begin{vmatrix} 4 & 0 & -1 \\ 1 & 2 & 1 \\ -3 & 6 & 5 \end{vmatrix}$, compute the following minors and cofactors:

a. m_{11} f. c_{11}

b. m_{21} g. c_{21}

c. m_{31} h. c_{31}

d. m_{22} i. c_{22}

e. m_{12} j. c_{12}

2.22 Compute $|\mathbf{A}|$ in Exercise 2.21.

2.23 Compute the inverse of the following matrices, if one exists:

a. $\begin{bmatrix} 1 & 0 \\ 0 & 1 \end{bmatrix}$

c. $\begin{bmatrix} 1 & 0 & 0 \\ 2 & 1 & 3 \\ 1 & 1 & 2 \end{bmatrix}$

b. $\begin{bmatrix} 3 & -1 & 2 \\ 1 & 2 & 1 \\ -2 & 1 & 3 \end{bmatrix}$

d. $\begin{bmatrix} 3 & -1 & 2 \\ 1 & 2 & 1 \\ 3 & -1 & 2 \end{bmatrix}$

2.24 Find the eigenvalues and eigenvectors of the following matrices:

a. $\begin{bmatrix} 3 & 5 \\ 4 & 5 \end{bmatrix}$ b. $\begin{bmatrix} 1 & 2 \\ -2 & 5 \end{bmatrix}$

2.25 Find the characteristic equation, eigenvalues, and corresponding eigenvectors for the following matrices:

a. $\begin{bmatrix} 1 & 2 \\ 4 & 3 \end{bmatrix}$ b. $\begin{bmatrix} 2 & 0 & 0 \\ 0 & 1 & 0 \\ 0 & 0 & 3 \end{bmatrix}$

CHAPTER 3

TRANSFORMATIONS

The geometers of antiquity took two different paths toward understanding the shapes of objects and the figures they drew to represent them. One path led them to study properties that they could measure, such as angles, distance, area, and volume. The other led them to study logical relationships among shapes, such as equivalence, similarity, and constructability. Both paths led them to the conclusion that certain geometric properties of an object do not change when the object is translated or rotated, expanded or twisted. Geometers discovered that different kinds of changes, *transformations*, left different properties unchanged, or *invariant*, and this led them to identify different kinds of geometry.

This chapter explores geometric transformations and their invariant properties. It introduces systems of equations that produce linear transformations and develops the algebra and geometry of translations, rotations, reflection, and scaling, including how to combine them. Vectors and matrices are put to good use here, in ways that clearly demonstrate their effectiveness.

3.1 The Geometries of Transformations

Here we will discover what geometric properties are preserved when we impose transformations on figures. Each transformation changes some properties and leaves others unchanged; for example, angles may be preserved but not distance.

Congruent geometry applies only to figures of identical size and shape. We can translate and rotate a figure without changing its congruent properties. *Conformal geometry* applies to *similar* figures, those whose size is different but whose corresponding angles are equal. Euclidean geometry studies the properties of both congruent and similar figures.

Affine geometry allows transformations where distances only are preserved between points on the same or parallel lines, and angles are not preserved at all. Parallelism is an important invariant of this geometry.

Projective geometry allows transformations where parallel lines are not preserved, but straight lines remain straight. A projective transformation transforms circles, ellipses, hyperbolas, and parabolas into one another.

47

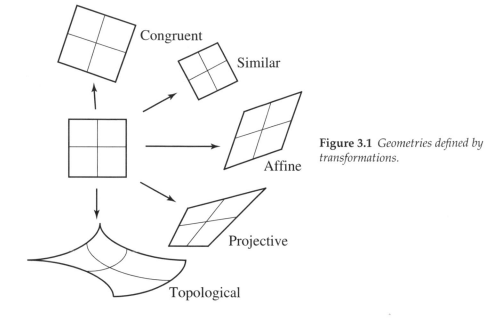

Figure 3.1 *Geometries defined by transformations.*

Topology is the least restrictive of all the geometries. It permits transformations that stretch, bend, and variously distort a figure, so long as no cutting or gluing takes place. Fewer geometric properties are preserved under topological transformations, but connectivity and order are preserved.

Figure 3.1 shows examples of the transformations that lead to these geometries. Each example begins with a square, which is then transformed in a way that is appropriate to the particular geometry.

Invariant Geometric Properties

A point is a *fixed point* if a transformation maps it onto itself: its position is unchanged. For example, a rotation in the plane fixes the point at the center of the rotation.

We let T denote some transformation (rotation, translation, scaling, and so on). If L denotes a line, then $T(L)$ denotes the result, or product, of the transformation operating on the line. If $T(L) = L'$, another line, then T is a *collineation*, which means that T preserves lines (Figure 3.2). For any point P on L, $T(P)$ is on $T(L)$.

Figure 3.2 *Collineation transformation.*

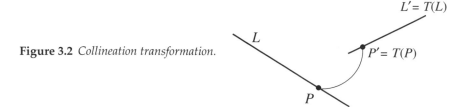

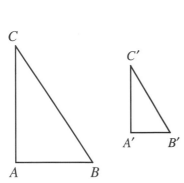

Figure 3.3 *Conformal transformation.*

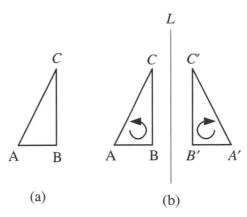

(a)   (b)

Figure 3.4 *An orientation-reversing transformation.*

Distance-preserving transformations also preserve angles, area, and volume. It is the most restrictive Euclidean geometry, preserving both size and shape. Translation and rotation are distance-preserving transformations. A square remains a square under these transformations, and retains all of its initial geometric properties related to size and shape.

Conformal transformations preserve the angle measures of a figure but not distances (size). Trigonometry and Euclidean geometry of similar figures are examples of conformal transformations in operation, because they are concerned with properties of similar, not necessarily congruent, figures (Figure 3.3).

Given triangle *ABC* in Figure 3.4a, we can sequentially visit its vertices in two different ways: clockwise or counterclockwise. We say that the triangle has two possible *orientations*. Any transformation that does not change the orientation of the figure is an *orientation-preserving transformation*, or a *direct transformation*. A transformation that reverses orientation is an *opposite transformation*. Reflection is an *orientation-reversing transformation* (Figure 3.4b).

Isometries

An *isometry* is a rigid-body motion of a figure. It preserves the distances between all points of a figure. (The word *isometry* is from the Greek words *isos* and *metron*, meaning equal measure.) The algebraic equations describing general isometries in the plane, not including translations, are

$$x' = ax + by$$
$$y' = cx + dy$$
(3.1)

with the restriction that

$$\begin{vmatrix} a & b \\ c & d \end{vmatrix} = \pm 1$$
(3.2)

and where x and y are the initial coordinates of a point and x' and y' are the new, transformed, coordinates. If the determinant is equal to minus one, then the transformation is orientation reversing. A similar set of equations and restrictions describes isometries in three dimensions.

Four kinds of isometries in the plane are possible:

1. Reflection
2. Rotation
3. Translation
4. Glide reflection

Similarities

A *similarity transformation* changes the size of a figure but does not change its shape. It is a *conformal transformation*. This is the first step in generalizing the isometries and working toward less restrictive transformations and the resulting geometries. A similarity transformation can be a combination of a scaling transformation (such as dilation) and an isometry. This means that figures related by a similarity transformation can occupy different spatial locations and have different rotational orientations (Figure 3.5).

The equations for a similarity transformation in the plane are

$$x' = ax - by$$
$$y' = \pm(bx + ay) \tag{3.3}$$

where

$$\sqrt{a^2 + b^2} = k \tag{3.4}$$

and

$$k \neq 0 \begin{cases} k > 0 & \text{for a direct similarity} \\ k < 0 & \text{for an opposite similarity} \end{cases} \tag{3.5}$$

The term k is the ratio of expansion or contraction.

Figure 3.5 *Similarity transformation.*

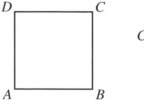

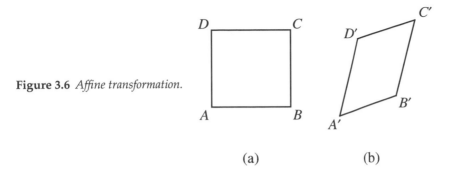

Figure 3.6 *Affine transformation.*

(a)   (b)

Affinities

An *affine* property of a geometric figure is one that is preserved by all affine transformations. Collinearity and parallelism are important invariant properties of affine transformations. But in affine transformations we must give up both distance and angle invariance. For example, an affinity transforms the square in Figure 3.6a into the trapezoid shape of Figure 3.6b, where $A'B' \neq AB$ and $\angle A' \neq \angle A$, but $A'D' \| B'C'$, and all points on line AD map onto line $A'D'$.

Algebraically, this means that we must relax most of the restrictions on the coefficients of the transformation equations (Equations 3.2–3.5), so that

$$x' = ax + by$$
$$y' = cx + dy$$

(3.6)

where the only restriction is that

$$\begin{vmatrix} a & b \\ c & d \end{vmatrix} \neq 0$$

(3.7)

Table 3.1 summarizes the possible affine transformations in the plane.

Transformations in Three Dimensions

The transformations and their algebraic representations we have just seen apply to geometric figures in the plane. We can readily extend them to points, lines, and arbitrary shapes in three-dimensional space. (We will generalize the notation scheme somewhat to make it more convenient to work with later.) So a transformed point in three dimensions is given by the following equations (again not including translations):

$$x' = a_{11}x + a_{12}y + a_{13}z$$
$$y' = a_{21}x + a_{22}y + a_{23}z$$
$$z' = a_{31}x + a_{32}y + a_{33}z$$

(3.8)

Table 3.1 Affine transformations

Transformation	Restriction on Coefficients
General (all)	$ad - bc \neq 0$
Identity	$a = 1, b = c = 0, d = 1$
Isometry	$\begin{cases} a^2 + c^2 = 1 \\ b^2 + d^2 = 1 \\ ab + cd = 0 \end{cases}$
Similarity	$\begin{cases} a^2 + c^2 = b^2 + d^2 \\ ab + cd = 0 \end{cases}$
Equiareal	$\lvert ad - bc \rvert = 1$
Shear: x direction	$a = 1, b = k_y, c = 0, d = 1$
y direction	$a = 1, b = 0, c = k_x, d = 1$
Strain: x direction	$a = k_x, b = c = 0, d = 1$
y direction	$a = 1, b = c = 0, d = k_y$
Dilation (uniform)	$a = k, b = c = 0, d = k$
Inversion	$a = -1, b = c = 0, d = -1$

For isometries the coefficients are restricted by

$$\lvert a_{ij} \rvert = \pm 1 \tag{3.9}$$

and for general affine transformations the only restriction is

$$\lvert a_{ij} \rvert \neq 0 \tag{3.10}$$

3.2 Linear Transformations

Equations like those in Equations 3.8 are homogeneous linear equations describing the transformation of a point x, y, z *onto* the point x', y', z'. (There is a technical distinction between the terms *into* and *onto*. *Onto* implies a one-to-one correspondence between all the points in the initial system and all the points in the transformed system.) If we account for translations (a rigid-body transformation), then Equations 3.8 become

$$
\begin{aligned}
x' &= a_{11}x + a_{12}y + a_{13}z + a_{14} \\
y' &= a_{21}x + a_{22}y + a_{23}z + a_{24} \\
z' &= a_{31}x + a_{32}y + a_{33}z + a_{34}
\end{aligned}
\tag{3.11}
$$

These equations are not homogeneous. For now, we will focus on the homogeneous equations.

We can assert that for every linear transformation from point x, y, z to point x', y', z' there is a corresponding unique linear transformation that brings point x', y', z' back to x, y, z. This means that we must be able to solve Equations 3.8 for x, y, z in terms of x', y', z'; that is,

$$x = a'_{11}x' + a'_{12}y' + a'_{13}z'$$
$$y = a'_{21}x' + a'_{22}y' + a'_{23}z' \tag{3.12}$$
$$z = a'_{31}x' + a'_{32}y' + a'_{33}z'$$

For this set of equations to have a unique solution, the determinant of the coefficients must not equal zero:

$$\begin{vmatrix} a'_{11} & a'_{12} & a'_{13} \\ a'_{21} & a'_{22} & a'_{23} \\ a'_{31} & a'_{32} & a'_{33} \end{vmatrix} \neq 0 \tag{3.13}$$

In the plane, Equations 3.12 become

$$x = a'_{11}x' + a'_{12}y'$$
$$y = a'_{21}x' + a'_{22}y' \tag{3.14}$$

Affine Transformation of a Line

The equation of a line in the plane is

$$Ax + By + C = 0 \tag{3.15}$$

Let's see how an affine transformation affects the form of this equation, and therefore the line itself. Substituting from Equations 3.14 we find that

$$A\left(a'_{11}x' + a'_{12}y'\right) + B\left(a'_{21}x' + a'_{22}y'\right) + C = 0 \tag{3.16}$$

or

$$\left(Aa'_{11} + Ba'_{21}\right)x' + \left(Aa'_{12} + Ba'_{22}\right)y' + C = 0 \tag{3.17}$$

To simplify things somewhat, we make the following substitutions:

$$A' = \left(Aa'_{11} + Ba'_{21}\right)$$
$$B' = \left(Aa'_{12} + Ba'_{22}\right) \tag{3.18}$$
$$C' = C$$

Then Equation 3.17 becomes

$$A'x' + B'y' + C' = 0 \tag{3.19}$$

This has the same form as Equation 3.15, and we conclude that straight lines transform into straight lines under affine transformations.

Affine Transformation of a Conic

Now let's apply a general affine transformation to a conic curve. A conic is represented by the second-degree equation

$$Ax^2 + 2Bxy + Cy^2 + 2Dx + 2Ey + F = 0 \tag{3.20}$$

Substituting from Equation 3.14 produces

$$A\left(a'_{11}x' + a'_{12}y'\right)^2 + 2B\left(a'_{11}x' + a'_{12}y'\right)\left(a'_{21}x' + a'_{22}y'\right)$$
$$+C\left(a'_{21}x' + a'_{22}y'\right)^2 + 2D\left(a'_{11}x' + a'_{12}y'\right) + 2E\left(a'_{21}x' + a'_{22}y'\right) + F = 0 \tag{3.21}$$

Expanding and then rearranging this awkward equation yields

$$A'x'^2 + 2B'x'y' + C'y'^2 + 2D'x' + 2E'y' + F' = 0 \tag{3.22}$$

where $A', B', \ldots, F'$ are combinations of $A, B, \ldots, F$ and the coefficients a'_{ij}. The form of Equation 3.22 is identical to Equation 3.20. This means that an affine transformation transforms one second-degree curve into another second-degree curve.

We can now make the following generalization: An affine transformation changes any curve of degree n into another curve of the same degree.

Matrix Form

Matrices are short-cuts to writing and manipulating transformation equations (Chapter 2). In expanded matrix form Equations 3.8 become

$$\begin{bmatrix} x' \\ y' \\ z' \end{bmatrix} = \begin{bmatrix} a_{11} & a_{12} & a_{13} \\ a_{21} & a_{22} & a_{23} \\ a_{31} & a_{32} & a_{33} \end{bmatrix} \begin{bmatrix} x \\ y \\ z \end{bmatrix} \tag{3.23}$$

which we can quickly simplify to

$$P' = AP \tag{3.24}$$

To find x, y, z in terms of x', y', z', we solve the matrix equation for P:

$$P = A^{-1}P' \tag{3.25}$$

where A^{-1} is the matrix inverse of A.

3.3 Translation

Rigid-body motions are produced by the isometries of *translation* and *rotation*. Algebraically, they are not really true motions in the sense that there is a definite path or trajectory connecting some initial position to some new transformed position. There are no intermediate positions. But we can think of these two isometries as motions within the plane or in space. A translation, then, is simply the movement of a set of points or a figure such that every point of it moves in the same direction and the same distance.

The translation of a point in three dimensions is given by

$$x' = x + a$$
$$y' = y + b \qquad\qquad (3.26)$$
$$z' = z + c$$

In the plane these equations simply become

$$x' = x + a$$
$$y' = y + b \qquad\qquad (3.27)$$

Admittedly, these are pretty dull equations, almost too simple to bother with. But consider this: They seriously complicate Equations 3.8 for reasons explained below, because if we want them to include translations we must rewrite them as (see Equation 3.11)

$$x' = a_{11}x + a_{12}y + a_{13}z + a_{14}$$
$$y' = a_{21}x + a_{22}y + a_{23}z + a_{24} \qquad\qquad (3.28)$$
$$z' = a_{31}x + a_{32}y + a_{33}z + a_{34}$$

These are not homogenous equations: The matrices that represent them are not square and have no *multiplicative inverse*. (This just means that in this form we have no way to calculate an inverse relationship between x, y, z and x', y', z'.) It also complicates the computation of combined transformations, that is, when we must combine translations and rotations to find their equivalent net effect. We will resolve both of these problems in the last section.

Two Points

Given two points P_1 and P_2 we can find a unique translation that takes P_1 into P_1 (Figure 3.7). We have $P_1 = x_1, y_1$ and $P_2 = x_2, y_2$. There are unique numbers a and b

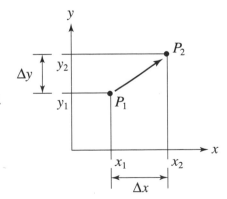

Figure 3.7 *Translation given by two points.*

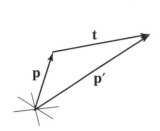

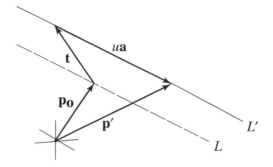

Figure 3.8 *Vector description of a translation.*

Figure 3.9 *Vectors and translation of a line.*

such that $x_2 = x_1 + a$ and $y_2 = y_1 + b$. A little simple algebra followed by substitution into Equations 3.27 produces

$$x' = x + (x_2 - x_1)$$
$$y' = y + (y_2 - y_1)$$
(3.29)

We can also write this as

$$x' = x + \Delta x$$
$$y' = y + \Delta y$$
(3.30)

where $\Delta x = x_2 - x_1$ and $\Delta y = y_2 - y_1$.

Vectors and Translations

Vectors produce a very concise description of a translation (Chapter 1). If we define a point by the vector $\mathbf{p}$, and an arbitrary translation by another vector $\mathbf{t}$, then we can express the translation of $\mathbf{p}$ by $\mathbf{t}$ (Figure 3.8) as

$$\mathbf{p}' = \mathbf{p} + \mathbf{t}$$
(3.31)

We can use vectors to translate the straight line L defined by $\mathbf{p} = \mathbf{p}_0 + u\mathbf{a}$. Because each point on the line must be translated the same direction and distance, we have (Figure 3.9)

$$\mathbf{p}' = \mathbf{p}_0 + u\mathbf{a} + \mathbf{t}$$
(3.32)

Note that the translation does not change the direction of the line, so L and L' are parallel.

A succession of translations $\mathbf{t}_1, \mathbf{t}_2, \ldots, \mathbf{t}_n$ is simply equivalent to their vector sum: $\mathbf{t} = \mathbf{t}_1 + \mathbf{t}_2 + \cdots + \mathbf{t}_n$.

3.4 Rotations in the Plane

A rotation in the plane about the origin is the simplest kind of rotation. It is so simple that it obscures many characteristics of rotations in three or more dimensions. We

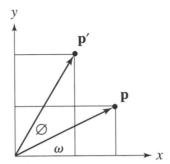

Figure 3.10 *Rotation in the plane.*

will use a right-hand coordinate system and the convention that positive rotations are counterclockwise. A rotation is an isometry, so the rotation transformations will be in the form of Equations 3.6. The coordinates of the transformed point $\mathbf{p}'$ must be functions of the coordinates of $\mathbf{p}$ and some rotation angle ϕ.

The derivation of these functions is easy, and Figure 3.10 shows some of the details. In the figure we see that

$$x' = |\mathbf{p}'| \cos(\omega + \phi)$$
$$y' = |\mathbf{p}'| \sin(\omega + \phi)$$

(3.33)

and

$$\cos \omega = \frac{x}{|\mathbf{p}|} \qquad \sin \omega = \frac{y}{|\mathbf{p}|}$$

(3.34)

Because an isometry preserves distance between points (in this case the origin and point $\mathbf{p}$), we see that $|\mathbf{p}'| = |\mathbf{p}|$. These two points are on the circumference of a circle whose center is at the origin. From elementary trigonometry identities, we know that

$$\cos(\omega + \phi) = \cos \omega \cos \phi - \sin \omega \sin \phi$$
$$\sin(\omega + \phi) = \sin \omega \cos \phi + \cos \omega \sin \phi$$

(3.35)

Substituting appropriately into Equations 3.31 from Equations 3.34 and 3.35 produces

$$x' = x \cos \phi - y \sin \phi$$
$$y' = x \sin \phi + y \cos \phi$$

(3.36)

This is the set of rotation transformation equations we are after.

Writing Equations 3.36 in matrix form produces

$$\begin{bmatrix} x' \\ y' \end{bmatrix} = \begin{bmatrix} \cos \phi & -\sin \phi \\ \sin \phi & \cos \phi \end{bmatrix} \begin{bmatrix} x \\ y \end{bmatrix}$$

(3.37)

or

$$\mathbf{P}' = \mathbf{R}_\phi \mathbf{P}$$

(3.38)

where R is the *rotation matrix* and

$$\mathbf{R}_\phi = \begin{bmatrix} \cos\phi & -\sin\phi \\ \sin\phi & \cos\phi \end{bmatrix} \tag{3.39}$$

Note that $|\mathbf{R}_\phi| = 1$, conforming to the restriction on the coefficients for an isometry transformation.

To reverse it, we apply the rotation $-\phi$ and obtain

$$\mathbf{R}_{-\phi} = \begin{bmatrix} \cos\phi & \sin\phi \\ -\sin\phi & \cos\phi \end{bmatrix} \tag{3.40}$$

Using matrix algebra, we find that

$$\mathbf{R}_{-\phi}\mathbf{R}_\phi = \begin{bmatrix} 1 & 0 \\ 0 & 1 \end{bmatrix} \tag{3.41}$$

or

$$\mathbf{R}_{-\phi}\mathbf{R}_\phi = \mathbf{I} \tag{3.42}$$

which means that

$$\mathbf{R}_{-\phi} = \mathbf{R}_\phi^{-1} \tag{3.43}$$

Successive Rotations in the Plane

We can compute the product of two successive rotations in the plane and about the origin as follows: First, we rotate $\mathbf{p}$ by ϕ_1 to obtain $\mathbf{p}'$:

$$\mathbf{p}' = \mathbf{R}_{\phi_1}\mathbf{p} \tag{3.44}$$

Second, we rotate $\mathbf{p}'$ by ϕ_2 to obtain $\mathbf{p}''$, where

$$\mathbf{p}'' = \mathbf{R}_{\phi_2}\mathbf{p}' \tag{3.45}$$

or

$$\mathbf{p}'' = \mathbf{R}_{\phi_2}\mathbf{R}_{\phi_1}\mathbf{p} \tag{3.46}$$

The product of the two rotation matrices yields

$$\mathbf{R}_{\phi_2}\mathbf{R}_{\phi_1} = \begin{bmatrix} \cos\phi_1\cos\phi_2 - \sin\phi_1\sin\phi_2 & -\sin\phi_1\cos\phi_2 - \cos\phi_1\sin\phi_2 \\ \cos\phi_1\sin\phi_2 + \sin\phi_1\cos\phi_2 & -\sin\phi_1\sin\phi_2 + \cos\phi_1\cos\phi_2 \end{bmatrix} \tag{3.47}$$

Again, we use common trigonometric identities to obtain

$$\mathbf{R}_{\phi_2}\mathbf{R}_{\phi_1} = \begin{bmatrix} \cos(\phi_1 + \phi_2) & -\sin(\phi_1 + \phi_2) \\ \sin(\phi_1 + \phi_2) & \cos(\phi_1 + \phi_2) \end{bmatrix} \tag{3.48}$$

This tells us that for two successive rotations in the plane about the origin, we merely add the two rotation angles ϕ_1 and ϕ_2 and use the resulting sum in the rotation matrix. Of course, this extends to the product of n successive rotations, so that

$$\mathbf{R} = \begin{bmatrix} \cos(\phi_1 + \phi_2 + \cdots + \phi_n) & -\sin(\phi_1 + \phi_2 + \cdots + \phi_n) \\ \sin(\phi_1 + \phi_2 + \cdots + \phi_n) & \cos(\phi_1 + \phi_2 + \cdots + \phi_n) \end{bmatrix} \qquad (3.49)$$

Rotation about an Arbitrary Point

To execute a rotation of a point or figure about some point other than the origin, we proceed in three steps (Figure 3.11). We let $\mathbf{p}_c$ and α define the center and angle of rotation, respectively.

1. Translate $\mathbf{p}_c$ to the origin. Translate all points on the figure in the same direction and distance: $\mathbf{p} - \mathbf{p}_c$.
2. Rotate the results about the origin: $\mathbf{R}_\alpha(\mathbf{p} - \mathbf{p}_c)$.
3. Reverse step 1, returning the center of rotation to $\mathbf{p}_c$ and thus producing.

$$\mathbf{p}' = \mathbf{R}_\alpha(\mathbf{p} - \mathbf{p}_c) + \mathbf{p}_c \qquad (3.50)$$

Rotation of the Coordinate System

Rotating the coordinate system about the origin and through an angle ϕ_c is equivalent to an oppositely directed rotation of points in the original system (Figure 3.12).

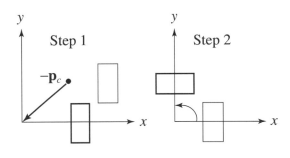

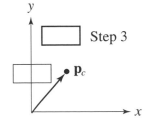

Figure 3.11 *Rotation about an arbitrary point.*

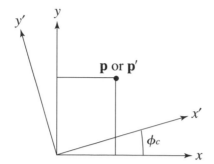

Figure 3.12 *Rotation of the x, y coordinate system.*

This means that $\phi = -\phi_c$, and the rotation matrix of Equation 3.39 becomes

$$\mathbf{R}_{-\phi_c} = \begin{bmatrix} \cos \phi_c & \sin \phi_c \\ -\sin \phi_c & \cos \phi_c \end{bmatrix} \tag{3.51}$$

3.5 Rotations in Space

The simplest rotations in space are those about the principal axes. We apply the right-hand rule to define the algebraic sign of a rotation. For example, if we look toward the origin from a point on the z axis, a positive rotation is counterclockwise; similarly for the x and y axes (Figure 3.13).

Rotation about the z axis produces a result similar to a rotation about the origin in the x, y plane. But we must account for the z coordinate. We do this by changing Equations 3.36 as follows:

$$\begin{aligned} x' &= x \cos \phi - y \sin \phi \\ y' &= x \sin \phi + y \cos \phi \\ z' &= z \end{aligned} \tag{3.52}$$

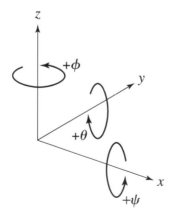

Figure 3.13 *Rotation convention.*

and the rotation matrix becomes

$$
\mathbf{R}_\phi = \begin{bmatrix} \cos\phi & -\sin\phi & 0 \\ \sin\phi & \cos\phi & 0 \\ 0 & 0 & 1 \end{bmatrix} \tag{3.53}
$$

Again, notice that $|\mathbf{R}_\phi| = 1$, as required by the rules of isometry.
For rotation about the x or y axis, the matrices are

$$
\mathbf{R}_\varphi = \begin{bmatrix} 1 & 0 & 0 \\ 0 & \cos\varphi & -\sin\varphi \\ 0 & \sin\varphi & \cos\varphi \end{bmatrix} \tag{3.54}
$$

and

$$
\mathbf{R}_\theta = \begin{bmatrix} \cos\theta & 0 & \sin\theta \\ 0 & 1 & 0 \\ -\sin\theta & 0 & \cos\theta \end{bmatrix} \tag{3.55}
$$

respectively, where, again, we see that $|\mathbf{R}_\varphi| = |\mathbf{R}_\theta| = 1$.

Successive Rotations

We can describe the rotation of a point or a point-defined figure in space as a product of successive rotations about each of the principal axes. First, we must establish a consistent sequence of rotations, because the order in which we execute rotations in space is important. Different orders produce very different results. What follows is the standard approach:

1. Perform the rotation about the z axis, $\mathbf{R}_\phi$.
2. Perform the rotation about the y axis, $\mathbf{R}_\theta$.
3. Perform the rotation about the x axis, $\mathbf{R}_\varphi$.

The matrix equation for this sequence of rotations is

$$
\mathbf{P}' = \mathbf{R}_\varphi \mathbf{R}_\theta \mathbf{R}_\phi \mathbf{P} \tag{3.56}
$$

If we let $\mathbf{R}_{\varphi\theta\phi} = \mathbf{R}_\varphi \mathbf{R}_\theta \mathbf{R}_\phi$, then

$$
\mathbf{P}' = \mathbf{R}_{\varphi\theta\phi} \mathbf{P} \tag{3.57}
$$

Multiplying the rotation matrices as we did in Equations 3.53, 3.54, and 3.55 produces

$$
\mathbf{R}_{\varphi\theta\phi} = \begin{bmatrix} \cos\theta\cos\phi & -\cos\theta\sin\phi & \sin\theta \\ \cos\varphi\sin\phi + \sin\varphi\sin\theta\cos\phi & \cos\varphi\cos\phi - \sin\varphi\sin\theta\sin\phi & -\sin\varphi\cos\theta \\ \sin\varphi\sin\phi - \cos\varphi\sin\theta\cos\phi & \sin\varphi\cos\phi + \cos\varphi\sin\theta\sin\phi & \cos\varphi\cos\theta \end{bmatrix}
$$

$$\tag{3.58}$$

We have already seen that $|\mathbf{R}_\varphi| = |\mathbf{R}_\theta| = |\mathbf{R}_\phi| = 1$, and from matrix algebra we can show that

$$|\mathbf{R}_{\varphi\theta\varphi}| = |\mathbf{R}_\varphi||\mathbf{R}_\theta||\mathbf{R}_\phi| = 1 \tag{3.59}$$

Rotation of the Coordinate System

Some problems in computer graphics or geometric modeling require that we find the coordinates of points in a new coordinate system, one that is rotated with respect to the system in which the points are originally specified, but sharing a common origin. We must account for three different angles of rotation: ϕ_c, θ_c, and φ_c (Figure 3.14). Under a general coordinate-system rotation there is a succession of intermediate axes.

Taking the rotations in order, we achieve the final position of the transformed coordinate axes in three steps:

1. Rotate the x, y, z system of axes through ϕ_c about the z axis. This moves the x axis to x_1 and the y axis to y_1. The z_1 axis is identical to the z axis.
2. Rotate the x_1, y_1, z_1 system of axes through θ_c about the y_1 axis. This moves the x_1 axis to x_2 and the z_1 axis to z_2. The y_2 axis is identical to the y_1 axis.
3. Rotate the x_2, y_2, z_2 system of axes through φ_c about the x_2 axis. This moves the y_2 axis to y_3 and the z_2 axis to z_3. The x_3 axis is identical to the x_2 axis.

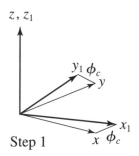

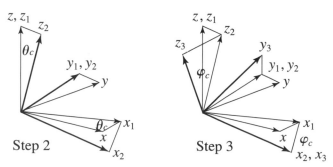

Figure 3.14 *Rotation of the x, y, z coordinate system.*

We use Equation 3.58, with $\phi = -\phi_c$, $\theta = -\theta_c$, and $\varphi = -\varphi_c$, to obtain

$$\mathbf{R}_{\varphi_c \theta_c \phi_c}$$

$$= \begin{bmatrix} \cos\theta_c \cos\phi_c & \cos\theta_c \sin\phi_c & -\sin\theta_c \\ -\cos\varphi_c \sin\phi_c + \sin\varphi_c \sin\theta_c \cos\phi_c & \cos\varphi_c \cos\phi_c + \sin\varphi_c \sin\theta_c \sin\phi_c & \sin\varphi_c \cos\theta_c \\ \sin\varphi_c \sin\phi_c + \cos\varphi_c \sin\theta_c \cos\phi_c & -\sin\varphi_c \cos\phi_c + \cos\varphi_c \sin\theta_c \sin\phi_c & \cos\varphi_c \cos\theta_c \end{bmatrix}$$

$$(3.60)$$

When we compare this to Equation 3.58, we see that some of the terms change sign, but the forms are identical. There is something else going on here, too. We note that $\mathbf{R}_{\varphi_c \theta_c \phi_c}$ is not the transpose or the inverse of $\mathbf{R}_{\varphi\theta\phi}$, that is,

$$\mathbf{R}_{\varphi_c \theta_c \phi_c} \neq \mathbf{R}_{\varphi\theta\phi}^T \tag{3.61}$$

Instead, we find that

$$\begin{aligned} \mathbf{R}_{\varphi_c \theta_c \phi_c} &= \mathbf{R}_\varphi^{-1} \mathbf{R}_\theta^{-1} \mathbf{R}_\phi^{-1} \\ &= \mathbf{R}_\varphi^T \mathbf{R}_\theta^T \mathbf{R}_\phi^T \\ &= [\mathbf{R}_\phi \mathbf{R}_\theta \mathbf{R}_\varphi]^T \end{aligned} \tag{3.62}$$

Rotation about an Arbitrary Axis

The most general rotation transformation of points and figures is through some angle α about some arbitrary axis $\mathbf{a}$ in space that passes through the origin of the coordinate system (Figure 3.15). We do this by a sequence of rotations with the intention of bringing the axis $\mathbf{a}$ into coincidence with the z axis.

1. Rotate about the z axis through an angle ϕ so that the axis $\mathbf{a}$ lies in the y, z plane. Rotate points similarly so that $\mathbf{P}' = \mathbf{R}_\phi \mathbf{P}$.
2. Rotate about the x axis through an angle ϕ so that the axis $\mathbf{a}$ is coincident with the z axis. Again, rotate points similarly, so that now $\mathbf{P}' = \mathbf{R}_\varphi \mathbf{R}_\phi \mathbf{P}$.
3. Rotate about the axis $\mathbf{a}$ (same as a rotation about the now coincident z axis) through α, yielding $\mathbf{P}' = \mathbf{R}_\alpha \mathbf{R}_\varphi \mathbf{R}_\phi \mathbf{P}$.

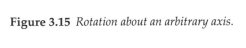

Figure 3.15 *Rotation about an arbitrary axis.*

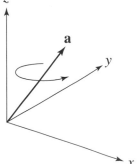

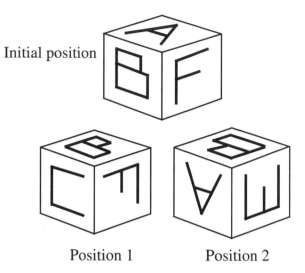

Initial position

Figure 3.16 *Equivalent rotations.*

Position 1   Position 2

4. Reverse the first rotation about the x axis: $\mathbf{P}' = \mathbf{R}_{\varphi}^{-1}\mathbf{R}_{\alpha}\mathbf{R}_{\varphi}\mathbf{R}_{\phi}P$.

5. Reverse the first rotation about the z axis, finally returning the axis $\mathbf{a}$ to its original position and producing the resulting net rotation of any point about some arbitrary axis and angle.

$$\mathbf{P}' = \mathbf{R}_{\phi}^{-1}\mathbf{R}_{\varphi}^{-1}\mathbf{R}_{\alpha}\mathbf{R}_{\varphi}\mathbf{R}_{\phi}\mathbf{P} \tag{3.63}$$

Equivalent Rotations

Euler (in 1752) proved that no matter how many times we twist and turn an object with respect to a reference position, we can always reach every possible new position with a single equivalent rotation. For example, try to visualize rotating the lettered block shown in Figure 3.16 through 90° about an axis perpendicular to the center of face F, placing the block in position 1. Next, rotate it 180° about an axis perpendicular to face B, placing it in position 2. We can put the block into position 2 with a single rotation of 180° about an axis passing through the centers of opposite edges AB and CD.

3.6 Reflection

We will begin with a synthetic definition of a reflection in the plane and then extend this to three dimensions. Given any point A and a line m not through A, a reflection of A in m produces its image point A'. The line segment AA' is perpendicular to m and bisected by it (Figure 3.17). We easily demonstrate the following characteristics of a reflection.

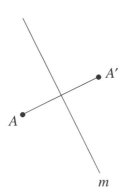

Figure 3.17 *Reflection across a line* m *in a plane.*

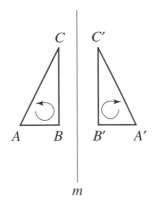

Figure 3.18 *Reflection across a line in a plane reverses orientation.*

1. A reflection in the plane reverses the orientation of a figure (Figure 3.18).
2. If a line is parallel to the axis of reflection, then so is its image. Similarly, if a line is perpendicular to the axis of reflection, then so is its image, and the line and image are collinear.
3. A reflection is an isometry, preserving distance, angle, parallelness, perpendicularity, betweenness, and midpoints.

Inversion in a Point in the Plane

To construct the inversion of a figure ABC through a point P, all lying in the plane, we prolong line AP so that $AP = PA'$, and similarly to obtain $BP = PB'$ (Figure 3.19). $A'B'C'$ is the inversion of ABC. By comparing the vertex sequence of these two triangles, we see that this transformation preserves orientation.

$A'B'C'$ is identical to the image produced if we rotate ABC 180° about P. For this reason, we often call an inversion a *half-turn*. Two successive half-turns about the same point is an *identity* transformation because it brings the figure back to its initial position. Two successive half-turns about two different points produces a translation (Figure 3.20).

Two Reflections in the Plane

Given a figure ABC and two intersecting lines m and n in the plane, the successive reflections of ABC about m and then n produce an image $A''B''C''$ equivalent to a rotation about the point of intersection of the lines and through an angle equal to twice the angle between the lines (Figure 3.21).

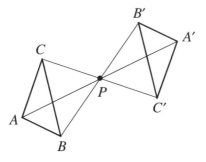

Figure 3.19 *Inversion in a point.*

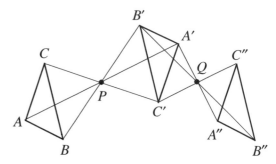

Figure 3.20 *Two successive half-turns.*

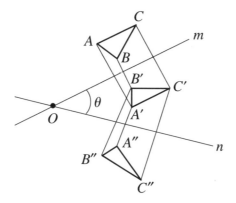

Figure 3.21 *Two successive reflections.*

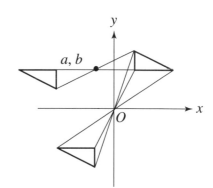

Figure 3.22 *Inversion in a point preserves orientation.*

Transformation Equations for Inversions and Reflections in the Plane

Inversion in the origin or in an arbitrary point a, b preserves orientation (Figure 3.22) and is given respectively by

$$\begin{aligned} x' &= -x \\ y' &= -y \end{aligned} \tag{3.64}$$

and

$$\begin{aligned} x' &= -x + 2a \\ y' &= -y + 2b \end{aligned} \tag{3.65}$$

Reflection in the x or y axis reverses orientation (Figure 3.23) and is given respectively by

$$\begin{aligned} x' &= x \\ y' &= -y \end{aligned} \tag{3.66}$$

and

$$\begin{aligned} x' &= -x \\ y' &= y \end{aligned} \tag{3.67}$$

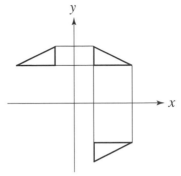

Figure 3.23 *Reflection in the x or y axis reverses orientation.*

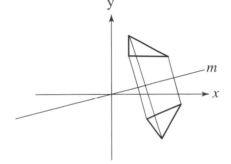

Figure 3.24 *Reflection in an arbitrary line.*

Reflection in an arbitrary line through the origin reverses orientation (Figure 3.24) and is given by

$$x' = x \cos 2\alpha + y \sin 2\alpha$$
$$y' = x \sin 2\alpha - y \cos 2\alpha$$

(3.68)

This transformation consists of a rotation through $-\alpha$ that brings the reflection axis m into coincidence with the x axis, followed by a reflection across the x axis, and completed by a rotation that returns line m to its original position.

Transformation Equations for Inversions and Reflections in Space

Inversion through the origin or an arbitrary point a, b, c in space reverses orientation (Figure 3.25) and is given respectively by

$$x' = -x$$
$$y' = -y$$
$$z' = -z$$

(3.69)

and

$$x' = -x + 2a$$
$$y' = -y + 2b$$
$$z' = -z + 2c$$

(3.70)

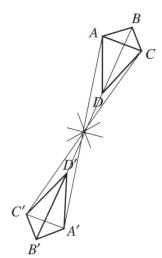

Figure 3.25 *Inversion through a point in space reverses orientation.*

Note that inversion through a point in spaces of even dimension preserves orientation, while inversion through a point in spaces of odd dimension reverses orientation.

Reflection across the x, y plane ($z = 0$ plane) reverses orientation (Figure 3.26) and is given by

$$
\begin{aligned}
x' &= x \\
y' &= y \\
z' &= -z
\end{aligned}
\tag{3.71}
$$

and similarly for the $x = 0$ and $y = 0$ planes.

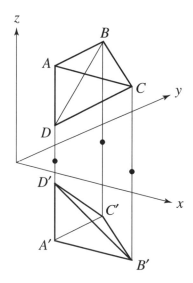

Figure 3.26 *Reflection across a plane in space reverses orientation.*

3.7 Homogeneous Coordinates

When we put a geometric figure through a sequence of translations, rotations, and other transformations, we discovered that a translation destroys the homogeneous character of the transformation equations and that we are required to use a mixture of matrix multiplications and additions (e.g., Equation 3.11). This means that the matrix equation describing the translation and rotation of a point is either

$$\mathbf{P}' = \mathbf{RP} + \mathbf{T} \tag{3.72}$$

or

$$\mathbf{P}' = \mathbf{R}(\mathbf{P} + \mathbf{T}) \tag{3.73}$$

depending on which occurs first. Additional translations and rotations quickly expand these equations into a nested computational mess.

If, on the other hand, a series of transformations contains only rotations, for example, then the net transformation can be expressed as a simple product of matrices, like

$$\mathbf{P}' = \mathbf{R}_\theta \mathbf{R}_\phi \mathbf{P} + \mathbf{T} \tag{3.74}$$

We can reestablish the homogeneous character of the equations describing a series of transformations that includes translations by introducing homogeneous co-ordinates. The theory of homogeneous coordinates is a part of the study of projective geometry and is not included here. Homogeneous coordinates are the coordinates of points in projective space, but we can use them to make our transformation mathematics easier. They allow us to combine several transformations into one matrix, and a long sequence of transformations that includes translations reduces to a series of matrix multiplications. No matrix addition is required.

We use a four-component vector to represent a point in three-dimensional space:

$$\mathbf{p}_h = \begin{bmatrix} hx & hy & hz & h \end{bmatrix} \tag{3.75}$$

where h is some scalar. These are the homogeneous coordinates of the point. The subscript h on $\mathbf{p}_h$ signifies that $\mathbf{p}$ is given by homogeneous coordinates, and is only used as a reminder here.

Homogeneous coordinates easily convert to ordinary coordinates by dividing by h, so that

$$x = \frac{hx}{h}, \quad y = \frac{hy}{h}, \quad z = \frac{hz}{h} \tag{3.76}$$

We notice that there is no unique homogeneous coordinate representation of a point in three- dimensional space. For example, the homogeneous coordinates [12 8 4 4], [6 4 2 2], and [3 2 1 1] all represent the point [3 2 1] in ordinary three-dimensional space. The computations are easier if we choose $h = 1$, so that $\begin{bmatrix} x & y & z & 1 \end{bmatrix}$ in homogeneous coordinates represents the point $\begin{bmatrix} x & y & z \end{bmatrix}$.

We make all transformations of homogeneous points by means of a 4×4 transformation matrix. To use this technique, we must represent curves or surfaces by their control point matrices. We add the fourth or homogeneous coordinate h to each control point. We make the required transformations and then convert the resulting homogeneous coordinates into ordinary three-dimensional coordinates by means of Equation 3.76.

Now let us investigate some of the possible transformations, expressing them as

$$\mathbf{P}'_h = \mathbf{T}_h \mathbf{P}_h \tag{3.77}$$

We translate points using the following matrix product:

$$\begin{bmatrix} x' & y' & z' & 1 \end{bmatrix} = \begin{bmatrix} 1 & 0 & 0 & t_x \\ 0 & 1 & 0 & t_y \\ 0 & 0 & 1 & t_z \\ 0 & 0 & 0 & 1 \end{bmatrix} \begin{bmatrix} x & y & z & 1 \end{bmatrix} \tag{3.78}$$

where t_x, t_y, t_z are components of the vector $\mathbf{t}$ that describes the transformation $\mathbf{p}' = \mathbf{p} + \mathbf{t}$. The matrix product yields

$$\begin{bmatrix} x' & y' & z' & 1 \end{bmatrix} = \begin{bmatrix} x + t_x & y + t_y & z + t_z & 1 \end{bmatrix} \tag{3.79}$$

from which we readily compute the ordinary transformed coordinates

$$x' = x + t_x, \quad y' = y + t_y, \quad z' = z + t_z \tag{3.80}$$

To rotate a point around a principal axis we use one of the three following matrix products. For rotation around the z axis:

$$\begin{bmatrix} x' & y' & z' & 1 \end{bmatrix} = \begin{bmatrix} \cos\theta & -\sin\theta & 0 & 0 \\ \sin\theta & \cos\theta & 0 & 0 \\ 0 & 0 & 1 & 0 \\ 0 & 0 & 0 & 1 \end{bmatrix} \begin{bmatrix} x & y & z & 1 \end{bmatrix} \tag{3.81}$$

For rotation around the y axis:

$$\begin{bmatrix} x' & y' & z' & 1 \end{bmatrix} = \begin{bmatrix} \cos\beta & 0 & \sin\beta & 0 \\ 0 & 1 & 0 & 0 \\ -\sin\beta & 0 & \cos\beta & 0 \\ 0 & 0 & 0 & 1 \end{bmatrix} \begin{bmatrix} x & y & z & 1 \end{bmatrix} \tag{3.82}$$

For rotation around the x axis:

$$\begin{bmatrix} x' & y' & z' & 1 \end{bmatrix} = \begin{bmatrix} 1 & 0 & 0 & 0 \\ 0 & \cos\gamma & -\sin\gamma & 0 \\ 0 & \sin\gamma & \cos\gamma & 0 \\ 0 & 0 & 0 & 1 \end{bmatrix} \begin{bmatrix} x & y & z & 1 \end{bmatrix} \tag{3.83}$$

The differential scaling equation is

$$[x' \quad y' \quad z' \quad 1] = \begin{bmatrix} s_x & 0 & 0 & 0 \\ 0 & s_y & 0 & 0 \\ 0 & 0 & s_z & 0 \\ 0 & 0 & 0 & 1 \end{bmatrix} [x \quad y \quad z \quad 1] \tag{3.84}$$

And finally for projections we have

$$[x' \quad y' \quad z' \quad 1] = \begin{bmatrix} 1 & 0 & 0 & 0 \\ 0 & 1 & 0 & 0 \\ 0 & 0 & 1 & 0 \\ p & q & r & 1 \end{bmatrix} [x \quad y \quad z \quad 1] \tag{3.85}$$

which we use in Chapter 18.

Thus, by what only amounts to temporarily adding an extra coordinate to the **P** and **P'** matrices and adjusting **R** and **T** accordingly, we can write an arbitrary sequence of transformations as a single equivalent matrix product.

Let's see how this works in two dimensions, for rotation and translation in the x, y plane. If we must first translate a point by **T** and then rotate the result by **R**, we write this as

$$\mathbf{P}' = \mathbf{RTP} \tag{3.86}$$

where

$$\mathbf{RT} = \begin{bmatrix} \cos\theta & -\sin\theta & t_x \cos\theta - t_y \sin\theta \\ \sin\theta & \cos\theta & t_x \sin\theta + t_y \cos\theta \\ 0 & 0 & 1 \end{bmatrix} \tag{3.87}$$

Or, reversing the order and translating the point before rotating it, we write

$$\mathbf{P}' = \mathbf{TRP} \tag{3.88}$$

where

$$\mathbf{TR} = \begin{bmatrix} \cos\theta & -\sin\theta & t_x \\ \sin\theta & \cos\theta & t_y \\ 0 & 0 & 1 \end{bmatrix} \tag{3.89}$$

This also shows us that

$$\mathbf{TR} \neq \mathbf{RT} \tag{3.90}$$

A very simple and obvious extension of these ideas applies to rotations and translations in space.

Exercises

3.1 Describe the following linear transformations:
   a. $x' = x,\ y' = ay$
   b. $x' = ax,\ y' = by$
   c. $x' = ax,\ y' = ay$
   d. $x' = (x - y)/\sqrt{2},\ y' = (x + y)\sqrt{2}$

3.2 How far is a point moved under the translation given by Equation 3.26?

3.3 Show that the product of the following two reflection matrices is a rotation:

$$\begin{bmatrix} 0 & 1 \\ 1 & 0 \end{bmatrix} \text{ and } \begin{bmatrix} 0 & -1 \\ -1 & 0 \end{bmatrix}$$

3.4 Reflections in space are summarized by the following matrix. Write out all the possible combinations as individual matrices, and give their corresponding linear equations, describing what each does.

$$\mathbf{R}_f = \begin{bmatrix} \pm 1 & 0 & 0 \\ 0 & \pm 1 & 0 \\ 0 & 0 & \pm 1 \end{bmatrix}$$

3.5 Describe the product of reflection of a point in two mutually perpendicular lines in the plane.

3.6 Describe the product of reflection of a point in three mutually perpendicular planes.

3.7 Show that the product of sequential reflections in two parallel lines, $x_l = a$ and $x_m = b$, of a point P_0 is the equivalent of a single translation. Find the translation.

3.8 Show that, in the plane, translating two points $\mathbf{p}_1$ and $\mathbf{p}_2$, each by $\mathbf{t}$, preserves the distance between them.

3.9 Compare $\mathbf{R}_{\theta\phi}$ and $\mathbf{R}_{\phi\theta}$ to show that different orders of rotation produce different results.

3.10 Show that the translation of a point P to P'' produced by two successive half-turns about different points A and B is equal to twice the directed distance from A to B. Draw a sketch to support your answer.

3.11 Derive Equation 3.65.

3.12 Derive Equation 3.68.

CHAPTER 4

SYMMETRY AND GROUPS

Symmetry brings to mind decorative patterns, tilings, and repetitive mirror-images. As a subject of study, it was once thought to be unrelated to serious mathematics and scientific thinking. Not until the 19th century did mathematicians discover a close relationship between analytical geometry and symmetry. Then they invented *group theory* to study the symmetry of mathematical objects.

Aside from the obvious application of symmetry to the decorative arts and engineered constructions, the understanding of symmetry is now central to mathematical physics and the study of elementary particles and cosmology. In the equations of quantum mechanics and general relativity, the presence or absence of certain algebraic symmetries profoundly affects how we interpret them. The highly intuitive geometric nature of the elementary principles of symmetry and groups makes this subject easy to teach and learn.

This chapter takes an analytical approach to symmetry. It defines a *group* and the more specialized *symmetry group*, including cyclic, dihedral, and finite symmetry groups. The symmetry of *ornamental groups* (*frieze groups* and *wallpaper groups*) reveals a surprising quantitative aspect to these aesthetic forms. The *crystallographic restriction* shows how symmetry depends on the nature of space itself. This is also true of plane lattices and tilings. The chapter concludes with a look at the rotational symmetry of polyhedra.

4.1 Introduction

When we say that a figure is *symmetric*, we mean that certain motions or rearrangements of its parts leave it unchanged as a whole. A figure is symmetric if it is congruent to itself in more than one way. For example, if we rotate a square 90° about its center, we see no difference between the *original* figure and its rotated *image*. They coincide exactly. So we can call this motion a *symmetry transformation*. Reflections, rotations, and inversions, subject to certain conditions, can produce symmetry transformations. Any rotation of a circle about its center, or any reflection about a diameter, is a *symmetry* of a circle.

We begin with an investigation of the six symmetries of an equilateral triangle (Figure 4.1). We will look for its symmetries by finding different ways of transforming

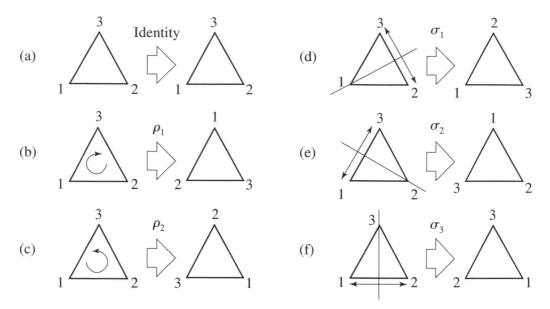

Figure 4.1 *Symmetry transformations. (a) Identity transformation. (b) and (c) Rotational symmetry. (d), (e), and (f) Reflection symmetry.*

it that do not change its overall appearance, orientation, or measurable properties. This means that a symmetry transformation must preserve the distances between the vertices and the angles between edges.

Rotating the triangle $0°$ or $360°$ does not change it (Figure 4.1a). Mathematicians call this the *identity transformation* and denote it with the Greek letter *iota*, ι. As simple and seemingly superfluous as it appears, we will see that the identity transformation is very important in the study of symmetry groups.

We can rotate an equilateral triangle about its center $120°$ clockwise or counterclockwise and leave it unchanged (Figures 4.1b and 4.1c). We will denote rotations by the Greek letter *rho*, and in these two cases by ρ_1 and ρ_2.

Finally, we can reflect an equilateral triangle across each of the three lines bisecting its vertex angles, also leaving it unchanged (Figures 4.1d, 4.1e, and 4.1f). We will use the Greek letter *sigma* to denote reflections, and specifically σ_1, σ_2, and σ_3 for this example.

The rotation ρ_1 moves vertex 1 into the position that was occupied by vertex 3, 3 moves to 2, and 2 moves to 1. For the reflection σ_1, vertices 2 and 3 interchange positions. If we ignore the vertex labels, then the original and transformed, or *image*, triangles are identical in position and orientation, as well as in all measurable properties.

The inversion of an equilateral triangle through its center is not a symmetry transformation (Figure 4.2). An inversion produces an image equivalent to that of a $180°$ rotation, which does not have the same relative orientation as the original.

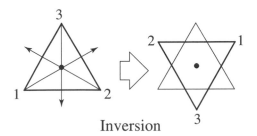

Inversion

Figure 4.2 *Inversion of an equilateral triangle.*

Figure 4.3 *Binary product of a symmetry transformation.*

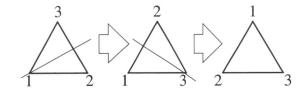

Products of Transformations

Now let's consider *binary products* of symmetry transformations of the equilateral triangle. By *product* we mean the result of two successive transformations. For example, $\sigma_2\sigma_1$ means reflection 1 followed by reflection 2, first interchanging vertices 2 and 3 (lower right vertex with upper vertex) and then interchanging vertices 1 and 2 (lower left vertex with upper vertex) (Figure 4.3). We see that this product is the equivalent of the single symmetry transformation ρ_1, so we can write an equation for it:

$$\rho_1 = \sigma_2\sigma_1 \tag{4.1}$$

We can construct a table of all the possible products of the symmetry transformations of the equilateral triangle (Table 4.1). Mathematicians call it a *Cayley table* to honor the English mathematician Arthur Cayley (1821–1895), who first used it. For our example, there are six symmetry transformations ($\iota, \rho_1, \rho_2, \sigma_1, \sigma_2,$ and σ_3), so we have a 6×6 Cayley table.

To find the product $\rho_2\sigma_1$, we enter the row whose first element is ρ_2 and move across to the entry at the intersection of this row with the column headed by σ_1. The

Table 4.1 Symmetry products of an equilateral triangle

	ι	ρ_1	ρ_2	σ_1	σ_2	σ_3
ι	ι	ρ_1	ρ_2	σ_1	σ_2	σ_3
ρ_1	ρ_1	ρ_2	ι	σ_3	σ_1	σ_2
ρ_2	ρ_2	ι	ρ_1	σ_2	σ_3	σ_1
σ_1	σ_1	σ_2	σ_3	ι	ρ_1	ρ_2
σ_2	σ_2	σ_3	σ_1	ρ_2	ι	ρ_1
σ_3	σ_3	σ_1	σ_2	ρ_1	ρ_2	ι

entry at the intersection is σ_2, so we see that

$$\rho_2\sigma_1 = \sigma_2 \tag{4.2}$$

The product of any two symmetries in this group is always another symmetry that is also a member of the group. Order is important; for example, we see that

$$\rho_2\sigma_1 \neq \sigma_1\rho_2 \tag{4.3}$$

Points, Lines, and Planes of Symmetry

For any symmetric figure, we find that certain sets of points, lines, or planes are *fixed*, or *invariant*, under a symmetry transformation of the figure. They are easy to identify in simple symmetric figures (Figure 4.4). A hexagon is symmetric under inversion through its center point or under rotations that are integer multiples of 60°. The hexagon also has six different lines fixed under reflections (Figure 4.4a). The cube has many lines and planes of symmetry, as we will see later, but only one of them is shown here (Figure 4.4b).

We can also build up a symmetric figure by a series of rotations and reflections. For example, beginning with an arbitrary figure, we reflect it across some line m (Figure 4.5). The image, together with its original, forms a new symmetric figure.

Analysis of Symmetry

Purely analytical techniques are also available to us for discovering the symmetries of forms defined by algebraic equations. For curves in the plane, we find:

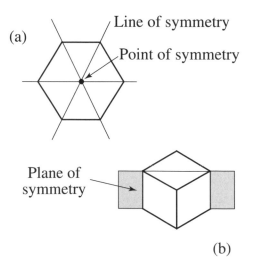

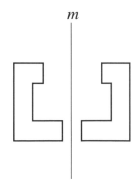

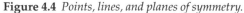

Figure 4.4 *Points, lines, and planes of symmetry.*

Figure 4.5 *Symmetric figure.*

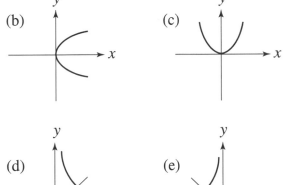
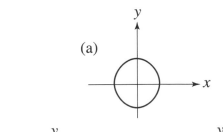

Figure 4.6 *Symmetries of conic curves.*

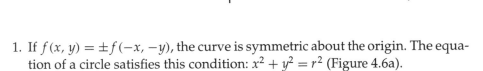

1. If $f(x, y) = \pm f(-x, -y)$, the curve is symmetric about the origin. The equation of a circle satisfies this condition: $x^2 + y^2 = r^2$ (Figure 4.6a).
2. If $f(x, y) = \pm f(x, -y)$, the curve is symmetric about the x axis. The equation of a parabola satisfies this condition: $x = y^2$ (Figure 4.6b).
3. If $f(x, y) = \pm f(-x, y)$, the curve is symmetric about the y axis. The equation of a parabola satisfies this condition: $y = x^2$ (Figure 4.6c).
4. If $f(x, y) = \pm f(y, x)$, the curve is symmetric about the line $x = y$. The equation of a hyper-bola satisfies this condition: $xy = k^2$ (Figure 4.6d).
5. If $f(x, y) = \pm f(-y, -x)$, the curve is symmetric about the line $x = -y$. The equation of a hyperbola satisfies this condition: $xy = -k^2$ (Figure 4.6e).

4.2 Groups

Group theory gives us a way to study symmetry in a mathematically rigorous way. A *group* is a special kind of set. It is a set S consisting of elements and an operator that combines these elements two at a time. We will use the symbol $\circ$ to denote the operator. A group must have the following four properties:

1. *Closure*: The product of any two elements of S is another element that is also a member of S. In other words, if A and B are elements of S and the product $A \circ B = C$, then C is also a member of S.

2. *Identity*: The set S must have an identity element I such that $A \circ I = I \circ A = A$ for all elements of S.

3. *Inverse*: Every element A of S must have an inverse element A^{-1} that is also a member of S and that satisfies $A \circ A^{-1} = A^{-1} \circ A = I$.

4. *Associativity*: An ordered sequence of operations must be associative; that is, $(A \circ B) \circ C = A \circ (B \circ C)$.

5. *Commutativity* (optional): If $A \circ B = B \circ A$ for any pair of elements in a group, then the group is commutative. Such groups are also called *Abelian groups* after the Norwegian mathematician Neils Henrik Abel (1802–1829). Most symmetry groups are not commutative.

Here are two examples of nongeometric groups:

1. A set consisting of all positive and negative numbers forms a group under the operation of addition. The identity element is because $n + 0 = n$. The inverse of any element n is $-n$.

2. All nonzero rational numbers form a group under the operation of multiplication. The identity element is because $n \times 1 = n$. The inverse of any element n is $1/n$.

Here is a geometric example: Let's consider two rotations of a circle about its center. The set of the two rotations $0°$ and π forms a group. Each rotation is its own inverse. The product of each with itself is a rotation of $0°$, and the product of the two is a rotation through π. On the other hand, the set of rotations $0°$ and $\pi/2$ is not a group because it does not contain the inverse of $\pi/2$ nor the product of $\pi/2$ with itself.

We can characterize groups in several ways. If a group has exactly n elements, then it is a *finite group* and has *order n*; otherwise a group is infinite. If α is some transformation and if there is a smallest positive integer n such that $\alpha^n = \iota$, then α has order n; otherwise the order is infinite. If every element of a group containing α is a power of α, then the group is cyclic and α is the generator (see *Cyclic Groups*).

4.3 Symmetry Groups

The symmetry transformations of a figure always form a group. Group theory organizes the study of symmetry and provides the analytical tools we need to classify the various kinds of symmetry.

Two properties of all symmetry transformations greatly simplify our studies. First, the product of two symmetry transformations is always another symmetry transformation. Second, a transformation that reverses a previous symmetry transformation is itself another symmetry transformation. These two properties correspond to the closure and inversion properties of a group. Thus, symmetry transformations themselves form a group.

Cyclic Groups

If every element of a group containing the transformation α is a power of α, then the group is *cyclic* with *generator* α. We denote a cyclic group as $\langle \alpha \rangle$. For example, if ρ is a rotation of 45°, then $\langle \rho \rangle$ is a cyclic group of order 8. If ρ is a rotation of 36°, then $\langle \rho \rangle$ is a cyclic group of order 10.

For simple cyclic groups, C_n denotes the cyclic group of order n generated by ρ, where ρ is a rotation of $360°/n$. This notation does not give us a way to specify a center of rotation, so we assume it to be the geometric center of the figure in question. Symmetry group C_1 contains only one element, the identity element. C_2 contains $\rho = 0°$ and $\rho = 180°$.

If $n > 2$, then C_n is the symmetry group of an interesting class of $2n$-gon (a polygon with $2n$ edges). These polygons have no lines of reflection symmetry. We can easily construct them by breaking the reflection symmetry of a regular polygon in a way that preserves rotational symmetry. Figure 4.7 shows some simple examples.

Dihedral Groups

A unique symmetry group exists for every regular polygon and polyhedron. We will study the symmetry of a square to understand dihedral symmetry (Figure 4.8). Four rotations, based on a cyclic subgroup of order 4 with generator ρ, where $\rho = 90°$, leave the square invariant. These rotations are: ρ, ρ^2, ρ^3, and ρ^4 (or 90°, 180°, 270°, 360°). Four reflections, σ, also leave the square invariant: σ_k, σ_l, σ_m, and σ_n.

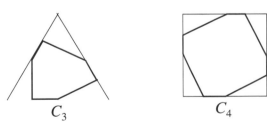

C_3 $\qquad\qquad$ C_4

Figure 4.7 *Examples of cyclic symmetry.*

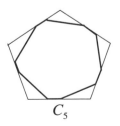
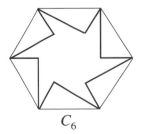

C_5 $\qquad\qquad$ C_6

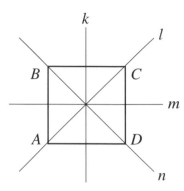

Figure 4.8 *Dihedral symmetry.*

Under a symmetry transformation, including the identity transformation, A may go to any one of four vertices, and B must go to one of the two adjacent vertices. This means that the images of the remaining vertices are thereby determined. There are eight permutations of symmetry for the square, and ρ and σ generate all of them. They are: ρ, ρ^2, ρ^3, and $\rho^4, \rho\sigma, \rho^2\sigma, \rho^3\sigma$, and $\rho^4\sigma$. When a symmetry group consists of rotations and reflections, we denote it as $\langle \rho, \sigma \rangle$ and call it a *dihedral group*. (*Dihedral* means having two faces. Here it indicates the presence of reflections in the symmetry group.)

4.4 Ornamental Groups

The so-called *ornamental* symmetry groups describe three patterns in the plane. One of them is the set of seven *frieze groups*, each of which exhibits symmetry along a line. Another is the set of 17 *wallpaper groups*, each of which exhibits symmetry in a repeated pattern that fills the plane. In other words, symmetry is the ordered repetition of a basic pattern in both frieze and wallpaper groups. Finally, a third ornamental symmetry group consists of the *rosette groups*, which in turn consist of cyclic and dihedral groups C_n and D_n. All frieze groups contain a subgroup of translations generated by one translation. All wallpaper groups contain a subgroup of translations generated by two nonparallel translations.

Frieze Groups

The word *frieze* is an architectural term and refers to a decorative band along a wall, usually with a repetitive pattern. For our purposes, a *frieze group* is the symmetry group of a repeated pattern along a line in a plane. The pattern is invariant under a sequence of translations along the line. Mathematicians have discovered that there are surprisingly, only seven possible frieze groups. Figure 4.9 shows an example of each one. The light lines are lines of reflection and not part of the pattern. (They are merely helpful artifacts of the explanation.) So are the small circles, which represent the centers of $180°$ rotations. We use them occasionally to create and describe a pattern. The F-notation is explained later. Finally, in all seven groups, we must imagine the pattern repeating indefinitely in both directions (right and left).

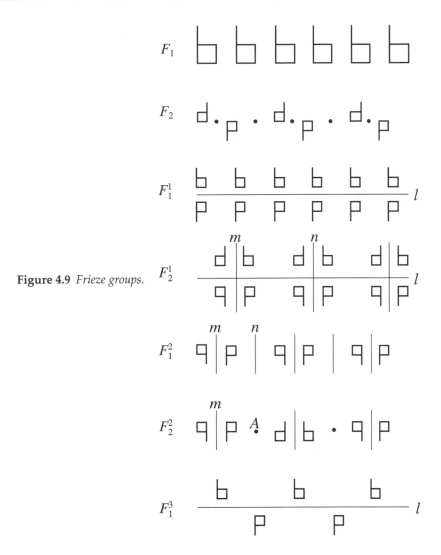

Figure 4.9 *Frieze groups.*

We generate the F_1 symmetry group by a repeated translation, $F_1 = \langle \tau \rangle$. This group has no point or line of symmetry and is not fixed by a glide reflection (a combination of translation and reflection relative to a fixed line).

The group F_2 has two points of symmetry, A and B, about which half-turns are executed. There are no lines of symmetry in this frieze pattern. We denote it as $F_2 = \langle \sigma_A, \sigma_B \rangle$.

F_1^1 has no point of symmetry, but l is a line of symmetry, so that $F_1^1 = \langle \tau, \sigma_l \rangle$.

F_2^1 has three lines of reflection symmetry l, m, and n, and $F_2^1 = \langle \sigma_l, \sigma_m, \sigma_n \rangle$.

F_1^2 has no point of symmetry but has two lines of symmetry, m and n, so that $F_1^2 = \langle \sigma_m, \sigma_n \rangle$.

F_2^2 has both a point and a line of symmetry, and $F_2^2 = \langle \sigma_m, \sigma_A \rangle$.

F_1^3 has no points or lines of symmetry but is generated by a glide reflection: $F_1^3 = \langle \gamma \rangle$.

No other symmetry groups will produce a frieze pattern.

As for the *F*-notation, the subscript indicates whether or not a half-turn is present in the pattern. The superscript 1 indicates that the centerline of the strip is a line of symmetry—a reflection axis. If the superscript is 2, then the center is not a line of symmetry, but there is a line of symmetry perpendicular to the centerline. Superscript 3 indicates that the symmetry group is generated by a glide reflection. The Hungarian mathematician Fejes-Tóth developed this notation.

Wallpaper Groups

Wallpaper groups are symmetry groups in the plane containing two nonparallel translations. Surprisingly, there are only 17 of these groups. All two-dimensional repetitive patterns in wallpaper, textiles, brickwork, or the arrangement of atoms in a plane of a crystal are but variations on one of these 17 groups. Thirteen of these groups include some kind of rotational symmetry, and four do not. Five groups have two-fold (180°) centers of rotational symmetry, three groups have three-fold (120°) centers of rotational symmetry, and two have six-fold (60°) centers of rotational symmetry. Each wallpaper group is a group of *isometries* (rigid-body transformations) in the plane whose translations are those in $\langle \tau_1, \tau_2 \rangle$, where τ_1 and τ_2 are not parallel and whose rotations (if any) are restricted to two-, three-, four-, and six-fold centers of rotation (see the Crystallographic Restriction section, following).

We identify a wallpaper group by the translation and rotation symmetries it has. Using the translations of a group, we describe a unit cell within the overall wallpaper pattern, which is the smallest area containing all of the elements of the particular symmetry group but without repetition. The unit cells form a plane lattice (Figure 4.10).

We construct the lattice for a wallpaper group by choosing any point P in the pattern, and a translation τ_1 that sends P to Q, an equivalent point in the pattern that has the smallest distance possible from P. Translations parallel to τ_1 generate a row of points equivalent to P on the line containing P and Q. We assume that there are other points in the pattern that are equivalent to P but not on the line containing P and Q. We find such a point R as close to P as possible. The translation that sends P into R we denote τ_2. If S is the image of Q under τ_2, then $PQRS$ is a parallelogram. We discover that all translations of a symmetry group are combinations of τ_1 and τ_2. In vector form

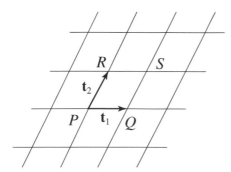

Figure 4.10 *Unit cell of a plane lattice.*

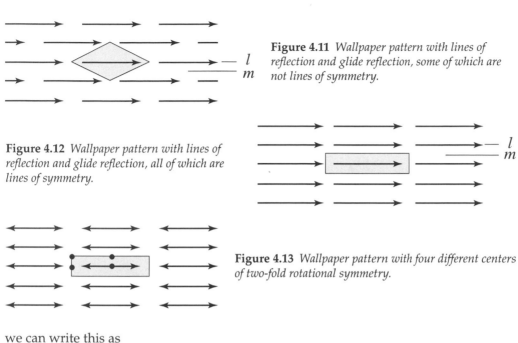

Figure 4.11 *Wallpaper pattern with lines of reflection and glide reflection, some of which are not lines of symmetry.*

Figure 4.12 *Wallpaper pattern with lines of reflection and glide reflection, all of which are lines of symmetry.*

Figure 4.13 *Wallpaper pattern with four different centers of two-fold rotational symmetry.*

we can write this as

$$\mathbf{t} = m\mathbf{t}_1 + n\mathbf{t}_2 \tag{4.4}$$

where m and n are positive or negative integers. They generate a lattice that contains all of the points equivalent to P at the intersections of the lattice grid.

We will discuss briefly three of the wallpaper groups. You can find complete descriptions of all 17 in several texts (see annotated Bibliography). The patterns in Figures 4.11, 4.12, and 4.13 are simple representative examples of these three groups.

The wallpaper pattern in Figure 4.11 has lines of reflection and glide reflection but has no centers of rotational symmetry. Some of the glide reflection lines are not lines of symmetry—line m, for example.

The pattern in Figure 4.12 also has lines of reflection and glide reflection, all of which are lines of symmetry. It has no centers of rotational symmetry.

The pattern in Figure 4.13 has four different centers of two-fold rotational symmetry (shown as small circles) and no reflection symmetry.

4.5 The Crystallographic Restriction

Thirteen wallpaper symmetry groups have centers of rotational symmetry. A center of rotational symmetry may belong to one of four possible types, according to the angles of rotation of the group. All possible angles are of the form $2\pi/n$, where n is an integer. The only allowable values of n are 1, 2, 3, 4, and 6. The proof of this restriction follows.

We choose the smallest possible translation of the group and denote it as $\mathbf{t}$ (Figure 4.14). We then assume that point A is an n-fold center of rotation and that

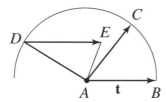

Figure 4.14 *The crystallograpic restriction.*

t translates A to B. A rotation of $2\pi/n$ about A sends B to some point C. It follows from group properties that a translation can send A to C as well. Because $BC \geq AB$, it follows that $\angle BAC \geq \pi/3$, or

$$\frac{2\pi}{n} \geq \frac{\pi}{3} \tag{4.5}$$

which means that

$$n \leq 6 \tag{4.6}$$

Now we must prove that $n \neq 5$. To do this we use any multiple of $2\pi/5$ and evaluate the results. We rotate B about A through an angle of $2 \cdot 2\pi/5$ to D, then translate D by **t** to E. It is obvious that $AE < AB$, which contradicts our requirement that AB must be the smallest possible translation. Therefore, $n = 5$ is not allowed, and we are, indeed, restricted to 2-, 3-, 4-, and 6-fold centers of rotation. The very nature of space itself dictates the number and kind of symmetry groups allowed in the plane, just as it dictates that there can be only five regular polyhedra in space.

4.6 Plane Lattices

Any regular system of points in the plane or in space forms a *lattice*. Lattices are a way to study the structure of space, crystal structures (how atoms organize themselves into regular repeating patterns in a solid), the ornamental symmetry groups, and tilings in the plane. We can generate a plane lattice as follows: Start from any fixed point O. All of the possible images of O produced by the set of translations $m\mathbf{a} + n\mathbf{b}$ form a lattice (Figure 4.15). Integers m and n are positive or negative, and $\mathbf{a}$ and $\mathbf{b}$ are nonparallel vectors.

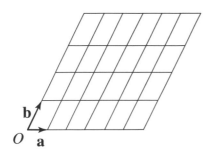

Figure 4.15 *Lattice formed by a set of translations.*

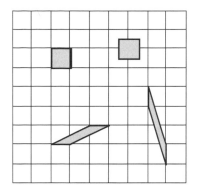

Figure 4.16 *Examples of a unit cell.*

As we have discussed previously, the *unit cell* of a lattice is the smallest part that can generate the whole lattice pattern by repetitive translations of itself. Another way to define a unit cell is as a parallelogram formed by two nonparallel line segments with lattice points at the vertices and no lattice points within it, except as a single central point (Figure 4.16). Note that all possible (allowable) unit cells in a square lattice enclose an equal area. Furthermore, depending on the magnitudes and relative directions of the vectors **a** and **b**, a lattice may be rectangular or triangular.

4.7 Tilings

A *tiling* of the plane is a set of polygonal shapes that are joined edge-to-edge to cover the plane without gaps or overlaps. A tiling is *monohedral* if a single size and shape will cover the plane. Square tiles are the simplest example of a monohedral tiling. A tiling is dihedral (Figure 4.17), trihedral, or *r*-hedral if two, three, or *r* different tiles are needed to cover the plane.

Some tilings may not exhibit symmetry, while others will have one of the wallpaper symmetries. The polygons forming the individual tiles may be without symmetry themselves.

There are only three *regular tilings* of the plane. They are formed by equilateral triangles, squares, and hexagons (Figure 4.18). Each vertex of a regular tiling is surrounded identically with congruent faces. The regular pentagon cannot tile the plane without gaps or overlaps, because three at a vertex leave a gap, and four must necessarily overlap. The proof that there are only three regular tilings of the plane follows:

A *p*-sided regular polygon's interior angles are each

$$\theta = \frac{180(p-2)}{p} \tag{4.7}$$

If the typical vertex of a regular tiling is surrounded by *q* regular *p*-sided polygons, then the sum of the interior angles around a vertex is

$$\frac{180q(p-2)}{p} = 360 \tag{4.8}$$

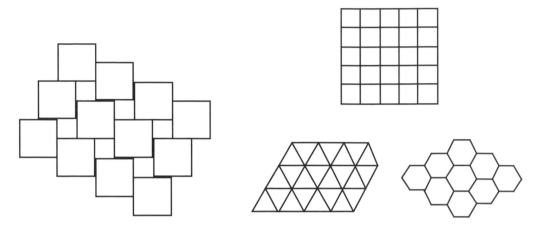

Figure 4.17 *Dihedral tiling.*   **Figure 4.18** *The only three regular tilings of a plane.*

which we can write as

$$(p-2)(q-2) = 4 \tag{4.9}$$

The positive-integer solutions to this equation are:

$$\{p, q\} = \{3, 6\}, \{4, 4\}, \text{ and } \{6, 3\} \tag{4.10}$$

These are also the only regular edge-to-edge tilings possible with congruent, but not necessarily regular, polygons of any kind.

4.8 Polyhedral Symmetry

A rigorous study of the symmetries of polyhedra was first done in the field of crystallography, and mathematicians soon discovered that all of the regular, semi-regular, and dual polyhedra have one of only three special rotational symmetry groups. We cannot generate these symmetries by the cyclic or dihedral groups. We denote them as the *3.2 symmetry group*, the *4.3.2 symmetry group*, and the *5.3.2 symmetry group*. The number designations mean combinations of rotational symmetry with 2-, 3-, 4-, and 5-fold axes.

The 3.2 Symmetry Group

A tetrahedron has four 3-fold axes of rotational symmetry. Each axis passes through a vertex and the center of the opposite face. One 3-fold axis is shown in two different views in Figures 4.19a and 4.19b. The view in Figure 4.19b is along the axis through vertex A. It is easy to see that 120° rotations leave the figure unchanged.

A tetrahedron also has three 2-fold axes of rotational symmetry. Each axis passes through the center points of opposite edges. Only one of these axes is shown in Figure 4.19c and 4.19d.

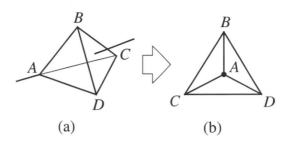

Figure 4.19 *The tetrahedron's axes of rotational symmetry.*

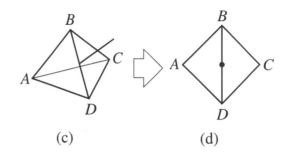

The 4.3.2 Symmetry Group

A cube belongs to the 4.3.2 rotational symmetry group. A cube is invariant under rotations of 90°, 180°, 270°, and 360° about any axis through the centers of opposite faces (Figure 4.20a). There are three of these axes.

A cube is invariant under rotations of 120° and 240° about any axis through opposite vertices, a body diagonal (Figure 4.20b). There are four of these axes.

A cube is also invariant under 180° rotations about axes passing through the centers of opposite edges (Figure 4.20c). There are six 2-fold axes.

Including the identity transformation, a cube has 24 rotations that leave it invariant. It has other symmetry transformations (reflection and inversion) that increase the total number of symmetry transformations to 48. Other polyhedra in the 4.3.2 symmetry group include the octahedron, the dual of the cube, the cuboctahedron, the great rhombicuboctahedron, and the small rhombicuboctahedron.

The 5.3.2 Symmetry Group

Two regular polyhedra belong to the 5.3.2 symmetry group: the icosahedron and its dual, the dodecahedron. They each have 5-fold, 3-fold, and 2-fold axes of rotational symmetry. Although these axes are easy to describe, they are difficult to draw with much clarity, so no figures are included here.

For the icosahedron, the 5-fold axes of symmetry pass through the center of each of its regular pentagonal faces and the center of the icosahedron. The 3-fold axes pass through each vertex and the center. The 2-fold axes pass through the center of each edge and the center of the icosahedron. Many other polyhedra belong to the 5.3.2 group.

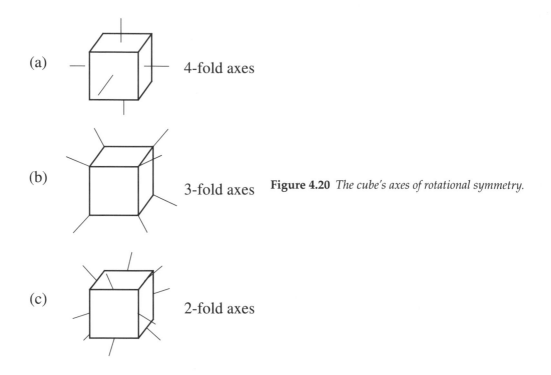

(a)   4-fold axes

(b)   3-fold axes   **Figure 4.20** *The cube's axes of rotational symmetry.*

(c)   2-fold axes

Symmetry Breaking

We can easily create, alter, or destroy the symmetry of a figure. For example, if we truncate one corner of a cube, we preserve one of its 3-fold axes of rotational symmetry and destroy the others (Figure 4.21). All 4-fold and 2-fold symmetries are also destroyed. This truncation preserves only three of the planes of reflection symmetry, destroying the other six. If we now truncate the opposite corner, as well, we restore the three 2-fold axes that are perpendicular to the surviving 3-fold axis.

If we truncate four nonadjacent corners of a cube, then the remaining symmetry is that of a tetrahedron (Figure 4.22). This is what the truncation did: The three 4-fold axes of rotational symmetry of the cube collapse into the tetrahedron's three 2-fold axes. The symmetry of the cube's 2-fold axes is destroyed, and the symmetry of all four 3-fold axes is preserved. The reflection symmetry of the six planes, through opposite edges of

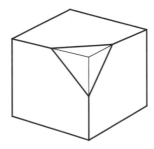

Figure 4.21 *Broken symmetry.*

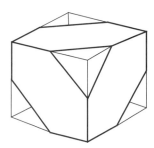

Figure 4.22 *Creating the 3.2 symmetry group by truncating a cube.*

the cube, is preserved, but the reflection symmetry defined by the three planes parallel to and midway between opposite pairs of the cube's faces is destroyed.

Exercises

4.1 Describe the relationship of the lines $x = -1$ and $y = 3$ to the curve $(x + 1)^2 + (y - 3)^2 = 10$.

4.2 Given that all points in the x, y plane are reflected in some line m, describe what points and lines are invariant.

4.3 For the curve $x = y^3$, is the origin a center of symmetry?

4.4 For the curve $x = y^4$, what are its lines of symmetry?

4.5 How many planes of symmetry does a cube have? Describe each one.

4.6 Describe the symmetries of the curve $y = x^3$.

4.7 Describe the symmetries of the curve $2x^2 + 3y^2 = 1$.

4.8 Describe the symmetries of the curve $xy = 3$.

4.9 Describe the symmetries of the curve $x^3 + x^2y + xy^2 + y^3 = 7$.

4.10 Which of the following sets are groups?
   a. Rotations of a circle through O, $2\pi/3$, and $4\pi/3$.
   b. Rotations of a circle through $\pi/2$, π, $3\pi/2$, and 2π.
   c. Reflection of a square about its diagonal, rotation about its center by π, and the identity element.
   d. The reflection of a circle in its center and the identity element.
   e. The reflection (inversion) of a circle in its center and the identity element.

4.11 Give the F number for each of the following frieze patterns:
   a. ...SSSSSSS...   e. ...MWMWMWM...
   b. ...AAAAAAA...   f. ...IIIIIII...
   c. ...FFFFFFF...   g. ...DDDDDDD...
   d. ...DWDMDWDM...

4.12 Describe the axes of the 4.3.2 symmetry group of the octahedron.

LIMIT AND CONTINUITY

Soon after geometry became a separate branch of mathematics, geometers began to study problems they could best solve by processes that they imagined could be continued indefinitely. These were problems related to vanishingly small distances, and problems related to sets containing infinitely many elements. For example, a circle begins to coincide with the limiting figure of a regular polygon inscribed in it as we increase the number of sides, giving rise to the concept of a nonending process. Or, we can successively halve the distance between two points AB on a straight line segment to produce lengths $1/2, 1/4, 1/8, \ldots, 1/2^n$ that of the initial length, again setting off a nonending process and introducing vanishingly small distances. Furthermore, the set of points on a line segment is an infinite set. All of these processes are related to the continuity of line segments and to the concept of limit. These, in turn, provide the basis for the calculus.

Limit and continuity processes are at work in computer graphics applications to ensure the display of smooth, continuous-looking curves and surfaces. They are an important consideration in the construction and rendering of shapes for computer-aided design and manufacturing, scientific visualization, and similar applications.

This chapter explains limit and continuity. It begins with the Greek method of exhaustion, which appeals strongly to our intuition and sets the stage for a more rigorous discussion of the limit process to come. The section on sequences and series extends this appeal to numbers and algebra. The chapter then reviews the definition of a function and follows this with discussions of the limit of a function, limit theorems and rules, and limit and the definite integral. The next two topics again appeal to our intuition, but with strong support from the more quantitative and analytical side of mathematics: these are the tangent to a curve, and rate of change. A discussion of intervals supports the concluding two sections on continuity and continuous functions.

Although this chapter reviews much of the material usually introduced in first-year calculus, its purpose is to highlight and re-emphasize these concepts, because they occur over and over again in more advanced studies.

5.1 The Greek Method of Exhaustion

Behind the innocent word "limit" is a subtle process, simple to describe, but awesome in its implications. It is the cornerstone of the calculus. The story of the calculus and the concept of limit are inseparable and must begin with the ancient Greek *method of exhaustion* and the great Archimedes. Although other Greeks, Eudoxus among them, had planted the seed of the idea earlier, it was Archimedes who nurtured it and reaped the harvest, applying this method to compute certain lengths, areas, and volumes. He developed rigorous logical proofs using the method of exhaustion to back up his calculations.

We can see the idea of a variable approaching a limit in a method for approximating the area of a circle. We begin by inscribing an equilateral triangle in a circle (Figure 5.1). Admittedly, the triangle is only the crudest approximation of the circle. Then, using simple constructions, we double the number of sides to create an inscribed hexagon. This approximation is marginally better. Double the number of sides again to create a regular dodecagon, and again to obtain a regular polygon of 24 sides. Now we are getting somewhere.

One more doubling and from across the room we can't tell them apart. If we continue this process indefinitely and study the resulting sequence of inscribed regular polygons with 3, 6, 12, 24, 48, 96, 192, 384, . . . sides, we are led to an intuitive conclusion that ultimately the polygonal area "exhausts" the area within the circumference of the circle. We would say that there is a "limiting" value to which all polygonal areas after, say, the thousandth in the sequence are very, very close. If such a limit exists, we are justified in asserting that this is also the exact measure of the area of the circle.

Here's a somewhat more algebraic description of what is happening as the number of sides of the inscribed polygon increases indefinitely: The area, A_P, of this polygon approaches a limit which it never exceeds, and that limit is the area of the circle. A_P is the dependent variable of some equation that ultimately is equal to the true area of a circle, A_C. As A_P approaches this limit, the difference $A_C - A_P$ decreases and eventually becomes less than any preassigned number, however small. We see that this process of increasing the number of sides of the inscribed polygon *exhausts* the difference in areas.

In general, any dependent variable v approaches a constant L as a limit when the successive values of v are such that the difference $v - L$ eventually becomes and remains less than any preassigned positive number, however small. We write this relationship as $\lim v = L$. It is the common practice of mathematicians, engineers, and scientists to use the notation $v \to L$, which we read as "v approaches L as a limit," or, more briefly, "v approaches L."

In the calculus of today, an "infinite process" of approximation is set up according to definite rules. If a *limit* exists for the computation and approximations in question,

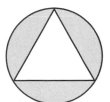

Figure 5.1 *Approximating the area of a circle.*

then the limiting number defines the length or area or volume, or whatever property that the application of the integral calculus is intended to measure. This tells us that if we are to advance beyond simple problems of length or area or volume, then we must master the concepts of *infinite process* and *limit* that are the basic notions of the calculus.

The techniques of early numerical integration let us approximate the length of a curve, like *PQ* in Figure 5.2, by measuring the length of a connected sequence of small straight chords *PA*, *AB*, *BC*, ... The closer together *A*, *B*, *C*, . . . are, and the shorter the chords, then the closer they will approximate the arcs *PA*, *AB*, *BC*, ... We see that the sum *PA* + *AB* + *BC* + . . . is the approximate length of the curve. The actual length of the curve is the *limit* of this sum as the chords become shorter and shorter but more and more numerous. Finding this limit, if it exists, is *integration*.

Here is a very similar procedure for finding the circumference of a circle. As the number of sides increases, the perimeter of the inscribed regular polygon gets closer and closer to the circumference of the circle. So to approximate the circumference of a circle, we could find the perimeter of an inscribed polygon with a very large number of sides. We can use this and the idea of a limit to define the circumference. Suppose we let *p* be the perimeter of an inscribed regular polygon with *n* sides and let *C* be the circumference of the circle. By choosing *n* large enough, we can find *p* as close to *C* as we want. We say *p* approaches *C* as a limit or *p* → *C*. The circumference of the circle is the limit of the perimeters of the sequence of inscribed regular polygons.

Archimedes credits Eudoxus as the first to state the lemma [a *lemma*: is a minor proposition or theorem helping to prove a more important or more general theorem] that now bears Archimedes' name—sometimes also known as the *axiom of continuity*. This served as the basis for the method of exhaustion, the Greek equivalent of the integral calculus. From this it is an easy step to prove the proposition that formed the basis of this method. As one early translation from the Greek rather awkwardly puts it:

> "If from any magnitude there be subtracted a part not less than its half, and if from the remainder one again subtracts not less than its half, and if this process of subtraction is continued, ultimately there will remain a magnitude less than any preassigned magnitude of the same kind."

This proposition is equivalent to the modern statement that if a) *M* is a given magnitude, b) ε is a preassigned and much smaller magnitude of the same kind, and c) *r* is a ratio such that $1/2 \leq r < 1$, then we can find a positive integer *N* such that $M(1 - r)^n < \varepsilon$ for all positive integers $n > N$. This means that the exhaustion property is equivalent to the modern statement that $\lim_{n \to \infty} M(1 - r)^n = 0$.

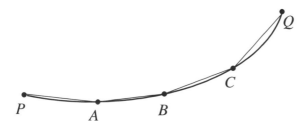

Figure 5.2 *Approximating a curve segment.*

5.2 Sequences and Series

The method of exhaustion and its variations shows us how to add smaller and smaller bits of area or length so that the accumulated total approaches some limiting area or length. Certain sets of real numbers may also approach some limiting value. We begin with some definitions.

A *sequence* is a succession of terms in a particular order and is formed according to some fixed rule. For example:

$$1, 4, 9, 16, 25, \ldots \quad \text{and} \quad 1, -x, \frac{x^2}{2}, -\frac{x^3}{3}, \frac{x^4}{4}, \ldots \tag{5.1}$$

The first, of course, is the sequence of squares of the natural numbers, and the second follows from a somewhat more complicated formula, soon explained.

A *series* is the sum of the terms of a sequence. From the above sequences we construct the following series:

$$1 + 4 + 9 + 16 + \cdots \quad \text{and} \quad 1 - x + \frac{x^2}{2} - \frac{x^3}{3} + \cdots \tag{5.2}$$

When the number of terms is fixed and countable, the sequence or series is *finite*. An *infinite* sequence has infinitely many terms, as many terms, in fact, as there are natural numbers. In these sequences, we are usually interested in what happens in the long run. We want to find out what happens to the nth term when n becomes very large. For example, in the sequence $1, 1/2, 1/3, 1/4, 1/5, \ldots, 1/n$, the terms get smaller and smaller, getting closer and closer to 0. This infinite sequence approaches zero as the limiting value. We indicate this by writing

$$\lim_{n \to \infty} \frac{1}{n} = 0 \tag{5.3}$$

again, where "lim" stands for *limit* and "$n \to \infty$" stands for n approaches infinity or, in other words n increases beyond any bound.

The *general*, or nth, term is an expression which describes how to construct the individual terms. In the first example above, the general, or nth, term is n^2. Here we construct the first term by setting $n = 1$, the second term by setting $n = 2$, and so on. In the second example, the nth term, except for $n = 1$, is

$$\frac{(-x)^{n-1}}{n-1} \tag{5.4}$$

To indicate that a sequence or series continues indefinitely, we use the ellipsis, a set of three dots, as in

$$1, 4, 9, \ldots, n^2 \tag{5.5}$$

Here is a geometric series of n terms:

$$S_n = a + ar + ar^2 + \cdots + ar^{n-1} \tag{5.6}$$

and if we apply some elementary algebra to this equation we will produce

$$S_n = \frac{a(1 - r^n)}{1 - r} \quad \text{or} \quad S_n = \frac{a(r^n - 1)}{r - 1} \tag{5.7}$$

The first form of Equation 5.7 is useful if $|r| < 1$, and the second form if $|r| > 1$.

If $|r| < 1$, then r^n decreases in numerical value as n increases indefinitely. In fact, we can make the value of r^n as close to zero as we want simply by choosing a large enough value for n. We write this as

$$\lim_{n \to \infty} r^n = 0 \tag{5.8}$$

and we see that the sum of the terms must be

$$\lim_{n \to \infty} S_n = \frac{a}{1 - r} \tag{5.9}$$

So for $|r| < 1$ the sum S_n of a geometric series approaches a limit as we increase the number of terms. In this case the series is *convergent*; it has a finite computable value.

On the other hand, if $|r| > 1$, then r^n becomes infinite as n increases indefinitely. So, from the second formula for S_n above, the sum S_n becomes infinite. In this case the series is *divergent*.

A peculiar situation arises if $r = -1$, for the series then becomes

$$a - a + a - a + \cdots \tag{5.10}$$

Here, if n is even, the sum is zero. If n is odd, the sum is a. As n increases indefinitely, the sum does not increase indefinitely, and it does not approach a limit. Such a series is an *oscillating series*.

Now we will consider the geometric series with $a = 1$ and $r = 1/2$. This expands to

$$S_n = 1 + \frac{1}{2} + \frac{1}{4} + \cdots + \frac{1}{2^{n-1}} \tag{5.11}$$

and using Equation 5.7 we find that

$$S_n = \frac{1 - \frac{1}{2^n}}{1 - \frac{1}{2}} = 2 - \frac{1}{2^{n-1}} \tag{5.12}$$

where $\lim_{n \to \infty} S_n = 2$.

We can interpret Equation 5.12 geometrically as follows: Lay off successive values of S_n on the x axis, as in Figure 5.3. Each succeeding point bisects the segment between the preceding point and the point at $x = 2$. Here the limit is obvious.

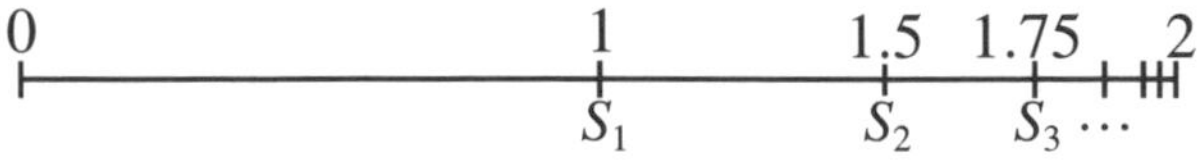

Figure 5.3 *Geometric interpretation of a convergent series.*

The limit of a sequence is a special type, because a sequence is a *function* whose *domain* is the set of natural numbers, $\{1, 2, 3, \ldots, n\}$, ordered according to size. The *range*, that is the set of function values, listed in the corresponding order, is $\{a_1, a_2, a_3, \ldots a_n\}$, the ordered set, which is a sequence.

Notice that subscripts will play a bookkeeping role for us, to help us keep track of similar terms and their order. At the very least, they serve as simple identification tags. Using numbers for subscripts usually, but not necessarily, indicates an ordering of some sort: We may read t_1 as "the first time" and t_2 as "the second time" of some sequence of events. Furthermore, mathematicians usually reserve the lowercase Roman italic letters i, j, k, m, and n for subscripts on a general or intermediate term of an expression or sequence of similar terms.

We can use the behavior of sequences and series to help us to define limits more carefully. Here are four different sequences, where we let $n \to \infty$:

1. $a_n = (n - 1)/(n + 1)$: For sufficiently large n we ignore the 1's and see that $a_n \to 1$.
2. $a_n = \frac{1}{2}a_{n-1} + 2$: No matter what value of a_1 we start with, we will always find that $a_n \to 4$.
3. a_n = the chance of living to the age of n years: Since the fountain of youth is as yet undiscovered, we can sadly but safely assume that $a_n \to 0$.
4. a_n = the proportion of zeros among the first n digits of π: Since the digits of π do not form a repeating sequence, we can expect that $a_n \to 0.10$.

Let us see what is required of a sequence of positive numbers for it to approach zero. In other words, what do we mean when we write $a_n \to 0$? Clearly, the numbers $\{a_1, a_2, a_3, \ldots a_n\}$ must somehow become smaller. In fact, for any small number we can specify the a_ns must eventually go below that number and stay below it.

We test for convergence to zero by choosing some small number—let's try 10^{-6} (that's one millionth or $1/1,000,000$). Then for some n, the a_n must be less than that number and stay below it. Before that point in the sequence, a_n may oscillate above and below our arbitrary threshold limit of 10^{-6}, but eventually all terms must go below this limit. Later in the sequence, maybe much later, all terms must be less than, say, 10^{-12}. Obviously, we can continue this process indefinitely. It does not matter how many terms must pass in a sequence before values of all succeeding terms fall below and remain below these ever decreasing and arbitrary threshold values. Because the sequence does eventually meet this test, we conclude that the sequence does indeed converge to zero, or approach zero as a limit.

The sequence $10^{-3}, 10^{-2}, 10^{-6}, 10^{-5}, 10^{-9}, 10^{-8}, \ldots$ converges to zero. Even though the values oscillate up and down, at some point all the succeeding numbers eventually stay below any arbitrarily small number.

However, we see that the sequence $10^{-4}, 10^{-6}, 10^{-4}, 10^{-8}, 10^{-4}, 10^{-10}, \ldots$ does not converge to zero, because the values do not remain less than 10^{-4}.

Mathematicians, engineers, and scientists often use the Greek lower case letter ε (epsilon) to represent an arbitrarily small positive number, and we will do the same.

The smaller ε is, the farther along the sequence we must go to reach an a_n whose value, and all succeeding values, are less than ε. Here is a more concise statement: For any ε there is an N such that $a_n < \varepsilon$ if $n > N$. Figure 5.4 shows examples of this.

Or consider the sequence: $\frac{1}{2}, \frac{4}{4}, \frac{9}{8}, \frac{16}{16}, \frac{25}{32}, \ldots, \frac{n^2}{2^n}$. Here the second and third terms are each greater than their preceding terms. However, the denominator 2^n grows faster than n^2, and the ratio goes below any ε we might choose.

Nothing prevents us from constructing sequences where the a_n may be negative as well as positive. The terms can converge to zero from the positive or negative side, or they can converge from both sides. For any of these conditions we require that for some a_n, and beyond, the values fall within any strip $\pm \varepsilon$ of zero, and stay there (Figure 5.5a).

Now that we allow a sequence to have members that are negative, we must define a member's distance from zero as its absolute value $|a_n|$. So $a_n \to 0$ really implies that $|a_n| \to 0$. We can revise our earlier statement to read: For any ε there is an N

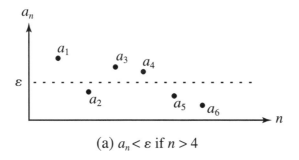

(a) $a_n < \varepsilon$ if $n > 4$

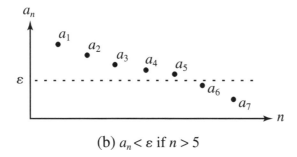

Figure 5.4 *Testing for convergence.*

(b) $a_n < \varepsilon$ if $n > 5$

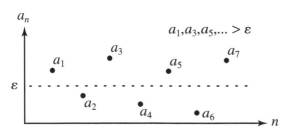

(c) Oscillating nonconvergence

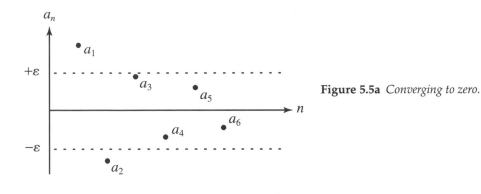

Figure 5.5a *Converging to zero.*

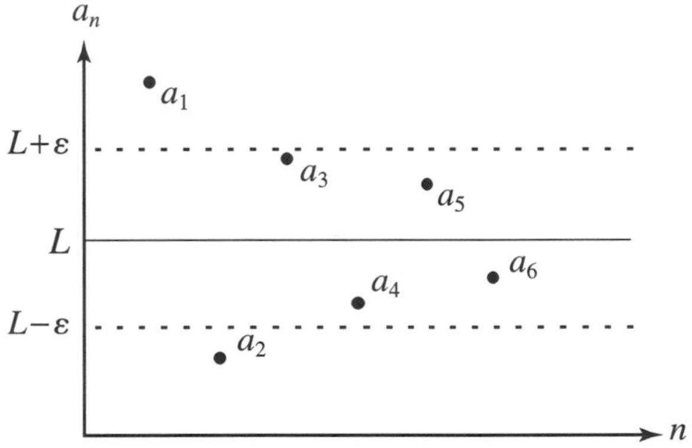

Figure 5.5b *Converging to L.*

such that

$$|a_n| < \varepsilon \text{ if } n > N \tag{5.13}$$

What about limits other than zero? If $a_n - L$ converges to zero, then the limit is L, and our test applies to the absolute value $|a_n - L|$. Now we are ready to further generalize the conditional statement defining convergence: For any ε there is an N such that

$$|a_n - L| < \varepsilon \text{ if } n > N \tag{5.14}$$

Notice that for a convergent sequence only a finite number of members are outside any ε strip around L (Figure 5.5b), and we can write $a_n \to L$ or $\lim a_n = L$. Both are equivalent to

$$\lim_{n\to\infty} a_n = L \tag{5.15}$$

Given two convergent sequences, $a_n \to L$ and $b_n \to M$, we can combine them to create other convergent sequences. We use the basic rules of arithmetic and algebra to do this.

Addition: If $c_n = a_n + b_n$, then $c_n \to (L + M)$.

Subtraction: If $c_n = a_n - b_n$, then $c_n \to (L - M)$.

Multiplication: If $c_n = a_n b_n$, then $c_n \to LM$.

Division: If $c_n = a_n/b_n$, then $c_n \to L/M$, for $M \neq 0$.

A useful special case is $ka_n \rightarrow kL$, where the constant k is factored out of the limit to give $\lim ka_n = k \lim a_n$.

5.3 Functions

We now shift our attention from sequences to functions. Instead of discrete values $a_1, a_2, a_3, \ldots, a_n$, we will explore a continuum of values represented by $f(x)$.

A *function* is one of the most important concepts in mathematics. It is usually an algebraic expression whose value depends on the value we assign to some variable. For example, $x^2 + 1$ is a function of x. If we set $x = 2$, then $x^2 + 1 = 5$. The symbol $f(x)$ is a short way to write the function of an independent variable x. (Sometimes just f is sufficient.) We read this as "f of x," and for this example we write $f(x) = x^2 + 1$. The number $f(x)$ is produced by the number x. Notice that $f(x)$ does not mean "f times x." It is a single, indivisible symbol standing for some function.

In the equation $y = f(x)$, the value of y depends on the value of x. The function $f(x)$ is some expression or rule that processes the values of x that we *input* and produces as *output* a value of y. We call x the independent variable and y the dependent variable.

We see that a function requires three things to describe its nature: two nonempty sets, A and B, and a rule associating each element of A with a corresponding element in B. The correspondence indicated by the arrows shown in Figure 5.6 is an example of a function *from A into B*.

Here, again, $f(x)$ denotes some function, and we say that $f(x)$ associates x with y. This means that y_1, for example, is the unique element of B that corresponds to the element x_1 of A according to the rule given by the function $f(x)$, where the subscripts on x and y mean that x_1 and y_1 are specific numbers. The independent variable x_1 is whatever we arbitrarily choose it to be, and the dependent variable y_1 is what the function $f(x)$ works out for us. It is also common to say that "$f(x)$ *maps* x to y," or "y is the *image* of x under control of $f(x)$," or "y is the value of $f(x)$, given x." In more concise mathematical notation, we have the familiar

$$y = f(x) \tag{5.16}$$

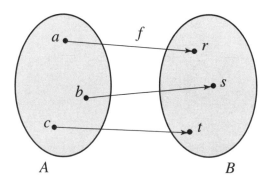

Figure 5.6 *A function as a mapping from A into B.*

and read this as "y equals f of x." The other mappings shown in the figure are expressed as $y_2 = f(x_2)$, and $y_3 = f(x_3)$. Notice that the terms *function, mapping,* and *transformation* are very nearly synonymous. (We will look at their subtle differences later.)

More about the f notation itself: There is nothing special about the letter f. It is obviously the first letter of the word "function," and that is probably why earlier mathematicians chose it to stand for the process of a function. We may use it and other letters to distinguish between functions operating simultaneously or in the same application. For example, $f(x) = x^2 - x + 1$, $g(x) = 4x + 3$, and $h(x) = -x$. Different letters may also indicate different variables, as with $f(x)$, $g(y)$, and $h(z)$. And, of course, we may have functions of several variables, as with $f(x, y)$, $\phi(x, y, z)$, and $g(r, \theta)$, for example.

Set A is the *domain* of the function $f(x)$, and the set of all values of $y = f(x)$ is the *range* of the function. We can denote these as $Dom(f)$ and $Ran(f)$, respectively. The domain of a function is the set of *inputs,* or x_is. The *range* is the set of outputs, or y_is.

The terms *into* and *onto* have very special meanings when we use them to describe the actions of functions. If f is a function from A *into* B, then $A = Dom(f)$ and $Ran(f)$ is a subset of B. However, if $Ran(f) = B$, then f is a function from A *onto* B, where the term "onto" implies that every element of B is an image of a corresponding element of A.

Let's look at some examples. If A is the set of all integers, $1, 2, 3, \ldots, n$, and B is the set of all even integers, $2, 4, 6, \ldots, 2n$, then $f(n) = 2n$ is a mapping from A onto B, because $Ran(f) = B$. For the same sets A and B, $f(n) = 4n$ is a mapping from A into B, because $Ran(f)$ in this case is clearly a subset of B. The function $f(n) = 2n$ generates all the elements of B, while the function $f(n) = 4n$ generates only integers divisible by four (forming a subset of B). Figure 5.7 illustrates this distinction.

Nothing prevents sets A and B from being identical, that is $A = B$. This lets us investigate mapping a set onto or into itself, an important notion in geometric transformations. The set of real numbers, R, is a good example. The domain of the function $f(x) = x^2$ is the set of all real numbers, but its range includes only positive numbers. So it is an example of mapping a set into itself. If $f(x) = x^3$, then it is a mapping of a set onto itself. Figure 5.8 gives a visual interpretation of this mapping. You can easily verify this by trying a few real numbers, positive and negative, as input to these functions. What do you conclude about the function $f(x) = \sin x$?

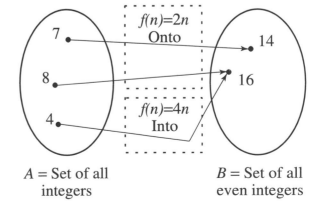

Figure 5.7 *Distinguishing between* onto *and* into.

$f(n)=2n$ Onto

$f(n)=4n$ Into

A = Set of all integers

B = Set of all even integers

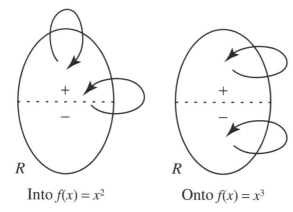

Figure 5.8 *Mapping a set* into *itself versus mapping a set* onto *itself.*

R

Into $f(x) = x^2$

R

Onto $f(x) = x^3$

Figure 5.8 illustrates the notion of mapping a set both into and onto itself. We choose to work with R, the set of real numbers. Mathematicians, particularly geometers, like to think of these numbers as points on a line, the so-called *real line* or *number line* extending indefinitely in both the positive and negative directions. This analogy gives us an excellent way to look at "onto" and "one-to-one" transformations.

Now, let's refine our notion of a function still further. If and only if we can pair off the elements of the two sets so that every member of either set has a unique mate in the other set, then we can say that the two sets are in a *one-to-one correspondence*. If f is a function such that for x_1 and x_2 in $Dom(f)$, then $x_1 \neq x_2$ implies that $f(x_1) \neq f(x_2)$, and we can call f a one-to-one function (Figure 5.9). In other words, if f is one-to-one, then $f(x_1) = f(x_2) \Rightarrow x_1 = x_2$. A discussion of mappings of the real line will help to explain these notions.

Let's begin with the function $f(x) = x+1$. This maps every real number exactly one unit to the right. For example, $x = -2.7$ transforms to $x' = -1.7$, $x = -0.2$ transforms to $x' = 0.8$, and $x = 1$ transforms to $x' = 2$. Every point in R has a unique image point, and every point in R is the unique image of some other point. No gaps or overlaps are produced by this transformation (Figure 5.10).

Now try $f(x) = -2x$. Do you see why it, too, is both onto and one-to-one (Figure 5.11)?

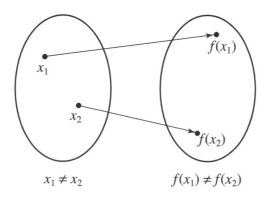

x_1

$f(x_1)$

x_2

$f(x_2)$

Figure 5.9 *A one-to-one* function.

$x_1 \neq x_2$

$f(x_1) \neq f(x_2)$

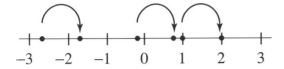

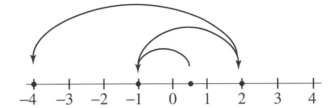

Figure 5.10 *A mapping on the real line.*

Figure 5.11 *Another mapping on the real line.*

The term "function," as we have used it here, is inclusive of the terms "transformation" and "mapping" and has a broader application. The terms "transformation" and "mapping" suggest geometric processes. The variables in a functional relation may be any kind of mathematical object. When we refer to mappings and geometric transformations, we usually mean to imply that the variables represent geometric objects. Point coordinates are a good example.

Think of a function as a machine with an input value to make it "go" and some sort of internal workings that produce an output (Figure 5.12). The input set, or domain, might be the set of real numbers, and the internal workings of the "machine" might be a simple mathematical expression, such as $3x^2$, or perhaps a complicated algorithm. The machine's output values constitute the range. So, simply insert a value from the set of values contained in the domain, and the machine cranks out the corresponding value in the range set. (Did you notice that in this example, the function produces the same output value for certain pairs of input values?)

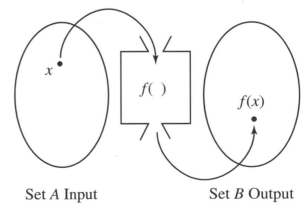

Figure 5.12 *A function as a "machine" for turning input values into output values.*

Set *A* Input   Set *B* Output

Let $f(x) = 3x^2$, then

$$f(-1) = 3$$
$$f(0) = 0$$
$$f(1) = 3$$
$$\vdots$$
$$f(a) = 3a^2$$
$$f(x_1) = 3x_1^2$$

A function contains instructions for generating all the possible values of the output. To know a function we must know something about its inputs and outputs, its domain and range. Throughout any problem statement and solution a particular function symbol, say $g(x)$, stands for the same rule. In simple cases, as we have seen, this rule takes the form of a series of algebraic operations on an independent variable. The function symbol says we must perform these same operations on whatever different values of the independent variable the problem demands. For example, if

$$f(x) = x^2 - 9x + 14 \tag{5.17}$$

then

$$f(y) = y^2 - 9y + 14 \tag{5.18}$$

and

$$f(b + 1) = (b + 1)^2 - 9(b + 1) + 14 = b^2 - 7b + 6 \tag{5.19}$$
$$f(0) = 0^2 - 9 \cdot 0 + 14 = 14 \tag{5.20}$$
$$f(-1) = (-1)^2 - 9(-1) + 14 = 24 \tag{5.21}$$
$$f(3) = 3^2 - 9 \cdot 3 + 14 = -4 \tag{5.22}$$

Let's see what happens to a function when we make some simple changes and create new functions. Figure 5.13a shows a graph of the function $f(x) = x^2 + 1$. We can add or subtract a constant from $f(x)$ in two different ways, $f(x) - 2$ and $f(x - 2)$, for example. However, it is not always easy to see at once the difference between these two

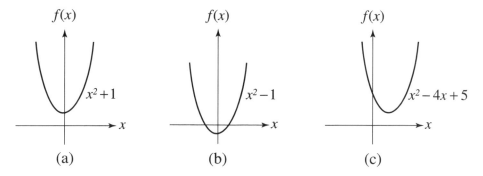

Figure 5.13 *Two different ways to change a function.*

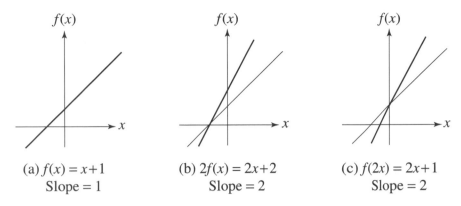

Figure 5.14 *More ways to change a function.*

new functions. In $f(x) - 2$, we subtract 2 from each value of the function $f(x)$. This moves the whole graph down (Figure 5.13b). In $f(x - 2)$, we subtract 2 from the dependent variable first and then compute the value of the function. This moves the graph over to the right (Figure 5.13c). The figure shows both movements. The new function $f(x) - 2$ is $x^2 - 1$, and the new function $f(x - 2)$, is $x^2 - 4x + 5$.

There are other ways to change a function. For example, instead of subtracting or adding, we can multiply the function or the independent variable by 2. This compresses the function's graph in the horizontal direction. Let's see what happens to the function $f(x) = x + 1$, Figure 5.14a. Figure 5.14b shows $2f(x)$, and Figure 5.14c shows $f(2x)$.

We can also combine two function by adding them, $f(x) + g(x)$, or even by multiplying them, $f(x) \cdot g(x)$.

5.4 Limit of a Function

Mathematicians base everything in calculus on two concepts: the limit of an infinite sequence, and the limit of a function. We have done the first. The second concept turns out to be easy to illustrate and easy to explain informally, but very difficult to define in a rigorous way.

We begin with some function of x, say simply $f(x)$, and some constant, a, such that $f(a) = L$. If we can make the value of $f(x)$ as close to L as we want by choosing values of x close enough to a but nonetheless different from a, then we say that the limit of $f(x)$, as x approaches a, is L. We write this as

$$\lim_{x \to a} f(x) = L \tag{5.23}$$

which we read as "the limit of f of x, as x approaches a, is L."

This means that if x is close to a, then $f(x)$ is close to L. If another value of x is even closer to a, then the resulting $f(x)$ is also even closer to L. If $x - a$ is small, then

$f(x) - L$ should be small. When we say "small" we mean arbitrarily small or below any ε. The difference $f(x) - L$ must become as small as we choose to make it by choosing values of x closer and closer to a. For example, as x approaches $a = 0$, the function $f(x) = 3 - x^3$ approaches $L = 3$. We can also express this idea by writing $f(x) \to L$ as $x \to a$, which we read "f of x approaches L as x approaches a."

The last statement is awkward and poses some problems because it involves two limits. We see that the limit $x \to a$ is forcing $f(x) \to L$. (Remember that for sequences, $n \to \infty$ forced $a_n \to L$.) We should not expect the same ε in both limits. That is, we cannot expect that $|x - a| < \varepsilon$ produces $|f(x) - L| < \varepsilon$. We may need to put x extremely close to a (closer than ε), and we must ensure that if x is close enough to a, then $|f(x) - L| < \varepsilon$.

This brings us to the so-called "epsilon-delta" definition of the limit of a function. First, we choose ε and show that $f(x)$ is within ε of L for every x near a. Then we choose some number δ (delta) that quantifies the meaning of "near a." Our goal is to get $f(x)$ within ε of L by keeping x within δ of a:

$$\text{If } 0 < |x - a| < \delta, \text{ then } |f(x) - L| < \varepsilon \qquad (5.24)$$

The concept of a limit is a mathematical refinement of the intuitive notion that the limit L is a number which $f(x)$ approaches as the value of x approaches a, which we interpret geometrically as a moving point, say x moving toward a, $f(x)$ moving toward L. This, of course, is a relic of the original Newtonian calculus, but it is still useful. It brings us to a graphical interpretation of the definition of a limit. Consider the graph of $f(x)$ near $x = a$. Figure 5.15 shows a function for which $\lim_{x \to a} f(x)$ exists, and Figure 5.16 shows a function for which $\lim_{x \to a} f(x)$ does not exist.

For the example in Figure 5.15, we force $f(x)$ to differ from L by less than a prescribed quantity ε, where $|f(x) - L| < \varepsilon$. Then all that we need to do is to choose x anywhere within a certain amount δ of the value $x = a$, where $0 < |x - a| < \delta$. That is, it is possible near $x = a$ on the curve in this figure to restrict the variation of $f(x)$ to as little as we want by sufficiently narrowing the vertical band around $x = a$. For this curve we have $\lim_{x \to a} f(x) = L$. In fact, here $f(a) = L$, and so in this special case we call $f(x)$ *continuous* (a property we will discuss in a later section). The functions in Figures 5.16 and 5.17 are not continuous.

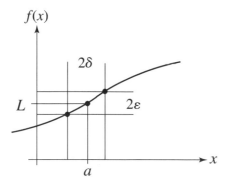

Figure 5.15 *Function approaching a limit.*

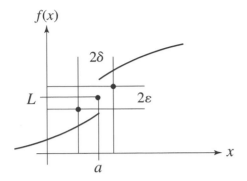

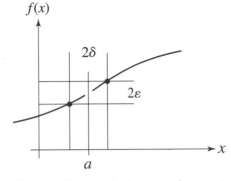

Figure 5.16 *Function has no unique limit at a. It is not continuous.*

Figure 5.17 *Function is not continuous at a.*

Next we consider the situation near $x = a$ for the curve in Figure 5.16. There, no matter how narrow a band we choose about $x = a$, the variation in $f(x)$ can never by made arbitrarily small because the $\lim_{x \to a} f(x)$ does not exist. Here $f(x)$ is a *step function* and has no limit as x approaches the step or jump in value.

It turns out that the existence or nonexistence of $f(a)$, the value of $f(x)$ at $x = a$, has nothing whatever to do with the existence or nonexistence of the limit of $f(x)$ as x *approaches a* (Figure 5.17). Here $f(x) \to L$, because when we apply the limit process we ignore the point $x - a$. The actual value $f(a)$ can be anything, with no effect on L.

Now let's look at a restricted version of the limit process. We can write

$$\lim_{x \to a^+} f(x) = L \tag{5.25}$$

and mean by $x \to a^+$ that each x involved is greater than a. This kind of limit is a *right-hand limit*. The independent variable x approaches a from the right.

A *left-hand limit*,

$$\lim_{x \to a^-} f(x) = M \tag{5.26}$$

with x remaining less than a, is also possible.

If the ordinary limit exists, then the right-hand and left-hand limits exist and all three have the same value. If the right- and left-hand limits exist and have the same value, the limit itself exists and has that value.

We can apply these ideas (and some new ones) to finding the limit of a very important function. We can show that if an angle α is measured in radians, then

$$\lim_{\alpha \to 0} \frac{\sin \alpha}{\alpha} \to 1 \tag{5.27}$$

Consider Figure 5.18 in which RV is a circular arc with radius r with its center at B. The angles RTB and SVB are right angles.

We see that the triangle RTB is contained in the sector RVB, and this sector is itself contained in the larger triangle SVB, so it follows that Area $\triangle RTB <$ Area sector $RVB <$ Area $\triangle SVB$.

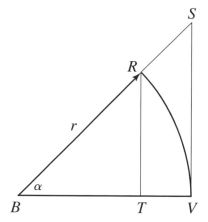

Figure 5.18 *The sine of an angle divided by the angle approaches 1 as the angle approaches zero.*

Since

$$\text{Area } \triangle RTB = \frac{1}{2}\overline{BT} \cdot \overline{RT} = \frac{1}{2}r\cos\alpha \cdot r\sin\alpha \qquad (5.28)$$

and

$$\text{Area sector } RVB = \frac{\alpha}{2\pi}(\pi r^2) = \frac{1}{2}\alpha r^2 \qquad (5.29)$$

and

$$\text{Area } \triangle SVB = \frac{1}{2}\overline{BV} \cdot \overline{SV} = \frac{1}{2}r \cdot r\tan\alpha \qquad (5.30)$$

we may conclude that, for $0 < \alpha < \frac{\pi}{2}$

$$\frac{1}{2}r^2\cos\alpha\sin\alpha < \frac{1}{2}\alpha r^2 < \frac{1}{2}r^2\tan\alpha \qquad (5.31)$$

Next, we divide each member of these inequalities by the positive quantity $\frac{1}{2}r^2\sin\alpha$ and obtain

$$\cos\alpha < \frac{\alpha}{\sin\alpha} < \frac{1}{\cos\alpha} \qquad (5.32)$$

We can rewrite this by inverting each member, so that

$$\frac{1}{\cos\alpha} > \frac{\sin\alpha}{\alpha} > \cos\alpha \qquad (5.33)$$

As α approaches zero, both $\cos\alpha$ and $1/\cos\alpha$ approach unity. Therefore, $\sin\alpha/\alpha$, which becomes squeezed in between them, must also approach unity. That is,

$$\lim_{\alpha\to0^+} \frac{\sin\alpha}{\alpha} = 1 \qquad (5.34)$$

This is also true for the left-hand limit, because

$$\frac{\sin \alpha}{\alpha} = \frac{\sin(-\alpha)}{(-\alpha)} \tag{5.35}$$

Notice that direct substitution of $\alpha = 0$ in the expression $\sin \alpha / \alpha$ produces the unacceptable result $0/0$. And this completes our proof of the assertion of Equation 5.27).

A careful application of this method lets us compute a variety of other important limits. For example, we can determine $\lim\limits_{\theta \to 0} \frac{\sin 3\theta}{\theta}$. Since

$$\frac{\sin 3\theta}{\theta} = 3 \cdot \frac{\sin 3\theta}{3\theta} \tag{5.36}$$

we simply take $\alpha = 3\theta$ in Equation 5.26 to produce

$$\lim_{\theta \to 0} 3 \cdot \frac{\sin 3\theta}{3\theta} = 3 \lim_{\theta \to 0} \frac{\sin 3\theta}{3\theta} = 3 \tag{5.37}$$

If you are tempted to think that Equation 5.27 justifies us in assigning a value to the nonsense symbol $\sin 0/0$, you should note that the value suggested by this equation would be in conflict with Equation 5.37.

5.5 Limit Theorems

How does the limit process apply to the algebraic combination of two or more functions? Sometimes the problems we must confront involve the sum or product of two or more functions, or the quotient of two functions. We will need the following three theorems on limits:

Theorem 1: The limit of the sum of two (or more) functions is equal to the sum of their limits:

$$\lim_{x \to a}[f(x) + g(x)] = \lim_{x \to a} f(x) + \lim_{x \to a} g(x) \tag{5.38}$$

Theorem 2: The limit of the product of two (or more) functions is equal to the product of their limits:

$$\lim_{x \to a}[f(x) \times g(x)] = \left[\lim_{x \to a} f(x)\right] \times \left[\lim_{x \to a} g(x)\right] \tag{5.39}$$

Theorem 3: The limit of the quotient of two functions is equal to the quotient of their limits, provided the limit of the denominator is not zero:

$$\lim_{x \to a} \frac{g(x)}{f(x)} = \frac{\lim\limits_{x \to a} g(x)}{\lim\limits_{x \to a} f(x)} \quad \text{if} \quad \lim_{x \to a} f(x) \neq 0 \tag{5.40}$$

These theorems assume that the limits of the two functions exist. However, even though it may be true that neither function separately approaches a limit, the sum, product, or quotient may do so.

The exceptional case of Theorem 3, in which the denominator approaches zero, requires further investigation. Thus, given any quotient $g(x)/f(x)$ in which $f(x)$ approaches zero, there are two possibilities: $g(x)$ does not approach zero or $g(x)$ also approaches zero.

In the first case, we can make the value of the quotient $g(x)/f(x)$ greater than any preassigned value by making $f(x)$ sufficiently small, so that the quotient will not approach a limit. In the second case, where the numerator and the denominator both approach zero, Theorem 3 does not apply, because the ratio of the limit is the meaningless expression $0/0$. However, the limit of the ratio $g(x)/f(x)$ may exist.

Here are some examples:

Example (a): Evaluate $\lim_{x \to 3}(x^3 + 2x)$. By Theorem 2 we have

$$\lim_{x \to 3}(x^3) = \lim_{x \to 3}(x \times x \times x) = 3 \times 3 \times 3 = 27 \tag{5.41}$$

and by Theorem 1:

$$\lim_{x \to 3}(x^3 + 2x) = \lim_{x \to 3}(x^3) + \lim_{x \to 3}(2x) = 27 + 6 = 33 \tag{5.42}$$

Example (b): Evaluate $\lim_{x \to 2} \frac{x^3 - 2x^2 - 3x + 6}{x - 2}$. By Theorems 1 and 2 we have

$$\lim_{x \to 2}(x^3 - 2x^2 - 3x + 6) = 8 - 8 - 6 + 6 = 0 \tag{5.43}$$

and

$$\lim_{x \to 2}(x - 2) = 0 \tag{5.44}$$

This gives us the exceptional case 2 under Theorem 3, but for all values of x except $x = 2$ we have

$$\frac{x^3 - 2x^2 - 3x + 6}{x - 2} = \frac{(x - 2)(x^2 - 3)}{x - 2} = x^2 - 3 \tag{5.45}$$

so that

$$\lim_{x \to 2} \frac{x^3 - 2x^2 - 3x + 6}{x - 2} = \lim_{x \to 2}(x^2 - 3) = 1 \tag{5.46}$$

Here is yet another way to look at the limit theorems of functions and their application: Suppose that the functions of $f(x)$, $g(x)$, and $h(x)$ each have the following limits,

$$\lim_{x \to a} f(x) = A \qquad \lim_{x \to a} g(x) = B \qquad \lim_{x \to a} h(x) = C \tag{5.47}$$

Then the following relationships hold:

$$\lim_{x \to a}[f(x) + g(x) - h(x)] = A + B - C \tag{5.48}$$

$$\lim_{x \to a}[f(x) \times g(x) \times h(x)] = ABC \tag{5.49}$$

$$\lim_{x \to a} \frac{f(x)}{g(x)} = \frac{A}{B} \text{ if } B \text{ is not zero} \tag{5.50}$$

Again we see that the limit of an algebraic sum, product, or quotient of functions of the same variable x is equal, respectively, to the same algebraic sum, product, or quotient of their respective limits, if for the quotient of two functions the limit of the denominator is not zero.

If c is a constant (independent of x) and A is not zero, then, from the previous discussion, we find

Theorem 4:

$$\lim_{x \to a}[f(x) + c] = A + c \tag{5.51}$$

$$\lim_{x \to a} cf(x) = cA \tag{5.52}$$

$$\lim_{x \to a} \frac{c}{f(x)} = \frac{c}{A} \tag{5.53}$$

Here are some examples:

Example (a): Show that $\lim_{x \to 2}(x^2 + 4x) - 12$. This function is the sum of two other functions, x^2 and $4x$. We begin by finding the limiting values of these two functions. From Theorem 2 we have

$$\lim_{x \to 2} x^2 = 4 \tag{5.54}$$

and from Theorem 4

$$\lim_{x \to 2} 4x = 4 \lim_{x \to 2} x = 8 \tag{5.55}$$

Using Theorem 1, we sum the two limits: $4 + 8 = 12$.

Example (b): Show that $\lim_{x \to 2} \frac{x^2 - 9}{x + 2} = -\frac{5}{4}$. Here we first consider the numerator, $\lim_{x \to 2}(x^2 - 9) = -5$ (from Theorems 2 and 4). For the denominator we have $\lim_{x \to 2}(x + 2) = 4$. So, from Theorem 3, we have the expected solution.

If the numerical value of a variable v ultimately becomes and remains greater than any preassigned positive number, however large, we say v becomes infinite. If v takes on only positive values, it becomes positively infinite. If v takes on only negative values, it becomes negatively infinite. The notation we use for these three cases is

$$\lim v = \infty \qquad \lim v = +\infty \qquad \lim v = -\infty \tag{5.56}$$

In these cases v does not approach a limit as we have defined it. We must understand the notation $\lim v = \infty$, or $v \to \infty$, as "v becomes infinite," and not as "v approaches infinity."

Because of the notation we use and for consistency, we sometimes read the expression $v \to +\infty$ as "v approaches the limit plus infinity." Similarly, $v \to -\infty$ is "v approaches the limit minus infinity." This expression is convenient, but we must not forget that infinity is not a limit, for infinity is not a number at all.

We can now write

$$\lim_{x \to 0} \frac{1}{x} = \infty \tag{5.57}$$

which means that $1/x$ becomes infinite when x approaches zero. So what happens if

$$\lim_{x \to a} f(x) = \infty \tag{5.58}$$

This tells us that if $f(x)$ becomes infinite as x approaches a as a limit, then $f(x)$ is discontinuous for $x = a$.

A function may have a limiting value when the independent variable becomes infinite. For example,

$$\lim_{x \to \infty} \frac{1}{x} = 0 \tag{5.59}$$

And, in general, if $f(x)$ approaches the constant value A as a limit when $x \to \infty$, we use the standard notation and write

$$\lim_{x \to \infty} f(x) = A \tag{5.60}$$

There are several special limits that can occur:

$$
\begin{aligned}
&\lim_{x \to 0} \frac{c}{x} = \infty \\[4pt]
&\lim_{x \to \infty} cx = \infty \\[4pt]
&\lim_{x \to \infty} \frac{x}{c} = \infty \\[4pt]
&\lim_{x \to \infty} \frac{c}{x} = 0
\end{aligned}
\tag{5.61}
$$

where $c \neq 0$. These special limits help us to find the limiting value of the quotient of two polynomials when the variable becomes infinite.

If f and g are functions of x, and if

$$\lim_{x \to a} f = A \quad \text{and} \quad \lim_{x \to a} g = 0 \tag{5.62}$$

and if A is not zero, then

$$\lim_{x \to a} \frac{f}{g} = \infty \tag{5.63}$$

5.6 Limit and the Definite Integral

Now we turn to the idea of integration, which gives us a way to add up infinitely many things, each of which is infinitesimally small. Actually doing the addition is not only not recommended, it is not even possible unless we use some approximations along the way. And that is the purpose of integration, to find the limits of infinite sums exactly and without the tedium of actually doing infinite sums. We saw a version of this earlier using the Greek method of exhaustion. (Computing the area of a shape that is not a rectangle is a good example of the adding up of nearly infinitely many nearly infinitesimally thin rectangles, or triangles in the case of polygons approximating a circle.)

We will need a good, efficient way to express sums of a large number of terms, usually an unspecified number n, such as

$$u_1 + u_2 + u_3 + \cdots + u_n \tag{5.64}$$

To avoid the trouble of writing out such expressions in full, we can use a single symbol. The symbol $\sum\limits_{i=1}^{n}$ means that we are to substitute $i = 1, 2, 3, \ldots, n$ successively in the expression following it and then add the results. Here are three examples,

$$\sum_{i=1}^{n} u_i = u_1 + u_2 + u_3 + \cdots + u_n \tag{5.65}$$

$$\sum_{j=1}^{n} a_j^2 = a_1^2 + a_2^2 + a_3^2 + \cdots + a_n^2 \tag{5.66}$$

$$\sum_{k=0}^{n} a_k x^{n-k} = a_0 x^n + a_1 x^{n-1} + a_2 x^{n-2} + \cdots + a_{n-1} x + a^n \tag{5.67}$$

The numbers or letters below and above Σ tell us where to start and stop summing. They are called the limits of the summation (not to be confused with the "limit" of a sequence or function). There is nothing special about the letters i, j, and k. They are "dummy" variables. We could just as well use them or others interchangeably as long as we are consistent within the context of a problem. Dummy variables are only on one side of the equation (the side with Σ), and they have no effect on the sum. The upper limit n is on both sides. Here are five more summations:

$$\sum_{k=1}^{n} k = 1 + 2 + 3 + \cdots + n \tag{5.68}$$

$$\sum_{j=1}^{4} (-1)^j = -1 + 1 - 1 + 1 = 0 \tag{5.69}$$

$$\sum_{j=1}^{5} (2j - 1) = 1 + 3 + 5 + 7 + 9 = 5^2 \tag{5.70}$$

$$\sum_{i=0}^{0} v_i = v_0 \tag{5.71}$$

$$\sum_{k=0}^{\infty} \frac{1}{2^k} = 1 + \frac{1}{2} + \frac{1}{4} + \cdots = 2 \text{[an infinite series]} \tag{5.72}$$

The numbers 1 and n, 1 and 4, 1 and 5, 0 and 0, and 0 and ∞ are the lower and upper limits of the sums. The dummy variable i, j, or k is the *index* of summation.

We can use the summation indices as exponents or scalar multipliers, so that

$$\sum_{i=0}^{3} a_i x^i = a_0 + a_1 x + a_2 x^2 + a_3 x^3 \tag{5.73}$$

Double summation provides a compact notation for even more complex expressions. We interpret the equation

$$\sum_{i=1}^{2} \sum_{j=1}^{3} a_i x_j = b \tag{5.74}$$

in this way: Call i the *outer index* and j the *inner index*. Start by setting the outer index to 1, and then exhaust the inner index. In other words, for $i = 1$, set $j = 1, 2$, and 3. This produces the following terms:

$$a_1 x_1 + a_1 x_2 + a_1 x_3 \tag{5.75a}$$

Next, increase the outer index i by 1, making $i = 2$, and again exhaust the inner index j, producing

$$a_2 x_1 + a_2 x_2 + a_2 x_3 \tag{5.75b}$$

This is the last step for this example (because the index i only takes on the values 1 and 2). The complete expansion of the summation of Equation 5.74 is

$$a_1 x_1 + a_1 x_2 + a_1 x_3 + a_2 x_1 + a_2 x_2 + a_2 x_3 = b \tag{5.76}$$

The integral calculus was invented as a way to calculate an area bounded by curves, by supposing the area to be divided into an "infinite number of infinitesimal parts called elements, the sum of all these elements being the area required." Historically, the integral sign $\int$ was merely an elongated S that early mathematicians used to indicate a sum. Here is a simple construction that demonstrates this:

First, divide the interval from $x = a$ and $x = b$ into any number of n subintervals, Δx_i, not necessarily equal. Next, construct the vertical ordinate lines at these points of division and make complete rectangles by drawing horizontal lines through the extremities of the ordinates, as in Figure 5.19.

It is clear that the sum of the areas of these n rectangles is an approximate value for the area under the curve between a and b. It is also clear that the limit of the sum of the areas of these rectangles, when their number n is indefinitely increased, will equal the

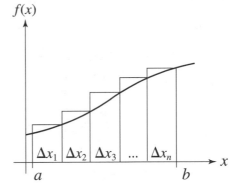

Figure 5.19 *Approximating the area under a curve.*

area under the curve. Now we will formulate this process into an explicit mathematical statement in the form of a summation:

1. Denote the lengths of the successive subintervals by

$$\Delta x_1, \Delta x_2, \Delta x_3, \ldots, \Delta x_n \tag{5.77}$$

2. Denote the x coordinates (the abscissas) of the points we choose, one in each subinterval, by

$$x_1, x_2, x_3, \ldots, x_n \tag{5.78}$$

3. The ordinates of the curve at these points are, respectively,

$$f(x_1), f(x_2), f(x_3), \ldots, f(x_n) \tag{5.79}$$

4. The area under the curve is approximately equal to

$$\text{Area} = f(x_1)\Delta x_1 + f(x_2)\Delta x_2 + \cdots + f(x_n)\Delta x_n \tag{5.80}$$

or

$$\text{Area} = \sum_{i=1}^{n} f(x_i)\Delta x_i \tag{5.81}$$

The limiting value of this sum when n increases without limit and when each subinterval approaches zero as a limit equals the value of the definite integral

$$\text{Area} = \int_a^b f(x)\,dx \tag{5.82}$$

so that

$$\int_a^b f(x)\,dx = \lim_{n\to\infty} \sum_{i=1}^{n} f(x_i)\Delta x_i \tag{5.83}$$

This is a very important result, because it shows us that we can calculate by integration a magnitude which is the limit of a summation.

Let's review all of this: The limit of the sum of the areas of more and more rectangles that approximate the area under a curve (the graph of a given function), and that become narrower and narrower, has a special symbol and also a special name. We denote it as $\int_a^b f(x)dx$, and refer to it as the *definite integral* of the given function from a to b. For any continuous function with the values $f(x)$, it defines the area of the region bounded by the x axis, the graph of $y = f(x)$, and the two vertical lines $x = a$ and $x = b$. The function whose values $f(x)$ are the heights of the rectangles is the *integrand*, and a and b are the *limits of integration*. They define two of the boundaries of the area. The term dx indicates that we divide the interval from a to b along the x axis into n parts, which we denote as Δx in the summation notation form.

5.7 Tangent to a Curve

In the calculus, the derivative of a function has a geometric interpretation. It is the slope of the tangent to the curve (the graph) of that function. This means that the value of the derivative at any point on the curve is equal to the slope of the tangent line to the curve at that point. One way to find the tangent is to find the limiting position of a secant. We then visualize the derivative as the limit of the slope of a variable secant as one point of a curve moves into coincidence with another. This process describes the *Fundamental Theorem* of the differential calculus and its application to geometry.

Now let's look at some details. We will find the tangent line to a curve at an arbitrary point P on it. Draw a secant through P and a neighboring point Q on the curve (Figure 5.20). Move Q along the curve so that it approaches closer and closer. This causes the secant line to revolve about P as it follows Q, until Q reaches its limiting position coincident with P. In this final position, the secant line lies on the tangent line at P.

Here, in four easy steps, is the algebra that corresponds to this geometry. Let $y = f(x)$ be the equation of the curve AB, that is, the graph of the function $f(x)$ in Figure 5.20. The coordinates of P are x, y, and the coordinates of Q, which is near P on the curve, are $x + \Delta x, y + \Delta y$. We want to find the value of the tangent of the angle ϕ, which of course is given by the ratio $\Delta y/\Delta x$.

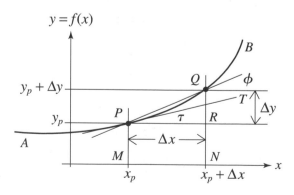

Figure 5.20 *Tangent to a curve.*

1. Evaluate this function at $x + \Delta x$, which yields $y + \Delta y$ (represented geometrically by NQ).

$$y + \Delta y = f(x + \Delta x) = NQ \qquad (5.84)$$

2. To find Δy compute $NQ - NR = RQ$, or

$$
\begin{aligned}
y + \Delta y &= f(x + \Delta x) & &= NQ \\
-y &= -f(x) & &= MP = NR \\
\hline
\Delta y &= f(x + \Delta x) - f(x) & &= RQ
\end{aligned}
\qquad (5.85)
$$

3. The ratio $\Delta y / \Delta x$ is simply

$$\frac{\Delta y}{\Delta x} = \frac{f(x + \Delta x) - f(x)}{\Delta x} = \frac{RQ}{MN} = \frac{RQ}{PR}$$

$$= \tan \angle RPQ = \tan \phi$$

$$= \text{slope of the secant line } PQ \qquad (5.86)$$

Now we have the ratio of the increments Δy and Δx, which equals the slope of the secant line drawn through the points P and Q on the graph of the function.

The geometry of the fourth and final step is as follows: We have fixed the value of x, so P is a fixed point on the graph. We allow Δx to vary and approach zero as a limit. The point Q moves along the curve and approaches P as a limiting position. The secant line drawn through P and Q then rotates about P and approaches the tangent line at P as its limiting position. In the figure we see that ϕ is the varying angle of the moving secant line PQ, and τ is the angle of the tangent line PT. Then it must be true that $\lim_{\Delta x \to 0} \phi = \tau$.

4. Assuming that $\tan \phi$ is a continuous function, we have the fourth and final step:

$$\frac{dy}{dx} = \lim_{\Delta x \to 0} \tan \phi = \tan \tau$$

$$= \text{slope of the tangent line at } P \qquad (5.87)$$

We see that as Δx approaches zero, Q approaches P along the curve, so that the slope of the secant approaches as its limit the slope of the curve at P. This limit is equal to the derivative of y with respect to x. This means that the derivative of a function is identical to the slope of the graph. And this means that if, in the formula for dy/dx, we substitute some value of x, the number we obtain is the slope of the curve at the point whose abscissa is the given value of x. It is at this point that the full story of the differential calculus begins.

5.8 Rate of Change

The idea of rate of change occurs all the time in our everyday experience. We are all familiar with expressions such as miles per hour, miles per gallon, pressure per square inch, price per pound, and so on. They all represent rates of change. An analogous idea in mathematics is that of the rate of change of a function.

Given some function $y = f(x)$, we can change x by some Δx. This, of course, causes a change in the dependent variable y, say Δy. If Δy is proportional to Δx, then the function changes uniformly, or at a uniform rate. This means that the change in y, corresponding to a given change in x, is constant, independent of the actual value of x. The rate is the change in the function divided by the change in the argument:

$$\frac{\Delta y}{\Delta x} = m \tag{5.88}$$

where m is a constant. The graph of this function is a straight line whose slope is m:

$$y = mx + k \tag{5.89}$$

When a function does not vary uniformly, the ratio $\Delta y / \Delta x$ is merely the average rate of change over the interval Δx. If we let Δx approach zero, this ratio approaches a definite limiting value, or the *instantaneous rate of change*:

$$\frac{dy}{dx} = \lim_{\Delta x \to 0} \frac{\Delta y}{\Delta x} \tag{5.90}$$

which is the rate of change of y with respect to x.

We need nothing more than our ordinary everyday experience to understand this idea. We can do a thought experiment to illustrate it. Suppose that we set up two posts beside a highway and measure the distance between them. A car drives past them. If Δx is the distance between the posts and if Δt is the time it takes the car to pass from one post to the other, then for a car traveling at a constant speed, that speed is simply $\Delta x / \Delta t$. However, if the speed varies, then this ratio is the average speed. If we wish to know the car's speed at a particular instant, say when it is passing the first post, common sense suggests what we place the posts close together. This decreases the time interval, Δt, so that there is less time for the speed to change very much, and the average speed will nearly equal the instantaneous speed.

In practice, we cannot use this procedure beyond a certain point, but it is clear that our intuitive idea of instantaneous speed is expressed exactly by $\Delta x / \Delta t$, the limit of $\Delta x / \Delta t$ as Δt approaches zero. We can now assert that the derivative of a function is identical to its rate of change. Now we realize that the three quantities derivative, slope, and rate of change are all equivalent.

The geometry shows us that when the slope of a curve (Figure 5.21) is positive (as on the curve segment AB), the ordinate y is increasing as x increases; when the slope is negative (as on the curve segment BC), the ordinate is decreasing. This simply tells us that a function increases or decreases according to whether its rate of change is positive or negative.

Calculus solves the following kinds of problems dealing with motion:

1. If the speed of an object is not constant but always changing, how do we compute the distance it travels?

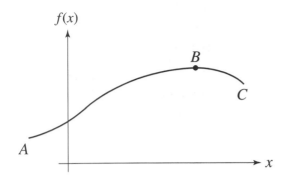

Figure 5.21 *Positive and negative rate of change of a function.*

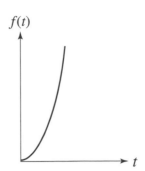

Figure 5.22 *Distance function.*

2. If the graph of distance traveled as a function of time, $f(t)$, is not a straight line, what is the slope at any point on it?

Each question is the converse of the other: find the distance from the speed, or find the speed from the distance.

For the very simplest cases, let's suppose that the speed at any time t is constant, that is, $s(t) = k$. For example, at $t = 1$ sec, the speed of an object is 4 meters/sec, and, in fact, no matter when we measure it, the speed is always the same: 4 meters/sec. The distance the object travels is simply $f(t) = kt$. But, with $s = 2t$, it is the acceleration that is constant. It equals 2, and is measured in, say, meters per second per second (or, m/sec²).

As a car accelerates, the speedometer goes steadily up. The distance traveled also goes up—faster and faster and each succeeding second it goes farther than in the previous second. If we measure t in seconds and s in meters per second, the distance f comes out in meters. After 14 seconds the speed is 28 meters per second, or just over 100 kilometers per hour (which is about 60 miles/hour). The speed and acceleration seem obvious. But how far has the car gone?

Let's look at a specific distance function. After an elapsed time t, the distance traveled is $f(t) = t^2$. Using this we can also find the speed $s(t)$ at that time t. The graph of $f(t) = t^2$ is a parabola (Figure 5.22). The curve starts at the beginning, with t set to zero. At $t = 5$, the distance traveled is $f(5) = 25$. After, say 10 seconds, f reaches 100. Speed is distance divided by time, but the speed is changing. Dividing $f = 100$ meters by $t = 10$ seconds gives us the speed $s = 10$ feet/sec, which is the average speed over 10 seconds. But how do we find the instantaneous speed, the reading on the speedometer when $t = 10$?

The car accelerates and its speed increases, the graph of t^2 gets steeper and it covers more distance in each second. The average speed between $t = 10$ and $t = 11$ is a better approximation, but it still is only an approximation of the exact speed at the time $t = 10$ seconds. But let's try it: The distance at $t = 10$ seconds is $f(10) = 10^2 = 100$ meters, and the distance at $t = 11$ seconds is $f(11) = 11^2 = 121$ feet. So between $t = 10$ and

$t = 11$ the car travels $121 - 100 = 21$ meters, and the average speed is

$$\frac{f(11) - f(10)}{11 - 10} = \frac{121 - 100}{11 - 10} = 21 \text{ meters/sec} \tag{5.91}$$

Notice that we divide the change in distance by the change in elapsed time. The car traveled 21 meters in that one second interval, so its average speed was 21 meters/second.

Now let's find the average over the half-second between $t = 10.0$ and $t = 10.5$. Again, we divide the change in distance by the change in time:

$$\frac{f(10.5) - f(10.0)}{10.5 - 10.0} = \frac{(10.5)^2 - (10.0)^2}{.5}$$
$$= 20.5 \text{ m/sec} \tag{5.92}$$

The average of 20.5 m/sec is closer to the speed at $t = 10.0$, but it is still not exact.

To find $s(10)$ we must continue to reduce the time interval. Geometrically this means that we find the slope between points that are closer and closer together on the curve. The limit is the slope at a single point. We know the average velocity between $t = 10.0$ and any later time $t = 10 + \Delta t$, because we know that the distance increases from 10^2 to $(10 + \Delta t)^2$. The change in time is Δt, so we have

$$s = \frac{(10 + \Delta t)^2 - 10^2}{\Delta t}$$
$$= \frac{100 + 20\Delta t + \Delta t^2 - 100}{\Delta t} \tag{5.93}$$
$$= 20 + \Delta t$$

This formula describes our previous calculations. The time interval from $t = 10.0$ to $t = 11.0$ is $\Delta t = 1$, and the average speed is $20 + \Delta t = 21$. When the time step was $\Delta t = .5$, the average speed was 20.5 m/sec. If the time interval is one millionth of a second, we see that the average speed is 20 plus $1/1,000,000$, or very nearly 20.

We may conclude that the speed at $t = 10.0$ sec is $s = 20$ m/sec, which is the slope of the curve, and we also conclude that if $f(t) = t^2$, then $s(t) = 2t$. We compute the distance at time $t + \Delta t$, subtract the distance at time t, and divide by Δt to find the average speed.

$$s = \frac{f(t + \Delta t)}{\Delta t}$$
$$= \frac{(t + \Delta t)^2 - t^2}{\Delta t} \tag{5.94}$$
$$= 2t + \Delta t$$

The average speed between time t and $t + \Delta t$ is $2t + \Delta t$. As the elapsed time approaches zero, the average speed approaches $s(t) = 2t$. If the distance is given by t^2, then the speed is $2t$. The speed is the derivative of $f(t)$, or $df(t)/dt$, and from our previous work we see that

$$\frac{df(t)}{dt} = \lim_{\Delta t \to 0} \frac{f(t + \Delta t) - f(t)}{\Delta t} \tag{5.95}$$

The average speed over a short time Δt is the ratio on the right side of this equation. The derivative, on the left side, is its limit as the time interval Δt approaches zero. Again, the distance at time t is $f(t)$, and the distance at time $t + \Delta t$ is $f(t + \Delta t)$. Subtraction gives us the change in distance during that time interval. We can sometimes use Δf to represent this difference; thus, $\Delta f = f(t + \Delta t) - f(t)$. The average speed is the ratio $\Delta f / \Delta t$, or change in distance divided by change in time.

The limit of the average speed is the derivative

$$\frac{df}{dt} = \lim_{\Delta t \to 0} \frac{\Delta f}{\Delta t} \tag{5.96}$$

if this limit exists.

Notice that Δf is the symbol for the change in f, and not Δ times f. Similarly Δt is not Δ times t. It is the time interval, which we can make as small as we choose. From the graph of $f(t)$, we see that $\Delta f / \Delta t$ is the average slope. When Δt and Δf approach zero in the limit, then the derivative $df(t)/dt$ gives us the slope at the instant t. For our example, $f(t) = t^2$, we have

$$\begin{aligned} \frac{\Delta f}{\Delta t} &= \frac{f(t + \Delta t) - f(t)}{\Delta t} \\[2mm] &= \frac{t^2 + 2t\Delta t + \Delta t^2 - t^2}{\Delta t} \\[2mm] &= 2t + \Delta t \end{aligned} \tag{5.97}$$

We must complete the algebra, including the division by Δt before we let Δt approach zero. Otherwise information is lost and we learn nothing from the ratio $\Delta f / \Delta t$. So we must expand $(t + \Delta t)^2$, subtract t^2, and then divide by Δt. Only then do we let $\Delta t \to 0$.

Here, again, we see the basis of the differential calculus at work. To get the instantaneous rate, we let Δt get smaller and smaller. Notice that Δf gets smaller too. The limiting value of $\Delta f / \Delta t$ is the *instantaneous rate of change* or derivative of $f(t)$ with respect to t. We calculate better and better average rates by taking smaller and smaller intervals. However, there is no "best" average rate when we do it this way, because we can always get a still better rate by decreasing the interval still further. The *limit* process gives the correct value because it is a number to which all "good" rates are close.

5.9 Intervals

There are two ways to define an *interval* between two points a and b on the *real number line*. Recall that the real number line is a straight line with one point on it, representing the origin or zero point, and two possible directions along it, one designated as positive and the other negative (Figure 5.23).

Each point on the figure corresponds to a real number whose absolute value is equal to the distance x of the point from the origin. The plus and minus signs indicate the direction to measure off this distance along the line.

We let the numbers a and b define the limiting or end points of an interval, where $a < b$.

An *open interval* does not contain the limit points, so we describe it by the expression

$$a < x < b \tag{5.98}$$

A *closed interval* does contain its limit points, so we describe it by the expression

$$a \leq x \leq b \tag{5.99}$$

Figure 5.24 shows both kinds of intervals.

A shorter way to describe a closed interval looks like this: $[a, b]$, using brackets, with the interval limits separated by a comma. For an open interval we use parentheses, with the interval limits here, too, separated by a comma, as in (a, b). We can also write for a closed interval

$$x \in [a, b] \tag{5.100}$$

which we read as "x is an element (or member) of the point set in the interval $[a, b]$." This notation is from set theory and applies to many algebraic and geometric situations. A similar notation for an open interval is

$$x \in (a, b) \tag{5.101}$$

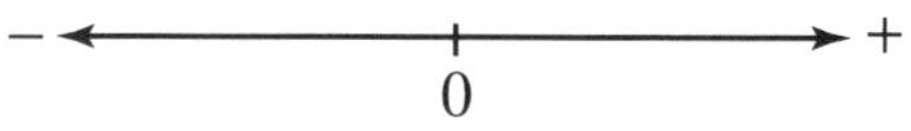

Figure 5.23 *The real number line.*

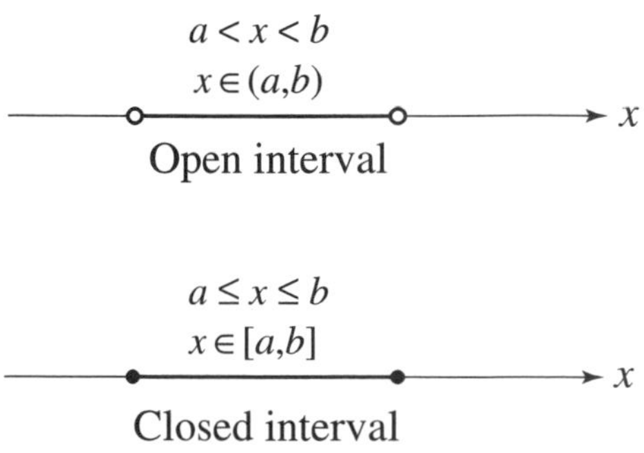

$$a < x < b$$
$$x \in (a,b)$$

Open interval

Figure 5.24 *Open and closed intervals on the real number line.*

$$a \leq x \leq b$$
$$x \in [a,b]$$

Closed interval

5.10 Continuity

As mathematicians refined the idea of a limit, they also came to a better understanding of *continuity*. They found that a big challenge for a successful theory of continuity was to put the real numbers into one-to-one correspondence with the points of the number line so that the numbers would form a continuum.

Many mathematicians, ancient and modern, worked hard to meet this challenge. Eudoxus, Dedekind, and Cantor all developed theoretical concepts that they hoped would fill the gaps in the ordered set of rational numbers. They wanted a geometric interpretation, one that produced a straight line that is continuous or unbroken. Their efforts were successful, and we may now speak of a "real number continuum."

But it was the discovery of irrational numbers over 2000 years ago by the Pythagoreans that first caused the need for an exact definition of the continuum. That produced a division in mathematical thinking that has lasted to the present. One school of thought has concentrated on working out the implication of Pythagoras' original belief that the framework of the universe was to be found in the natural numbers ... 1, 2, 3, 4, ... This sequence and the set of integers (including the negative versions of the natural numbers), ordered according to size, are the *discrete* numbers. We apply the term "discrete" to any sequence and to any ordered set in which every term (except the first, if any) has a unique predecessor, and every term (except the last, if any) has an immediate successor. For example, any finite set (arranged in any order) is discrete.

Because their focus is on the natural numbers, those inclined toward the Pythagorean view are called discretists. In contrast to this approach, there are others who would base science on the real-number continuum. The discretists have made their most important contributions to the classic theory of numbers, algebra, and logic. Those whose interest is the mathematics of continuity are likely to be geometers, analysts (calculus), and physicists. There are some important exceptions, of course, in physics. Since Democritus (ca 400 BC), matter has been thought of as discrete or atomistic. Modern quantum theory is the most striking example of the physics of the discrete or discontinuous.

We all have an intuitive idea of "continuous motion", like the path of a thrown ball or the flight path of an airplane made visible by a condensation trail. The motion is smooth and unbroken. As for the concept of continuity, the natural numbers 1, 2, 3, ... do not offer an appropriate mathematical model. For example, the points on a straight line do not have clear-cut identities like the natural numbers do, where the step from one number to the next is the same. No matter how close together we choose the points on this line, we can always find, or at least imagine, another point. There is no "shortest step" from one point to the next. In fact there is no "next" point at all. Choose any point and identify it by its coordinate on the real line. Then identify the point you believe to be the "next" point to it by giving its coordinate on the real line Whatever it is, you can always find another point in between them. The concept of continuity allows no "nextness."

5.11 Continuous Functions

We begin with an assertion: A function $f(x)$ is *continuous* within given limits if between these limits an infinitesimally small increment Δx in the variable x always produces an infinitesimally small increment in the function itself, so that $f(x + \Delta x) - f(x) \to 0$. We then look at its implications.

We are again concerned with the limit of $f(x)$ as $x \to a$, but now $f(a)$ becomes important. In the limit x approaches a but never reaches it, so we could ignore $f(a)$ itself. But, for a *continuous function*, this final number $f(a)$ must meet certain tests:

1. The number $f(a)$ must exist, that is, $f(x)$ is defined at $x = a$.
2. The limit of $f(x)$ must exist, that is, $x \to a$, $f(x) \to L$.
3. The limit L must equal $f(a)$.

If these conditions are met, then $f(x)$ is continuous at $x = a$. We can express these tests symbolically by $f(x) \to f(a)$ as $f(a)$.

Let's look at some functions that are not continuous at $x = 0$ (Figure 5.25):

1. In Figure 5.25a, $f(x)$ is a *step function*. It jumps from 1 to 2 at $x = 0$.
2. In Figure 5.25b, $f(x) = 1/x^2$ becomes larger without limit as $x \to 0$.
3. In Figure 5.25c, $f(x) = \sin(1/x)$, and the frequency of the oscillation approaches infinity as $x \to 0$.

The functions in these graphs fail to meet the limit test. The step function has a jump discontinuity. It does not have an ordinary two-sided limit but has, instead, one-sided limits, from both the left and right. The limit from the left, $x \to 0^-$, is 1. The limit from the right, $x \to 0^+$, is 2.

In the graph of $1/x^2$, the obvious limit is $L = +\infty$. But this is not acceptable, again because infinity is not a number. However, we can still express this situation in the usual way: $1/x^2 \to \infty$ as $x \to 0$, which means that $1/x^2$ goes above and stays above any L as $x \to 0$.

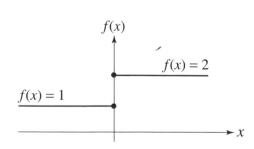

Figure 5.25a *Step function: discontinuous at $x = 0$.*

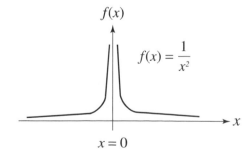

Figure 5.25b *Asymptotic function: increases without limit at $x = 0$.*

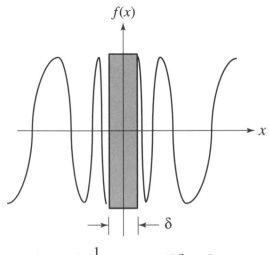

Figure 5.25c *Frequency of oscillation approaches infinity as x approaches zero.*

$$f(x) = \sin\frac{1}{x} > \varepsilon, \text{ even if } \delta \to 0$$

These two are the most common discontinuities. In Figure 5.25c, however, we see that $\sin(1/x)$ has no limit as $x \to 0$. This function does not even blow up. The value of the function never exceeds 1. As x gets smaller and $1/x$ gets larger, the sine function oscillates faster and faster. We cannot constrain the graph of this function within a small interval $\pm\varepsilon$, no matter how small we make δ.

There are two important properties of a continuous function on a closed interval. Given the interval $a \leq x \leq b$, or in short form $[a, b]$, at the endpoints a and b, $f(x)$ must approach $f(a)$ and $f(b)$, respectively.

First, we have the *extreme-value property*: A continuous function on the interval $[a, b]$ has a maximum value M and a minimum value N. At points x_{max} and x_{min} in the interval, it reaches those values, so that

$$f(x_{max}) = M \tag{5.102}$$

where $M \geq f(x) \geq [f(x_{min}) = N]$ for all x in $[a, b]$.

Next, we have the *intermediate-value property*: If the number F is between $f(a)$ and $f(b)$, there is a point c between a and b where $f(c) = F$. Thus, if F is between the minimum N and the maximum M, there is a point c between x_{max} and x_{min} where $f(c) = F$.

Here is an example to show why we need closed intervals and continuous functions: Given the function $f(x) = x$ and the interval constraint $0 < x \leq 1$, we see that the function never reaches its minimum $x = 0$. If we close the interval by defining $f(0) = 4$ (a discontinuity-producing jump), the function still does not reach the minimum. Finally, also because of the jump, the intermediate value $F = 2$ is not reached (Figure 5.26). A function $f(x)$ is continuous in an interval when it is continuous for all values of x in this interval, in which case it possesses both the extreme-value property and the intermediate-value property. Conversely, these properties may serve as tests of the continuity of a function.

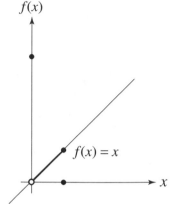

Figure 5.26 *Closed intervals and continuous functions.*

Now for a final review of continuity: A function $f(x)$ is continuous for the value $x = a$, or at the point $x = a$, if it satisfies the following three conditions:

1. The function is defined at $x = a$, that is, $f(a)$ exists.
2. The function approaches a limit as x approaches a, that is, $\lim_{x \to a} f(x)$.
3. The limit approached is equal to the value of the function at the point in question, that is

$$\lim_{x \to a} f(x) = f(a) \qquad (5.103)$$

Here are three examples:

1. Given $f(x) = x^3 + 2x$, we find that $\lim_{x \to 3} f(x) = 33$ and that $f(3) = 27 + 6 = 33$. Since conditions 1, 2, and 3 are satisfied, we see that this function is continuous at $x = 3$.
2. The function $f(x) = \frac{x^3 - 2x^2 - 3x + 6}{x - 2}$ is discontinuous at $x = 2$ because $f(2)$ is not defined.
3. The function $f(x) = \sqrt{x}$ is discontinuous at $x = 0$ because $\lim_{x \to 0} \sqrt{x}$ does not exist. For the function to be continuous, the limit must exist for any direction of approach, and in this case approach through negative values is excluded.

Given that a function is continuous in an interval if it is continuous at all points of the interval leads us to the following important conclusion: As x changes from any value a to any other value b, the function $f(x)$, if continuous in the interval $x = a$ to $x = b$, takes on every value between $f(a)$ and $f(b)$.

Exercises

In Exercises 5.1–5.5, write the first five terms of each series.

5.1 $\displaystyle\sum_{n=1}^{\infty} \frac{1}{n^2}$

5.4 $\displaystyle\sum_{n=1}^{\infty} \frac{1}{3n^4 - 30n^3 + 105n^2 - 149n + 72}$

5.2 $\displaystyle\sum_{n=1}^{\infty} \frac{(-1)^n}{n!}$

5.5 $\displaystyle\sum_{n=2}^{\infty} \frac{1 + (-1)^n}{n^2 + 1}$

5.3 $\displaystyle\sum_{n=1}^{\infty} \frac{2n^2 - 9n + 13}{6(n-1)!}$

In Exercises 5.6–5.11, write the next three terms and find the general term.

5.6 $2 + 4 + 8 + 16 + \cdots$

5.7 $1 - \dfrac{1}{2} + \dfrac{1}{3} - \dfrac{1}{4} + \cdots$

5.8 $-\dfrac{1}{2} + 0 + \dfrac{1}{4} + \dfrac{2}{5} + \dfrac{3}{6} + \cdots$

5.10 $\dfrac{\sqrt{x}}{2} + \dfrac{x}{2 \cdot 4} + \dfrac{x\sqrt{x}}{2 \cdot 4 \cdot 6} + \dfrac{x^2}{2 \cdot 4 \cdot 6 \cdot 8} + \cdots$

5.9 $x + x^2 + \dfrac{x^3}{2} + \dfrac{x^4}{6} + \cdots$

5.11 $\dfrac{x^2}{3} - \dfrac{x^3}{5} + \dfrac{x^4}{7} + \dfrac{x^5}{9} + \cdots$

In Exercises 5.12–5.16, write the first four terms of the series whose nth term is given.

5.12 $\dfrac{2^{n-1}}{\sqrt{n}}$

5.15 $\dfrac{x^{n-1}}{\sqrt{n}}$

5.13 $\dfrac{n+2}{2n-1}$

5.16 $\dfrac{(-1)^{n-1} x^{2n-1}}{(2n-1)!}$

5.14 $\dfrac{n}{3^{n-1}}$

In Exercises 5.17–5.24, find the limits if they exist.

5.17 $\displaystyle\lim_{x \to \infty} \frac{5 - 2x^2}{3x + 5x^2}$

5.21 $\displaystyle\lim_{x \to 2} \frac{x+3}{x^2 - 2}$

5.18 $\displaystyle\lim_{x \to \infty} \frac{4x + 5}{2x + 3}$

5.22 $\displaystyle\lim_{x \to 0} \frac{f(x + \Delta x) - f(x)}{\Delta x}$

5.19 $\displaystyle\lim_{x \to 0} \frac{4x^2 + 3x + 2}{x^3 + 2x - 6}$

5.23 $\displaystyle\lim_{\theta \to 0} \frac{\sin^2 \theta \cos^2 \theta}{\theta^2}$

5.20 $\displaystyle\lim_{h \to 0} \frac{x^2 h + 3xh^2 + h^3}{2xh + 5h^2}$

5.24 $\displaystyle\lim_{x \to 1} \frac{\sin x}{x}$

In Exercises 5.25–5.32, find the interval of convergence and test each series at its interval endpoints.

5.25 $\displaystyle\sum_{n=0}^{\infty}(-1)^n x^n$

5.29 $\displaystyle\sum_{n=0}^{\infty}(-1)^n n! x^n$

5.26 $\displaystyle\sum_{n=1}^{\infty}\frac{x^n}{2n-1}$

5.30 $\displaystyle\sum_{n=1}^{\infty}\frac{n! x^n}{2n-1}$

5.27 $\displaystyle\sum_{n=0}^{\infty}\frac{(-1)^n x^n}{(2n+1)^2 3^{n+1}}$

5.31 $\displaystyle\sum_{n=0}^{\infty}\frac{x^n}{n!}$

5.28 $\displaystyle\sum_{n=1}^{\infty}\frac{(n+1)x^{2n}}{5^n}$

5.32 $\displaystyle\sum_{n=0}^{\infty}\frac{(-1)^n x^{2n}}{(2n)!}$

5.33 Given $f(x) = x^3 - 5x^2 - 4x + 20$, find
    a. $f(1)$ c. $f(0)$
    b. $f(5)$ d. $f(a)$

5.34 Given $f(x) = 4 - 2x^2 + x^4$, find
    a. $f(0)$ d. $f(2)$
    b. $f(1)$ e. $f(-2)$
    c. $f(-1)$

5.35 Given $f(x) = x^3 - 5x^2 - 4x + 20$, find $f(t+1)$

5.36 Given $f(x) = x^3 + 3x$, find $f(x + \Delta x) - f(x)$

5.37 Given $f(x) = \frac{1}{x}$, find $f(x + \Delta x) - f(x)$.

In Exercises 5.38–5.42, evaluate the right- or left-hand limit, as indicated.

5.38 $\displaystyle\lim_{x\to 2^+}\frac{\sqrt{x-2}}{\sqrt{x^2-4}}$

5.41 $\displaystyle\lim_{x\to 2^+}\frac{(x^4 - 4x^3 + 5x^2 - 4x + 4)^{\frac{1}{4}}}{(x^2 - 3x + 2)^{\frac{1}{2}}}$

5.39 $\displaystyle\lim_{x\to 3^+}\frac{x-3}{\sqrt{x^2-9}}$

5.42 $\displaystyle\lim_{x\to 1^+}\frac{(x^2 + 4x - 5)^{\frac{1}{2}}}{(x^2 - 4x + 3)^{\frac{1}{3}}}$

5.40 $\displaystyle\lim_{x\to 1^-}\frac{\sqrt{1-x^3}}{\sqrt{1-x^2}}$

Use $\displaystyle\lim_{\phi\to 0}\frac{\sin\phi}{\phi} = 1$ to find the limits in Exercises 5.43–5.50.

5.43 $\lim\limits_{\theta \to 0} \dfrac{\sin k\theta}{\theta}$

5.47 $\lim\limits_{\theta \to 0} \dfrac{\sin \theta}{\theta^2}$

5.44 $\lim\limits_{\theta \to 0} \dfrac{\sin^2 \theta}{\theta}$

5.48 $\lim\limits_{x \to 0} x \csc x$

5.45 $\lim\limits_{\theta \to 0} \dfrac{\tan \theta}{\theta}$

5.49 $\lim\limits_{x \to 0} \dfrac{\tan a x}{\sin b x}$

5.46 $\lim\limits_{\theta \to 0} \dfrac{\sin \theta^2}{\theta}$

5.50 $\lim\limits_{x \to \pi/2} \dfrac{\cos x}{x - \pi/2}$

5.51 Find the number given by $\sum\limits_{j=1}^{4} 1/j$.

5.52 Find the number given by $\sum\limits_{j=2}^{5} (2j - 3)$.

5.53 Find the number given by $\sum\limits_{j=0}^{6} 2^j$

5.54 $\sum\limits_{j-0}^{n} 2^j$

5.55 Use the summation notation to express the sum $2 + 4 + 6 + \cdots + 100$, and compute the total sum.

5.56 Use the summation notation to express the sum $1 + 3 + 5 + \cdots + 199$, and compute the total sum.

5.57 Use the summation notation to express the sum $1 - \frac{1}{2} + \frac{1}{3} - \frac{1}{4}$, and compute the total sum.

5.58 Use the summation notation to express the total area under the curve of $f(x)$. Assume that this area is divided into n rectangles with base width of Δx and whose height just touches the curve at the points $x = \Delta x, 2\Delta x, \ldots, n\Delta x$.

In Exercises 5.59–5.64, find the points at which the function is discontinuous.

5.59 $\dfrac{x^2 + 5}{x^2 - 9}$

5.60 $\dfrac{2x + 1}{x^2 - 4x + 4}$

5.63 $\cot \phi$

5.61 $\dfrac{x^2 - 2x}{x^3 - 2x^2 + 2x}$

5.64 $\sqrt{x^2 - 2ax + a^2}$

5.62 $\sin \phi$

In Exercises 5.65–5.70, find the value of k that makes the function continuous.

5.65 $f(x) = \begin{cases} \sin x & x < 1 \\ k & x \geq 1 \end{cases}$

5.68 $f(x) = \begin{cases} k & x \neq 4 \\ 1/x^3 & x = 4 \end{cases}$

5.66 $f(x) = \begin{cases} k + x & x < 0 \\ k^2 + x^2 & x \geq 0 \end{cases}$

5.69 $f(x) = \begin{cases} (\tan x)/x & x \neq 0 \\ k & x = 0 \end{cases}$

5.67 $f(x) = \begin{cases} (\sin x)/x^2 & x \neq 0 \\ k & x = 0 \end{cases}$

5.70 $f(x) = \begin{cases} (\sin x - x)/x^k & x \neq 0 \\ 0 & x = 0 \end{cases}$

CHAPTER 6

TOPOLOGY

What makes a sphere different from a torus? Why are left and right not reliable directions on a Möbius strip? Topology answers questions like these. It is the study of continuity and connectivity and how to preserve them when geometric figures are deformed. Topology is a major branch of geometry, but it is a relative newcomer. Its first principles were discovered by the great German mathematician Georg Friedrich Bernhard Riemann (1826–1866), and further developed by the equally great French mathematician Henri Poincaré (1854–1912).

For many years mathematicians regarded topology as merely an interesting footnote to geometry. Studying the topology of the Möbius strip and Klein bottle, considered geometric curiosities and nothing more, was largely done as a mathematical recreation and not to be taken too seriously. Now, however, topology is an important part of many disciplines, including geometric modeling and mathematical physics. In fact, for the latter, the Möbius strip and Klein bottle both appear in superstring theory as aspects of Feynman diagrams representing the interactions of elementary particles.

Topological properties are not *metrical*. They are not such things as length, area, volume, or angles. They are concerned with connectivity and continuity. The properties of geometric shapes that are invariant under transformations that stretch, bend, twist, or compress a figure, without tearing, puncturing, or causing self-intersections, are topological properties. The property of being an open or closed curve or surface is a topological invariant. One-sidedness and two-sidedness are topologically invariant properties of surfaces. Lines, parabolas, and the branches of a hyperbola belong to the class of topologically equivalent figures we call *simple arcs*. This chapter discusses these subjects and also topological equivalence, the topology of closed paths, piecewise flat surfaces, closed curved surfaces, Euler operators, orientation, curvature, and intersections.

6.1 Topological Equivalence

Euclidean geometry is based on measurements and comparisons: Two figures are equal if the lengths and angles of corresponding parts are equal. Projective geometry is based on straight lines: Two figures are projectively equivalent if we can pass from one to another by a projective transformation. Projective properties can be represented analytically and measured, although the quantities are not quite the same as those in

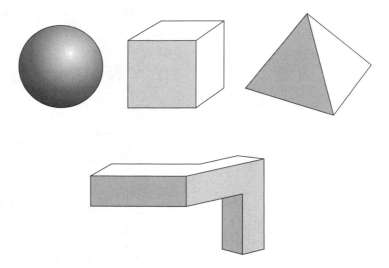

Figure 6.1 *Topologically equivalent shapes.*

Euclidean geometry. Other geometries are possible. One of them has no ordinary quantitative measure associated with it at all, and that geometry is *topology*.

Two geometric shapes may look quite different but turn out to be *topologically equivalent*. Two shapes are topologically equivalent when we can change one into the other by any continuous deformation as long as it preserves continuity. The four shapes in Figure 6.1 are topologically equal. We can easily imagine distorting any of the nonspherical polyhedral shapes into a sphere. For a physical representation of this distortion process, imagine constructing a cube using very thin and stretchy rubber sheets. Next simply inflate the "rubber" cube until it is more or less spherical. All of the shapes in Figure 6.2 are topologically equal, because we can distort each of them into the shape of a torus or donut.

Can the toruslike shapes be distorted into a spherical shape? No. So the shapes in Figure 6.2 are not topologically equal to those in Figure 6.1. The surfaces of these

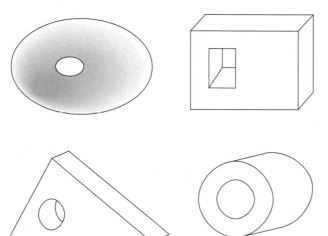

Figure 6.2 *More topologically equivalent shapes.*

two sets of shapes reflect their topological differences. For example, all simple non-self-intersecting closed curves that we can construct on the surface of a sphere are equivalent. Imagine moving several of these closed curves around on a sphere's surface, stretching, shrinking, and deforming them until they become congruent and superimposed, which is possible with curves 1 and 2 on the sphere in Figure 6.3. Next we see that there are three families of closed curves on a torus, curves 3, 4, and 5 on the torus in Figure 6.3, none of which can be brought into correspondence with a member of a different family.

Study the curves 1, 2, and 3 in Figure 6.3. Each of these curves we can shrink to a point. This cannot be done to curves 4 and 5 without cutting the surface of the torus. In fact, closed curves, like 1, 2, and 3 are what we call *topological disks* on the surfaces on which they lie, because they can be shrunk to a point. Closed curves 4 and 5 cannot be shrunk to a point and therefore do not define topological disks.

We can readily find other properties of curves on surfaces that undergo rubber sheet deformations. Look at the examples in Figure 6.4. Here the property of

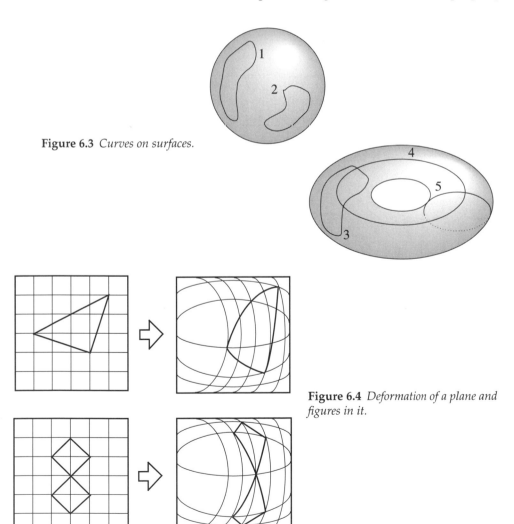

Figure 6.3 *Curves on surfaces.*

Figure 6.4 *Deformation of a plane and figures in it.*

self-intersection is preserved. What about metric properties, such as length, angle, area, and so forth? Are these preserved? No, they are not because, again, we can imagine stretching or compressing a surface, causing an increase or decrease in any or all of these metric properties. That is why they are not topological properties.

6.2 Topology of a Closed Path

The general *closed path theorem* states: The total turning along any closed path or curve is an integer multiple of 360°, denoted N_R, the *rotation number* of the path. It is an intrinsic property of the path, independent of where the path starts or how it is oriented. Figure 6.5 shows several examples, including loops. Closed, non-self-intersecting paths on a plane surface always appear to have total turning equal to ± 360°. (The algebraic sign depends on whatever convention we establish for the direction in which we traverse the path.) This is true for both the square and curved paths of Figure 6.5a and 6.5b. But self-intersecting closed paths always have total turning different from ± 360° (Figure 6.5c and 6.5d). Clearly, we must discriminate between the two classes of paths: simple and self-intersecting.

The *simple closed-path theorem* states that the total turning in a non-self-intersecting closed path is ± 360° (clockwise or counterclockwise). In other words, the rotation number N_R of any simple closed path is ±1. By studying some examples of simple closed paths we can see the truth of this theorem, which, as simple and obvious as it appears, is difficult to prove rigorously. The theorem implies that there is a relationship between the two properties of a closed path—the total turning and the crossings, or self-intersection points.

It is important to be clear about how we measure turning along a path. For example, the turning at a vertex of an equilateral triangle is 120° and not 60° (Figure 6.6). It is the exterior angle that we measure when we change direction.

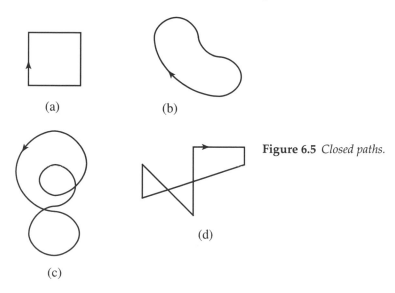

(a)

(b)

Figure 6.5 *Closed paths.*

(d)

(c)

Figure 6.6 *Measuring* turning *along a path.*

Curves defining closed paths that we can deform into one another are curves that are topologically equivalent. Total turning is a topological invariant for closed paths, and any two closed paths that are topologically equivalent must have the same total turning. The converse is also true: If two paths have the same total turning, they can be deformed into one another.

The simple closed-path theorem is more comprehensive than the closed-path theorem, and its proof is considerably more complicated. The simple closed-path theorem is a link between local and global information. Since crossing points on a curve are nonlocal phenomena, there is no direct way for a traversal algorithm (that is, an algorithm proceeding stepwise around the path) to detect a crossing point. To do it requires looking at the entire curve at once. But we can determine total turning with local computations.

A simple exercise shows how this theorem relates total turning to the existence of crossing points: We can walk around a closed path, perhaps as defined by a set of written instructions, carefully accumulating our total turning (taking clockwise as positive). If we complete the circuit and find on returning to the starting point that the total turning does not equal $\pm 360°$, then we can assert that somewhere the path must have at least one crossing point. Although we do not know where the crossing points are, and we were unable to observe them while traversing the path, by applying the theorem we can determine that one or more crossing points must exist. Thus, the simple closed-path theorem is an example of a powerful principle: It is often possible to determine global properties by accumulating local information.

Let's see what happens as we walk along a self-intersecting path with only 90° turns (Figure 6.7). Starting at O, we make our first turn at A, which is a left turn (or counterclockwise). Proceeding around the path and back to where we started at O, we count three left and three right 90° turns. This means that our total turning is 0°. Somewhere along the path is an intersection point.

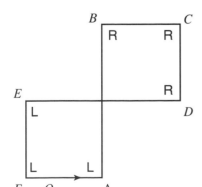

Figure 6.7 *Turning along a self-intersecting path.*

Deformation of Curves and Planes

To develop proofs of the simple closed-path theorem and related theorems about the topology of simple closed curves, we can look at what happens when we deform curves and the planes in which they lie. We must show that, given a simple closed path (a path with no crossing points), the total turning around the path is equal to $\pm 360°$. There are many paths for which this result is obvious; the simplest is the square. This suggests a way to prove the theorem—by showing that we can deform any simple closed path into a square. Because total turning is invariant under deformations, we can show that any simple closed path has the same total turning as a square, $\pm 360°$, depending on the direction we move around the square. This simplifies the problem of proving the theorem to that of showing how any simple closed curve can be deformed into a square. In fact, given any simple closed path, not only can the path itself be deformed into a square, but the entire plane can be deformed, pulled and stretched, so that the path becomes a square. We can even consider the plane itself as being deformed into a curved surface.

Deformations of paths and planes are closely related. Imagine that the plane is a stretchable rubber sheet. We allow any stretching or shrinking but not cutting or overlapping. If we draw a closed path on the rubber-sheet plane, then any deformation of the plane results in a deformation of the path. (Figure 6.4). Straight-line segments become curves under such deformations. Many of the changes that are legitimate for the deformation of curves in space, such as overlap, do not and cannot happen to curves in a deformed plane. Also, crossing points are neither created nor destroyed during such deformations of a plane.

Plane deformations are more conservative than path deformations. Every plane deformation is a path deformation, but a path deformation that introduces cross-overs is too violent to be a plane deformation. The mathematical term for a path deformation is *regular homotopy*, while a rubber-sheet deformation is an *ambient isotropy*.

We can deform any simple closed path into a square, and we can do this with a conservative deformation, that is, a plane deformation. So we state the deformation theorem for simple closed curves as follows: For any simple closed curve in the plane, there is a rubber-sheet deformation of the plane that reduces the curve to a square. Figure 6.8 shows a curve deformed into a rectilinear polygon, where we see that the total turning is reduced to summing 90° clockwise and counterclockwise turns.

The Jordan Curve Theorem

The *Jordan curve theorem* follows from the deformation theorem. It states: Any simple closed curve in the plane divides the plane into exactly two regions, an inside and an outside. Properties such as *dividing the plane into two regions* and *having an inside and an outside* are also invariant under rubber sheet deformations of the plane. If the undeformed curve has these properties, then the deformed curve must have them, too.

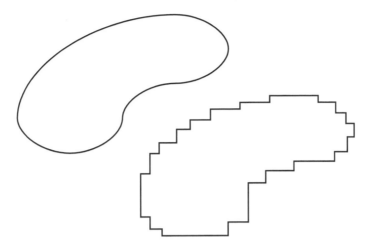

Figure 6.8 *A curve deformed into a rectilinear polygon.*

Consider a closed curve C in the plane (Figure 6.9). It divides the plane into two regions, A and B, where any two points within one of the regions (such as a and b or c and d) can be connected by a path that lies entirely within the region. If we take a point from each of the regions (such as e and f), then every path connecting them necessarily cuts C. The bounded region A is the *inside* of C, and B the *outside*. The first proof of the Jordan curve theorem, that every simple closed curve in the plane divides the plane into two regions, appeared in 1893 in the important book *Cours d'Analyse* by C. Jordan (1838–1922), although Jordan did not solve the problem completely. Even though Jordan's original proof has been greatly simplified, it is still not easy to prove this apparently obvious theorem.

Not only does a closed curve have an inside and outside, but the inside itself is deformable, in the rubber sheet sense, into the interior of a circle in the plane. A region that can be deformed in this way is a topological disk. A topological disk has no holes or isolated points in it.

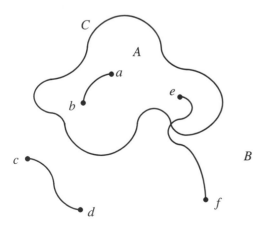

Figure 6.9 *A closed curve divides the plane into two separate regions.*

The Jordan curve theorem probably seems obvious, but simple closed curves can be complex and convoluted. The theorem is not true if we consider curves on surfaces other than the plane. For example, a simple closed curve can be drawn on a torus that does not divide the torus into two distinct regions if it loops through the torus's central hole (Figure 6.3).

Angle Excess

We will now define a new quantity called *angle excess*, or simply *excess*. It is the turning that a reference pointer undergoes when we carry it around a closed path. We associate angle excess with a closed path on a curved surface. We can restate the closed-path theorem so that it holds for simple closed paths on arbitrary surfaces:

$$T + E = 360° \tag{6.1}$$

where T is the total turning along the path and E is the angle excess along the path.

Figure 6.10 shows the angle excess associated with a closed path on a sphere: $\angle AOB$. A trip along this path starts at point O on the equator, proceeds to the pole along some meridian, then returns to the equator along a different meridian. The trip continues along the equator back to the starting point. Note that the reference pointer is always transported parallel to itself relative to the surface in which the path lies. For a fixed path, the excess is the same no matter where the trip begins.

Angle excess is additive, and we find that the following theorems apply:

1. If a triangle on an arbitrary surface is divided into two subtriangles, then the angle excess of the triangle is the sum of the angle excesses of the pieces.
2. The angle excess of any polygon on an arbitrary surface is the sum of the areas of the pieces in any polygonal subdivision.
3. For any topological disk on an arbitrary surface, the angle excess around the boundary is equal to the total curvature of the interior.

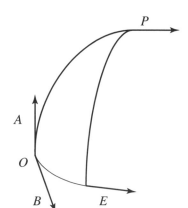

Figure 6.10 *Angle excess of a closed path on a sphere.*

6.3 Piecewise Flat Surfaces

A polyhedron is a good example of a piecewise flat surface. The surface of a polyhedron must be an arrangement of its face polygons such that two and only two of these polygons meet at an edge. It must be possible to traverse the surface of a polyhedron by crossing its edges and moving from one polygonal face to another, until all polygons have been traversed by this continuous path.

The term *simple polyhedra* refers to all polyhedra that we can deform into a sphere without cutting (see Sections 12.1 and 12.8). The five *regular polyhedra*, the Platonic solids, are a subset of the simple polyhedra. The regular polyhedra are convex, and every convex polyhedron is a simple polyhedron. For a counterexample, a toroidal polyhedron is not a simple polyhedron and is not convex. A nonsimple polyhedron is the topological equivalent of a solid object with holes in it, like a donut or torus. Figure 6.11 shows an example of a nonsimple polyhedron. Note that we cannot deform it into a sphere. However, we can deform any simple polyhedra into a sphere and any rectangular parallelepiped with a hole into a torus. Deformations let us replace more complicated structures with topologically equivalent polyhedra.

Connectivity Number

We assign a *connectivity number* N to a polyhedron (see Section 12.9). If the surface of a polyhedron is divided into two separate regions by every closed path (loop) defined by edges of the polygons making up its faces, then the polyhedron has connectivity $N = 0$. All simple polyhedra have $N = 0$, because the surface of a sphere is divided into two parts by any closed curve lying on it. Conversely, any polyhedron with $N = 0$ can be deformed into a sphere. In Figure 6.11 we see that there are closed loops of edges that do not divide the surface of a nonsimple polyhedron into two parts. We assign a polyhedron a connectivity number, N, greater than one, and define it as the maximum number of distinct loops with or without common points (intersections) that do not divide its surface into two separate regions.

We define any closed non-self-intersecting surface as a sphere with G handles. The number G is the *genus* of the surface, where $G = 0$ for a sphere, $G = 1$ for a torus, and $G = 2$ for the surface of a solid figure eight. We also define the genus G of a polyhedron

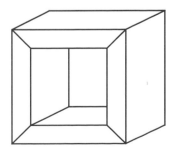

Figure 6.11 *Polyhedron with a hole through it.*

as the maximum number of nonintersecting loops to be found that do not divide its surface into two regions.

Euler's formula applies to polyhedra of any genus and connectivity. Thus,

$$V - E + F = 2(1 - G) = 2 - N \tag{6.2}$$

where V, E, and F are the number of vertices, edges, and faces, respectively. This is the *Euler-Poincaré* formula. These relationships are developed in more detail in Sections 12.8 and 12.9.

By counting its vertices, edges, and faces, we can determine the connectivity and genus of any polyhedron:

$$N = -V + E - F + 2$$
$$G = \frac{-V + E - F + 2}{2} \tag{6.3}$$

On any closed surface we can construct a network, called a *net*, of edges and vertices analogous to those defining polyhedra. The edges do not have to be straight lines. A net must have the following properties:

1. Each edge is terminated by two vertices.
2. Each edge subtends only two faces.
3. Each vertex subtends three or more edges.
4. Each face, each loop of edges, must be a topological disk.
5. The net must divide the entire surface into topological disks.

For a sphere and all topologically equivalent shapes, $N = 0$. For all toruslike shapes $N = 2$, and $N = 4$ for all shapes like a solid figure-eight. This means that we can use Equation 6.2 to determine N and thus determine the type of surface. Look at the three examples in Figure 6.12. In Figure 6.12a, the net on the sphere has the following characteristics: $V = 8$, $E = 16$, $F = 10$, and $N = 0$. These values do indeed satisfy Equation 6.2. In Figure 16.12b, we find the following characteristics of the net on the toruslike shape: $V = 16$, $E = 32$, $F = 16$, $N = 2$. Again, Equation 6.2 is satisfied. And, finally, for the solid figure-eight shape in Figure 6.12c, $V = 24$, $E = 44$, $F = 18$, and $N = 4$, which also satisfies the equation.

Take another look at Figure 6.12b. Do you see why it is necessary to have edges like 1–5, 2–6, 3–7, and 4–8? *Hint*: If these edges were missing, would the face contained between the loop of edges 1-2-3-4-1 and 5-6-7-8-5 be a topological disk?

Curvature of Piecewise Flat Surfaces

Now let's look at what seems to be a paradox: the curvature of piecewise flat surfaces. All of the curvature in a piecewise flat surface is concentrated at the vertices. This makes total curvature easy to compute. To do this we sum up the angle excesses of

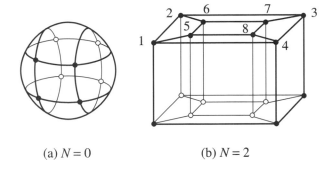

(a) $N = 0$ (b) $N = 2$

Figure 6.12 *Measuring connectivity, N.*

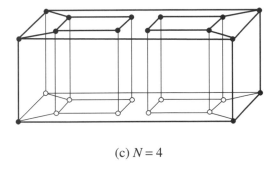

(c) $N = 4$

small paths around each of the vertices:

$$K = \sum_{i=1}^{V} E_i \tag{6.4}$$

where E_i is the excess of a path around the vertex i.

To transform this expression into the simplest, most meaningful terms, we recall that excess is equal to 2π minus the total turning, so we rewrite Equation 6.4 as

$$K = \sum_{i=1}^{V} (2\pi - T_i) \tag{6.5}$$

where T_i is the total turning of a path around the vertex i. By factoring out the 2π terms (one for each vertex) we obtain

$$K = 2\pi V - \sum_{i=1}^{V} T_i \tag{6.6}$$

where V is the number of vertices in the surface.

We can further clarify this expression if we use the fact that the total turning around a vertex is equal to the sum of all the interior angles meeting at that vertex. Summing these angles over all the vertices gives us all the interior angles of all the faces in the surface. Regrouping these angles according to the faces they lie in, we compute total curvature as

$$K = 2\pi V - \sum_{i=1}^{F} f_i \tag{6.7}$$

where f_i is the sum of the interior angles of the face i.

This is a surprising result, because it shows that we computed the second term without knowing how the edges are joined together. If we know all the individual faces of a piecewise flat surface and we know V, we can compute total curvature.

For piecewise square surfaces, the formula is even simpler, because the sum of interior angles of any face is 2π. Therefore, $K = 2\pi V - 2\pi F$ or $K = 2\pi(V - F)$. This equation for total curvature does not depend on angles at all! And we can do even better. Curiously enough, we can compute the total curvature of any closed piecewise flat surface without knowing any angles. All we need to know is the number of faces, vertices, and edges. For a closed piecewise flat surface with V vertices, E edges, and F faces, the total curvature is

$$K = 2\pi(V - E + F) \tag{6.8}$$

To prove Equation 6.8, we must show that we can express the summation term of Equation 13 independently of the values of the angles. We do not know much in general about the sum of the interior angles of a face, but we do know a closely related quantity: the sum of the exterior angles. The boundary of each face is a simple closed path, so we know that the sum of the exterior angles is 2π. To relate the exterior angles to the interior angles, we note that each exterior angle pairs off with an interior angle to sum to $180°$, and there are as many of these pairs as there are edges to a face. Therefore,

$$\begin{aligned} f_i &= \text{sum of } (\pi - \text{exterior angles}) \\ &= \pi \times e_i (\text{sum of exterior angles of the face}) \\ &= \pi \times e_i - 2\pi \end{aligned} \tag{6.9}$$

where e_i is the number of edges of the face i. Summing this quantity over all the faces yields

$$\sum_{i=1}^{F} f_i = \sum_{i=1}^{F} e_i - 2\pi F \tag{6.10}$$

Now we simplify the summation on e_i, keeping in mind that in a closed surface each edge is shared by precisely two faces, and that summing the number of edges in a face over all the separate faces counts each edge exactly twice. This means that

$$\sum_{i=1}^{F} e_i = 2E \tag{6.11}$$

Combining Equation 6.10 with 6.11 yields

$$\sum_{i=1}^{F} f_i = 2\pi (E - F) \tag{6.12}$$

Substituting Equation 6.12 into 6.7 yields

$$K = 2\pi V - \sum_{i=1}^{F} f_i \tag{6.13}$$

or

$$K = 2\pi \, (V - E + F) \tag{6.14}$$

The quantity $(V - E + F)$ is the *Euler characteristic* of a surface and we denote it by the Greek letter χ (chi). Using this notation, we rewrite the curvature formula as

$$K = 2\pi \chi \tag{6.15}$$

We can think of this equation as a convenient way to calculate the total curvature of a piecewise flat surface. However, it implies much more. As we have seen, K has special properties. First, K is defined for all surfaces, not only piecewise flat ones. Also, K is topological invariant. Therefore, the equation leads us to suspect that the Euler characteristic also has these properties. This suggests the following questions: Can we show directly that $(V - E + F)$ is a topological invariant? Can we define $(V - E + F)$ for all surfaces, not just piecewise flat ones? If we achieve these goals, we have gained considerable understanding of the topology of surfaces.

Here then is an important question about topological invariance: What happens to $(V - E + F)$ when we deform a piecewise flat surface? And we must answer: Absolutely nothing! After all, we readily see that $(V - E + F)$ is just a counting algorithm, and unless we cut a face in two, remove an edge, or perform some other nontopological transformation, none of the numbers can change.

6.4 Closed Curved Surfaces

Euler's characteristic, the sum $(V - E + F)$, works for an arbitrary surface. To see this, we begin with the image of a deformed piecewise flat surface, which suggests a new definition for the terms vertices, edges, and faces on a general closed, curved surface. We will now define faces on such surfaces not as flat polygons but as topological disks, and edges not as straight lines but as simple arcs with a vertex at either end. We will define a *net* on a closed, curved surface as an arbitrary collection of simple arcs, terminated at each end by a vertex, that divides the surface everywhere into topological disks. For convenience, we will continue to call the elements of a net *edges*, *vertices*, and *faces*. Given such a net, we can define its Euler characteristic by the same formula: $\chi = V - E + F$.

This definition of the Euler characteristic raises a question: If we can draw an infinite number of different nets on a surface, how do we decide which net to use in computing the Euler characteristic? The answer is that it does not matter which one we choose, because all valid nets on the same closed surface have the same Euler characteristic.

To prove this, we could imagine starting with a particular net on a surface, then transforming this net into a different one by adding or deleting vertices or edges.

We can do this in several ways, but we single out the following two elementary net transformations:

1. Adding (or deleting) a face by adding (or deleting) an edge between existing vertices, or
2. Adding (or deleting) a vertex.

How do these transformations affect the value of χ? If we add an edge, then E increases by 1. But F also increases by 1; hence, $F - E$ is unchanged. And because V is unchanged, $\chi = V - E + F$ is unchanged by the first type of transformation.

If we insert a new vertex into an edge, we produce not only a new vertex, but also a new edge. Thus, V and E each increase by 1, and F is unchanged; then χ is again unchanged. We conclude that χ is invariant under these two net transformations. To complete the proof that any two nets on the same surface will give the same value of χ, we assert that, given any two nets, we can always get from one to the other by some sequence of these transformations.

Every surface has an Euler characteristic, χ, and it is a topological invariant. Now we have two topological invariants for surfaces: the Euler characteristic χ and the total curvature K. We also know that they are related by $K = 2\pi\chi$ for piecewise flat surfaces. But does this hold for any surface? It does, and we can now assert the following theorem: For any closed surface, the total curvature and the Euler characteristic are related by $K = 2\pi\chi$. This is the *Gauss-Bonnet theorem*.

There are two proofs of this theorem. The first is based on the fact that K and χ are both topological invariants. We start with an arbitrary surface and deform it into a piecewise flat surface. Because K and χ are topological invariants, the deformation leaves them unchanged.

We know that $K = 2\pi\chi$ for the piecewise flat surface, so now we must show that we can deform any surface into a piecewise flat surface. Intuitively, this seems reasonable and our approach is to flatten the surface piece by piece, pushing all the curvature into the edges between the pieces. Then we flatten and straighten the edges piece by piece, pushing all the curvature into the vertices. The details of this procedure are difficult to express more rigorously, and they are not necessary here. A second proof that $K = 2\pi\chi$ is based on a direct computation of K for any surface and is similar to the way we proved the theorem for piecewise flat surfaces.

The Gauss-Bonnet theorem is important in the geometry of surfaces because it produces a relationship between quantities defined purely in terms of topology (such as the Euler characteristic) and quantities defined purely in terms of distances and angles (such as total curvature).

6.5 Euler Operators

Euler objects, such as polyhedra, always satisfy Euler's formula. The processes that add or delete faces, edges, and vertices to create a new Euler object are the *Euler operators*. These operators give us a rational way to construct solid, polyhedra-like objects and make sure that they are topologically valid.

The connectedness of the boundary surface of a solid is a property distinct from, and independent of, the enclosed interior points. Connectivity, orientation, and the characteristic of being non-self-intersecting are global properties of the surface and depend on all of its parts. Euler's formula describes a quantitative relationship between these parts that allows us to assign certain global characteristics: the number of handles or through holes, total curvature, connectivity, and so on.

Since we have just seen that Euler's formula is not restricted to plane-faced polyhedra, but also applies to any closed surface on which we can construct a proper net, the formula becomes a useful check on the topological validity of any solid whose surface we can express as a net of surface patches, curve segments, and vertices. To apply the formula, however, other conditions must be met:

1. All faces must be simply connected (that is, they are topological disks), with no holes in them, and bounded by a single ring of edges.
2. The solid object must be simply connected (that is, its complement is connected) with no holes through it.
3. Each edge must adjoin exactly two faces and terminate at a vertex at each end.
4. At least three edges must meet at each vertex.

It is easy to see that the objects in Figure 6.13 satisfy Euler's formula and that the nets are proper. Each net is a collection of simple arcs (edges), terminated at each end by a vertex. These arcs divide the surface into topological disks. Distorting the shapes by making the straight edges curves and the faces nonplanar surfaces does not change the applicability or validity of the formula.

If we add vertices, edges, and faces to a model, we must do it in a way that satisfies both the Euler formula and the four conditions listed above. In Figure 6.14a and 6.14b, a cube is correctly modified. In Figure 6.14c, the formula is certainly satisfied, but notice that edges (1, 5) and (2, 5) do not adjoin two faces. Also, only two edges meet at vertex 5, and edge (1, 2) adjoins three faces. So, as it stands, the object in Figure 6.14c is not a valid solid. We can remedy this situation by adding edges (3, 5) and (4, 5), resulting in a net gain of two edges and two faces. Edges (1, 2), (2, 3), (3, 4), and (4, 1) no longer define a face. An interesting modification of Euler's formula states that if a polyhedron is divided into C polyhedral cells, then the vertices, edges, faces, and cells

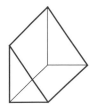
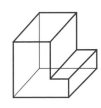

Figure 6.13 *Polyhedrons that satisfy Euler's formula.*

(a)   (b)

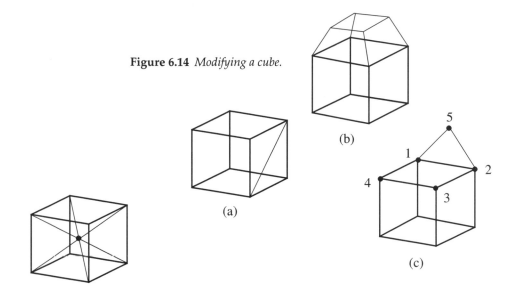

Figure 6.14 *Modifying a cube.*

(a)

(b)

(c)

Figure 6.15 *Changing a cube into a six-celled polyhedron.*

are related by

$$V - E + F - C = 1 \qquad\qquad (6.16)$$

For example, adding a vertex to the center of a cube (Figure 6.15) and joining it to each of the other eight vertices with edges creates a six-celled polyhedron. It is easy to verify the vertex, edge, face, and cell count.

6.6 Orientation

We have explored various topological properties of surfaces formed by taking a collection of planar pieces and gluing them together along their edges to create piecewise flat surfaces. We have learned that any surface formed in this way will be flat everywhere except possibly along the edges, where the pieces are glued together. In fact, if all the planar pieces have straight edges, then the piecewise flat surface has curvature only at the vertices. We have also learned that we can generalize our findings; the topological invariants apply equally well to curved surfaces.

Now let's explore another very important topological property—*orientation*. We will use a data structure called a *topological atlas* to do this. It is a way to keep track of which edges are adjoining. Figure 6.16 shows the atlas of a truncated pyramid. This topological atlas is similar to an ordinary road atlas, which is a collection of separate maps, each containing information directing the user to the next map.

Figure 6.17 shows two different ways of joining a pair of edges. The first, in Figure 6.17a, identifies an *orientation-preserving* construction. The other possibility,

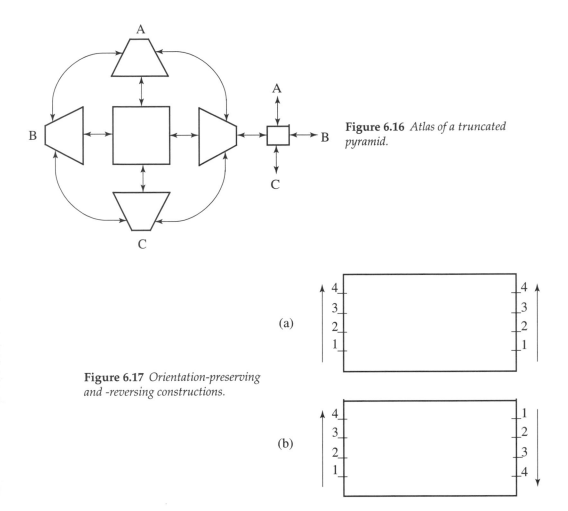

Figure 6.16 *Atlas of a truncated pyramid.*

(a)

Figure 6.17 *Orientation-preserving and -reversing constructions.*

(b)

in Figure 6.17b, identifies an *orientation-reversing* construction. We number each edge and use an arrow to indicate the ascending direction of the numbers. Next, we join opposite edges so that the numbers match and the direction of the two arrows is the same. We can use either numbers or arrows alone to identify the way the two edges are joined. Orientation-reversing constructions have both arrows pointing in the same rotational sense—either clockwise or counterclockwise—around the face. If one arrow points clockwise and the other counterclockwise, the construction is orientation preserving. Including orientation-reversing joints in an atlas lets us construct new kinds of surfaces, including some mathematical curiosities called *nonorientable surfaces*.

The atlas in Figure 6.18a defines a cylinder, while the atlas in 6.18b defines a torus. The atlas in 6.18c asks us to make a half twist before joining the edges together, resulting in a *Möbius strip*.

As inhabitants of a Möbius strip we would observe a curious phenomenon. If we start out at some point and take a trip all the way around it, when we return to our initial position, we would find that left and right are reversed. This happens because

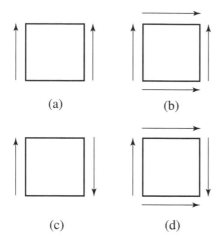

(a)   (b)

(c)   (d)

Figure 6.18 *A cylinder, torus, Möbius strip, and Klein bottle.*

right and left are not intrinsically defined. Externally, whether a turn appears left or right depends on the side of the surface we are looking from. It depends on our setting up an external reference. Thus, we must specify from which side we are to make this observation.

Right and left are characteristics of the Möbius strip inhabitant's motions. We can define right and left on the surface, but the definition works only locally. We cannot look at some point on the surface, then decide a priori that one direction is either right or left. We must first move around on the surface. There is nothing to ensure that, as we move around the surface, the commands to go left or right will generate the same orientation each time we return to a given point. This potential reversal and confusion is exactly what happens on the Möbius strip. A surface on which left and right are never reversed is *orientable*, and we can establish a consistent left and right definition on the surface. If we find a path that confuses left and right, the surface is *nonorientable*.

We can demonstrate the orientability of a surface in terms of another property. For example, an orientable surface is one on which we can define clockwise and counterclockwise rotations in a consistent way: We choose a point P on the surface S and imagine standing at that point. Next, we decide which of the two possible rotations to call clockwise; this is called an orientation at P. Now, we let Q be another point on S. (Note that P and Q may be coincident.) As we travel to Q along some path, we keep track of which rotation we defined as clockwise. This induces an orientation at Q, that is, a sense of clockwise or counterclockwise for rotations at Q.

There are obviously many paths from P to Q. There is not always a unique shortest path. Different paths may induce different orientations. If we obtain the same orientation no matter which path we take, then the surface is orientable. However, if we travel around the meridian of a Möbius strip, we end up with the opposite orientation from our original one. Therefore, again we conclude that the Möbius strip is nonorientable, and our new definition agrees with the old one.

Any closed surface that fits into three-dimensional space must be topologically the equivalent of a sphere with g handles, where g can equal zero. Now we add another condition, namely, that the surface must be orientable. A sphere is orientable, and it is not hard to show that adding a handle cannot change that.

The Möbius strip does fit into three-dimensional space, but it is not closed. To transform the Möbius strip into a closed surface, we make an orientation-preserving construction of the top and bottom edges on the atlas exactly as we did when we produced a torus from a cylinder (Figure 6.18d). The resulting surface is a _Klein bottle_, and, like the Möbius strip, it is nonorientable. Producing the Klein bottle involves pushing the surface through itself—a rather drastic operation, but the best we can do in three dimensions. The Klein bottle will not fit into a three-dimensional space without self-intersections.

6.7 Curvature

We can show that the total curvature for a sphere is 4π and that the angle excess is proportional to the area, or $E = kA$. Thus, for any polygon on a sphere of radius r, $E = k$, where $k = 1/r^2$ (if is measured in radians). For a simple closed path on the surface of a sphere, total turning in radians is expressed as $T = 2\pi - A/r^2$, where A is the area enclosed by the path and r is the radius of the sphere.

In general, for an arbitrary surface, k is a measurement taken at any point on the surface. The value of k at a point is the excess per unit area of a small patch of surface containing the point. We call k the _curvature density_ of the surface at a point. We use the term _density_ because k is "something per unit area," and we use the term _curvature density_ rather than _excess density_ because, although k is measured using excess, it tells us how curved the surface is at the point. To approximate a small patch of an arbitrary surface by a small patch of some surface we know very well, we observe that within a small enough area, almost any surface will appear planar.

We can make an even better approximation by using a small piece of a sphere. We will choose the approximating sphere to have the same excess per unit area as a small piece of the arbitrary surface. Therefore, if k does not change radically in the small piece, then all of the geometry there—angles, total turning, and so on—is very close to the geometry on a sphere whose radius is determined by $k = 1/r^2$. The smaller the radius of the approximating sphere (that is, the greater the curvature density), the more curved is the surface. A football, for example, is not very curved in the middle; k is small there. But at the pointed ends, the football is curved as much as a sphere of small radius. From the middle to the pointed ends, k gradually increases.

Curvature density is a local quantity that is measured in the vicinity of a point on an arbitrary surface. A global version of curvature density is called _total curvature_. We start with a region of an arbitrary surface and divide it into polygonal pieces. For each polygon we compute the excess and sum the excesses for all the pieces. The result is the total curvature K of the region. Of course, for this definition to make sense, we must be sure that if we subdivide a region into polygons in two different ways, the sum of the excess of the pieces is the same in both cases.

If the initial region is itself a polygon, the additivity theorem implies that K is precisely equal to the excess around the boundary of the polygon. Therefore, K is a kind of excess over the region, and it is true for any region on the surface, not just for polygons.

Handles on Spheres

To compute K for a region that has no boundary at all, the sphere, we divide the region into two pieces, for example, the northern and southern hemispheres. Each of these is bounded by the equator, which we know has excess 2π. This means that 4π is the total curvature of a sphere. Observe that the curvature density k of a sphere depends on the radius, but the total curvature K is the same for all spheres.

It turns out that the total curvature K is a topological invariant for closed surfaces. For surfaces with a boundary, the total curvature is unchanged by deformations that do not affect the vicinity of the boundary of the surface.

Any torus has zero total curvature. We are not asserting that a torus is flat like a cylinder or a plane, which would imply that the curvature density is zero everywhere. We are merely asserting that any torus has just as much negative curvature as positive curvature.

We now know the total curvature of two kinds of closed surfaces, the sphere and the torus. A sphere is not topologically the same as a torus, yet they are related. The torus can be described as a sphere with a handle attached. The process of "handle attaching," though not a deformation, is a practical way of making new closed surfaces from old ones because it is easy to keep track of how total curvature changes in the process. Let's explore this process a little more.

We start with a sphere and flatten out two regions on it. From each region we cut out a flat disk and attach a handle in the holes left by the disks. The handle is topologically a cylinder, but the edges of the cylinder must be flared out to blend with the flat regions around the disks. The result is topologically a torus. In other words,

$$Sphere - 2Disks + Handle = Torus \tag{6.17}$$

Next, we compute the total curvature on each side of the equation. The sphere has $K = 4\pi$. The disks were flat and thus have $K = 0$. The torus, as we just asserted, also has $K = 0$. Then, for the equation to be balanced, the handle must have a total curvature of -4π.

By flattening, cutting, and gluing, we can attach a handle to any surface. We now know the curvature in the handle, and we see that it always decreases the total curvature by 4π. We find that the total curvature of a two-holed torus is the same as that of a torus with a handle attached or a sphere with two handles attached. This is expressed as

$$K(2 \text{ hole torus}) = 4\pi - 4\pi - 4\pi = -4\pi \tag{6.18}$$

In general, for a surface that is topologically the same as a sphere with g handles attached, the total curvature is

$$K(\text{sphere} + g \text{ handles}) = 4\pi(1 - g) \tag{6.19}$$

The important fact about spheres with handles is this: Any closed surface in three-dimensional space is topologically equivalent to a sphere with some number of handles attached. We also see from Equation 6.19 that the total curvature of any closed surface in three-dimensional space is an integer multiple of 4π.

Recall that total turning for a closed curve path is always an integer multiple of 2π; so now we have discovered an analogous property for closed surfaces in three-dimensional space—all of that arbitrary denting and bending and all of those excess angles must somehow combine to give precisely an integer multiple of 4π. Remarkably, in changing the total curvature of an ordinary closed surface, we must change it by a multiple of 4π or not at all.

There are closed surfaces that are too twisted to fit into three-dimensional space, and, as a consequence, do not have a total curvature that is a multiple of 4π. As it turns out, any closed surface, including these twisted ones, must have total curvature equal to a multiple of 2π.

6.8 Intersections

Consider the intersections of the closed curves and straight lines in Figure 6.19. The following properties are true if these elements all lie in a plane:

1. An infinitely long (unbounded) straight line or open curve always intersects a closed curve an even number of times, if they intersect at all.
2. Two closed curves intersect an even number of times, if they intersect at all.
3. Either do not count tangent points, or count each as two intersections.

We can use these properties to decide if a point is inside or outside a closed curve or closed surface. In Figure 6.20 we see that:

1. A point is outside a closed curve or surface if a semi-infinite line drawn from it intersects the closed curve or surface an even number of times.
2. A point is inside a closed curve or surface if a semi-infinite line drawn from it intersects the closed curve or surface an odd number of times.
3. Either do not count tangent points, or count each as two intersections.

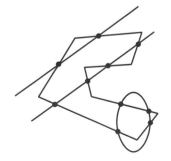

Figure 6.19 *Intersection of unbounded lines and closed curves.*

Figure 6.20 *Points inside and outside a closed curve.*

Here is another property to consider: If an unbounded plane intersects a closed surface, then the intersection is one or more closed curves. (A closed curve can have straight or "curved" sides. Thus, a square is a closed curve.)

The properties of intersections on more topologically complicated surfaces produce some surprises. For example, two closed curves may have only one point of intersection (see curves 4 and 5 in Figure 6.3).

Exercises

6.1 Sketch the atlas of a tetrahedron.

6.2 Find Euler's formula for the space of four dimensions (use Poincaré's generalization).

6.3 How many regular polytopes are there in four-dimensional space?

6.4 Show that every face of a polyhedron must be a topological disk.

6.5 Show that the total turning of a path around the vertex of a polyhedron is equal to the sum of the interior angles meeting at that vertex.

6.6 Show that $K = 2\pi(V - F)$ for a cube.

6.7 Compute K for each of the five regular polyhedra.

6.8 For piecewise square surfaces, show why $K = 2\pi(V - E + F)$ reduces to $K = 2\pi(V - F)$.

6.9 Construct a toroidal polyhedron, and compute its Euler characteristic and K. As a special challenge, construct a toroidal polyhedron that has the minimum number of vertices. Edges must be straight-line segments, and faces must be planar. (Remember that each face must be a topological disk.) Accurately sketch the polyhedron.

Are the following statements (Exercises 6.10–6.25) true or false? Briefly explain your answer.

6.10 The total turning along any closed path on a plane is an integer multiple of 360°.

6.11 Any two closed paths that are topologically equivalent must have the same total turning.

6.12 The Jordan curve theorem states that any simple closed curve in the plane divides the plane into two regions, an inside and an outside.

6.13 Every simple closed curve on the surface of a torus divides the surface into two separate regions.

6.14 On a closed path on a curved surface, the angle excess plus the total turning equals 360°.

6.15 For any topological disk on an arbitrary surface, the angle excess around the boundary is equal to the total curvature of the interior.

6.16 A torus has zero total curvature.

6.17 A two-holed torus has a total curvature equal to 4π.

6.18 Angle excess is always zero around a closed path on a plane surface.

6.19 Simple polyhedra can be continuously deformed into a sphere.

6.20 Euler's formula for simple polyhedra relates total curvature to V, E, and F.

6.21 All the curvature in a closed, piecewise flat surface is concentrated at the vertices.

6.22 The quantity $V - E + F - 2$ is called the *Euler characteristic* of a polyhedron.

6.23 Every face of a polyhedron must be a topological disk.

6.24 A net on a general surface is a collection of simple arcs that divide the surface everywhere into topological disks.

6.25 The surface of a sphere has a unique net.

CHAPTER 7

HALFSPACES

This chapter first reviews elementary set theory and its geometric interpretation in Venn diagrams, explaining the Boolean operators *union, intersection,* and *difference.* It reviews the important notion of a *function*, discusses two-dimensional halfspaces created by straight lines and open and closed curves in the plane, shows how to classify an arbitrary point with respect to containment in the halfspace, and how to create three-dimensional halfspaces using planes and surfaces. It shows how to combine halfspaces to create a finite, bounded shape and how to classify an arbitrary point with respect to that shape.

7.1 Definition

A straight line or open curve divides the Cartesian coordinate plane into two regions called *halfspaces* (Figure 7.1). One region we identify as the *inside half* (+) and the other as the *outside half* (−). A closed curve, such as a circle or ellipse, also divides the plane into two regions, one enclosed and finite and the other open and infinite. These are also examples of halfspaces.

In three dimensions, a plane or surface divides space into two regions. These, too, are halfspaces and are sometimes called *directed surfaces*. This spatial division also applies to spaces of higher dimensions. An n-dimensional space is divided into halfspaces

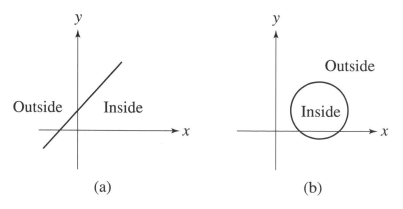

Figure 7.1 *Examples of halfspaces.*

by a hypersurface of $n - 1$ dimensions. We can combine two or more simple halfspaces using Boolean operators to create more complex shapes and to establish or limit the visibility of objects in a computer-graphics display. The mathematics and geometry of halfspaces makes it easy to do window and clipping operations and to solve many containment problems in computer graphics.

7.2 Set Theory

Set theory is the primary source of the algebra and operations that govern combining halfspaces to form complex shapes. A *set* is a well-defined collection of objects that are, in turn, *elements* or *members* of the set. Examples of a set are: all natural numbers, all even numbers, all white crows, all diamonds in a deck of cards, or the first ten prime numbers. In geometry, the point is a basic element, and in geometric modeling, computer graphics, and similar applications we may examine and operate on the set of points along a curve, the set of points inside a circle, or the set of points on the surface of some object, and so on.

A set that contains all of the elements of all of the sets in a particular problem is called the *universal set*. It is symbolized by E. In geometry, E is often taken to mean all the points in a two- or three-dimensional space. We use uppercase italic letters to denote a particular set. The *complement* of a set A with respect to a universal set E is the set of all elements of E that are not elements of A. The complement of A is written cA. The *null set*, denoted as $\varnothing$, is a set having no elements at all. It is also known as the *empty set*.

A set A is a subset of another set B if and only if every member of A is also a member of B. The symbol $\subset$ denotes the relationship between set and subset. $A \subset B$ says that A is a subset of B. For example, if B is the set of all natural numbers and A is the set of all even numbers, then $A \subset B$. A is a *proper subset* of B if every member of A is contained in B and if B has at least one element not in A. Every set is a subset of itself, but not a proper subset. We form new sets by combining the elements of two or more existing sets. The mathematics that describes this process is called *Boolean algebra*.

George Boole (1815–1864), one of England's most original mathematicians, invented an algebra for the study of logic. He did this by separating the symbols of mathematical operations from the objects upon which they operate. He studied these operators as independent mathematical objects and asked two important questions: Can these operations be combined? Are there algebra-like rules for doing this? The results of his investigations are formalized as Boolean algebra, where sets are treated as algebraic quantities by the operations of *union*, *intersection*, and *difference*.

The *union* operator $\cup$ combines two sets A and B to form a third set C whose members are all of the members of A plus all of the members of B. This is expressed as

$$C = A \cup B \tag{7.1}$$

For example, if A consists of four members a, b, c, and d and if B consists of three members c, d, and e, then if $A \cup B = C$, we find that C consists of the five members a, b, c, d, and e.

Another operator, the *intersection* operator ∩, combines two sets A and B to form a third set C whose members are only those common to both A and B. This is expressed as

$$C = A \cap B \tag{7.2}$$

Using the example of sets A and B, we find that if $C = A \cap B$, then the members of C are c and d.

The *difference* operator, $-$, combines two sets A and B to form a third set C whose members are only those of the first set that are not also members of the second. This is expressed as

$$C = A - B \tag{7.3}$$

Here, we find that if $C = A - B$, then the members of C are a and b.

The union and intersection operators are commutative:

$$A \cup B = B \cup A \tag{7.4}$$

and

$$A \cap B = B \cap A \tag{7.5}$$

But the difference operator is not commutative:

$$A - B \neq B - A \tag{7.6}$$

This is demonstrated in the example, where we find $A - B = \{a, b\}$ and $B - A = \{e\}$. The curly brackets { } are commonly used in set theory to enclose members of a set. They are part of what is called the *set-builder* notation.

Set operations obey the rules of Boolean algebra. They govern the ways we can combine sets and determine the results we should expect. The properties of this algebra are summarized in Table 7.1.

Venn diagrams, named after the logician John Venn (1834–1883), give us a visual, geometric interpretation of set theory and the Boolean operations. In many applications of geometry today, sets consist of points, where the universal set is often the set of points defining a Euclidean space of one, two, or three dimensions. Venn diagrams suggest ways to operate on subsets of those points that define geometric shapes and classify these points as being inside, outside, or on the boundary of the shape.

Figure 7.2 shows four different Venn diagrams. The outermost rectangle of each diagram encloses the universal set. The smaller inner rectangles are sets within the universal set: sets A, B, and C. In each diagram, the outcome of the indicated operation is shown as a shaded region.

7.3 Functions Revisited

Function notation is used to develop the mathematics of halfspaces. Recall that a function is a mathematical expression—not an equation, although a function may appear

Table 7.1 Properties of operations on sets

Union Properties
1. $A \cup B$ is a set (closure)
2. $A \cup B = B \cup A$ (commutativity)
3. $(A \cup B) \cup C = A \cup (B \cup C)$ (associativity)
4. $A \cup \emptyset = A$ (identity)
5. $A \cup A = A$ (Idempotency)
6. $A \cup cA = E$ (complementation)

Intersection Properties
1. $A \cap B$ is a set (closure)
2. $A \cap B = B \cap A$ (commutativity)
3. $(A \cap B) \cap C = A \cap (B \cap C)$ (associativity)
4. $A \cap E = A$ (identity)
5. $A \cap A = A$ (Idempotency)
6. $A \cap cA = \emptyset$ (complementation)

Distributive Properties
1. $A \cup (B \cap C) = (A \cup B) \cap (A \cup C)$
2. $A \cap (B \cup C) = (A \cap B) \cup (A \cap C)$

Complementation Properties
1. $cE = \emptyset$
2. $c\emptyset = E$
3. $c(cA) = A$
4. $c(A \cup B) = cA \cap cB$
5. $c(A \cap B) = cA \cup cB$

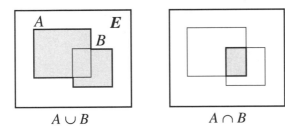

$A \cup B$ $A \cap B$

Figure 7.2 *Venn diagrams.*

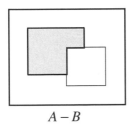

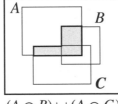

$A - B$ $(A \cap B) \cup (A \cap C)$

in an equation—given in terms of one or more variables (review Section 5.3) When we assign values to these variables, the function has a corresponding value. We say that the function is *evaluated* to produce a number. Think of a function as a mathematical machine (or simple algorithm). Variable values are the *input*; the function, or machine, generates a value—the *output*. The lowercase letters f, g, and h are usually used to denote a function, although other letters will do equally well.

Here is an example: The function f evaluated at x is written $f(x)$. If f represents the expression $x^2 + 2x - 1$, then $f(x) = x^2 + 2x - 1$. If we evaluate this function at t, we have $f(t) = t^2 + 2t - 1$. Or, if we evaluate it at $x = 3$, then $f(x = 3) = f(3) = 14$. Of course, we can use f to denote any other mathematical expression as well.

7.4 Halfspaces in the Plane

The equation of a straight line in the plane is

$$Ax + By + C = 0 \tag{7.7}$$

which divides the plane into two semi-infinite regions, or halfspaces, and serves as the boundary between them (Figure 7.3).

The halfspace bounded by the line is expressed as

$$h(x, y) = Ax + By + C \tag{7.8}$$

where h denotes a halfspace. This means that points in the halfspace are a function of the values x and y. If a particular pair of values of x and y yields $h(x, y) = 0$, then that pair represents a point on the line, or boundary, of the halfspace. Some pairs of values of x and y produce an inequality, either $h(x, y) > 0$ or $h(x, y) < 0$. The coordinates of all points on one side of the boundary produce values with the same algebraic sign. Points on the other side produce values with the opposite sign. The following conventions will help to classify a point with respect to a halfspace:

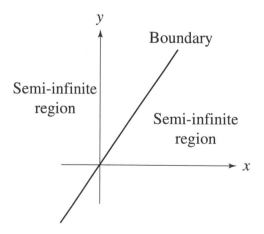

Figure 7.3 *Dividing the x, y plane into two halfspaces.*

1. If the coordinates of a point yield $h(x, y) = 0$, then the point is on the boundary of the halfspace.
2. If the coordinates of a point yield $h(x, y) > 0$, then the point is in the *inside* halfspace.
3. If the coordinates of a point yield $h(x, y) < 0$, then the point is in the *outside* halfspace; it is in the complement of $h(x, y)$.

Note that the conventions establish the semi-infinite region whose points yield positive values for $h(x, y)$ as the halfspace of interest.

In Figure 7.4a the line $2x - y + 1 = 0$ divides the x, y plane into two halfspaces. By convention, the halfspace of interest to us, $h_1(x, y) = 2x - y + 1$, is defined by the points that yield $h_1(x, y) > 0$. The origin lies inside this halfspace because $h_1(0, 0) = 1 > 0$. The point $(1, 3)$ is on the boundary because $h_1(1, 3) = 0$. The point $(-1, 3)$ is outside this halfspace because $h_1(-1, 3) = -4$.

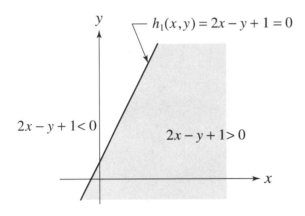

(a)

Figure 7.4 *Algebraic representation of halfspaces.*

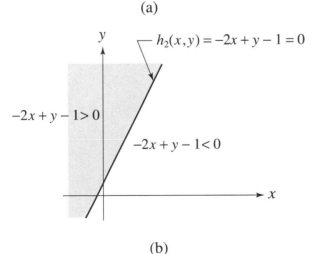

(b)

Changing the sign of $h_1(x, y)$ reverses the inside/outside point classification. This makes the complement of $h_1(x, y)$ the halfspace of interest (Figure 7.4b). Let

$$h_2(x, y) = -h_1(x, y) \tag{7.9}$$

so that

$$h_2(x, y) = -2x + y - 1 \tag{7.10}$$

The origin is now outside $h_2(x, y)$ because $h_2(0, 0) = -1 < 0$. The point $(1, 3)$ is still on the boundary because $h_2(1, 3) = 0$. The point $(-1, 3)$ is inside this halfspace because $h_2(-1, 3) = 4$.

The boundaries of halfspaces are not restricted to straight lines. For example, in Figure 7.5 the parabola $x - y^2 - 1 = 0$ divides the plane into two semi-infinite regions. The halfspace $h(x, y) = x - y^2 - 1$ is shown shaded.

This method works for closed curves, too. In Figure 7.6 the circle $-(x - 3)^2 - (y - 2)^2 + 1 = 0$ divides the plane into two regions. One region consists of points inside the circle, and the other consists of points outside of it. The complement of this function defines a circular hole in the plane.

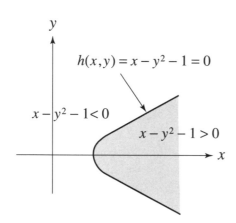

Figure 7.5 *Halfspace defined by a parabola.*

Figure 7.6 *Halfspace defined by a circle.*

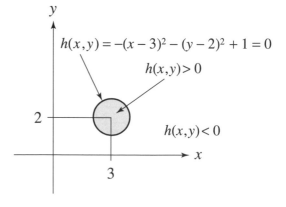

7.5 Halfspaces in Three Dimensions

The equation of a plane is

$$Ax + By + Cz + D = 0 \qquad (7.11)$$

It divides three-dimensional space into the two semi-infinite regions lying on either side of it. For a halfspace bounded by a plane, we write

$$h(x, y, z) = Ax + By + Cz + D \qquad (7.12)$$

The conventions for classifying a point in three-dimensional space with respect to a halfspace are analogous to the conventions for the plane, except that now we evaluate points with three coordinates instead of two. Thus, if $h(x, y, z) > 0$, the point is in the *inside* halfspace. If $h(x, y, z) = 0$, it is on the boundary. If $h(x, y, z) < 0$, the point is in the *outside* halfspace.

A simple example is that of the halfspaces formed by the $z = 0$ plane, where $h(x, y, z) = z$. The boundary is $h(x, y, z) < 0$, which is the x, y plane. A point $\mathbf{p}(x, y, z)$ is above this plane if $h(x, y, z) > 0$ and below it if $h(x, y, z) < 0$. The complement of the halfspace reverses the classification without changing the boundary.

Consider the following halfspace:

$$h(x, y, z) = 2x + 3y + 6z - 18 \qquad (7.13)$$

Its boundary is the plane $2x + 3y + 6z - 18 = 0$, part of which is shown shaded in Figure 7.7. The origin (and any other point on the same side of the plane) is outside this halfspace because $h(0, 0, 0) = -18$. The point $(1, 1, 4)$ is inside the halfspace because $h(1, 1, 4) = 11$.

We are not limited to using planes. For example, Figure 7.8 shows a halfspace defined by a cylindrical surface. We could use spheres, quadric surfaces, or any other implicitly defined surface to create a halfspace.

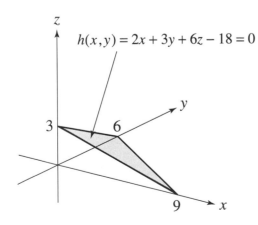

Figure 7.7 *Halfspace defined by a plane.*

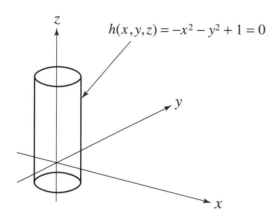

Figure 7.8 *Halfspace defined by a cylindrical surface.*

7.6 Combining Halfspaces

We can combine halfspaces using the Boolean intersection operator to create complex closed shapes. The halfspaces must always have the same number of dimensions. That is, we can combine a two-dimensional halfspace with another two-dimensional halfspace (but not with a three-dimensional halfspace). In doing this we preserve *dimensional homogeneity* (all elements have the same number of dimensions).

The shaded area in Figure 7.9 is the intersection of two halfspaces h_1 and h_2, where

$$h_1 = h_1(x, y) = -x + 30 \qquad\qquad (7.14)$$

and

$$h_2 = h_2(x, y) = y - 10 \qquad\qquad (7.15)$$

We can express this intersection as

$$g_1 = h_1 \cap h_2 \qquad\qquad (7.16)$$

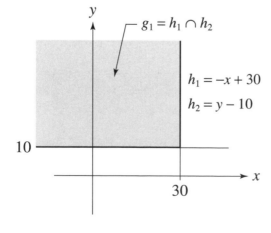

Figure 7.9 *Intersection of two halfspaces.*

(Note the abbreviated notation.) Any point outside either h_1 or h_2 is outside g_1. Any point inside both h_1 or h_2 and not on the boundary of either is inside g_1. Any point inside h_1 or h_2 and on the boundary of the other is on the boundary of g_1. This holds true for classifying a point with respect to the intersection of any two halfspaces.

If we take the intersection of g_1 and h_3, where

$$h_3 = x - y \tag{7.17}$$

we get a finite and closed triangular shape (Figure 7.10), expressed as

$$g_2 = h_1 \cap h_2 \cap h_3 \tag{7.18}$$

By including the halfspace

$$h_4 = (x - 30)^2 + (y - 10)^2 - 10 \tag{7.19}$$

we get a more complex shape (Figure 7.11), which is the result of the following series of intersections:

$$g_3 = h_1 \cap h_2 \cap h_3 \cap h_4 \tag{7.20}$$

Note that h_4 creates a hole.

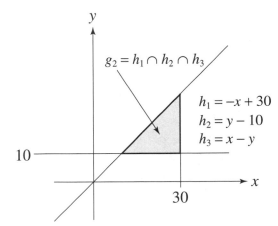

Figure 7.10 *Closed triangular shape.*

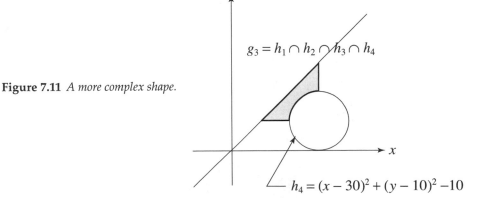

Figure 7.11 *A more complex shape.*

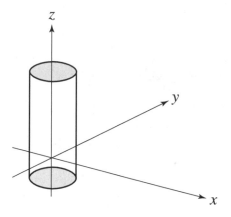

Figure 7.12 *A capped cylinder.*

In general, if we need n halfspaces to define the complex shape of some geometric object g, then

$$g = \bigcap_{i=1}^{n} h_i \tag{7.21}$$

where $\bigcap_{i=1}^{n}$ denotes the accumulated product of intersections of n halfspaces. This equation applies to both two- and three-dimensional halfspaces. Here is another way to express Equation 7.21:

$$g = h_1 \cap h_2 \cap \ldots \cap h_i \cap \ldots \cap h_n \tag{7.22}$$

This process by itself does not guarantee a finite g, which could be null if one of the h_i does not intersect the net intersection of the other halfspaces.

The capped cylinder in Figure 7.12 is

$$g = h_1 \cap h_2 \cap h_3 \tag{7.23}$$

where

$$\begin{aligned}
h_1 &= -x^2 - y^2 + 4 \\
h_2 &= z \\
h_3 &= -z + 8
\end{aligned} \tag{7.24}$$

This cylinder is eight units high, with the plane of one end at $z = 0$ and the other at $z = 8$. Its radius is two units, and its axis coincides with the z axis.

The object in Figure 7.13 is more complex. Let's change the notation somewhat to reflect this complexity. We will denote the final object as C, and we see that it is built up by combining two simpler objects A and B. Thus,

$$C = A \cup B \tag{7.25}$$

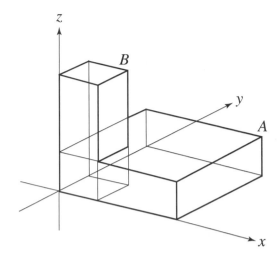

A   **Figure 7.13** *Union of two rectangular solids.*

where

$$A = \bigcap_{i=1}^{6} h_{A,i}$$

$$B = \bigcap_{i=1}^{6} h_{B,i}$$

(7.26)

The halfspaces are

$$
\begin{array}{ll}
h_{A,1} = x & h_{B,1} = x \\
h_{A,2} = -x + 4 & h_{B,2} = -x + 1 \\
h_{A,3} = y & h_{B,3} = y \\
h_{A,4} = -y + 4 & h_{B,4} = -y + 1 \\
h_{A,5} = z & h_{B,5} = z \\
h_{A,6} = -z + 2 & h_{B,6} = -z + 6
\end{array}
$$

Note that, as you can see in the figure, the two objects have three halfspaces in common: $h_{A,1} = h_{B,1}$, $h_{A,3} = h_{B,3}$, and $h_{A,5} = h_{B,5}$.

Three rules apply to classifying points with respect to any shapes created by Equation 7.21:

1. If and only if a point is inside all the halfspaces h_i, then it is inside the geometric object g.
2. If and only if a point is outside at least one of the halfspaces h_i, then it is outside the geometric object g.
3. If and only if a point is on the boundary of at least one halfspace h_i and inside the remaining halfspaces, then it is on the boundary of the geometric object g.

Once we have created two or more finite, enclosed shapes, we can combine them using an appropriate sequence of Boolean union, intersect, and difference operators.

Exercises

7.1 If the universal set E consists of the integers 1 through 24, a set A in E consists of the even integers, and B in E consists of multiples of 3, list the members of
a. $A \cup B$ c. $A - B$
b. $A \cap B$ d. $B - A$

7.2 Find the complement of $A \cup B$ (see Exercise 7.1).

7.3 Find the complement of $A - B$ (see Exercise 7.1).

7.4 Using Venn diagrams, demonstrate that for $A \subset B$ it is necessary and sufficient that $A \cup B = B$ or $A \cap B = A$.

7.5 Show that $A - (A - B) = A \cap B$.

7.6 Classify the following points in the plane with respect to $h_1 = y - 2$:
a. $(0, 0)$ f. $(-1, -2)$
b. $(4, -2)$ g. $(-4, 1)$
c. $(4, 4)$ h. $(-3, 1)$
d. $(2, 3)$ i. $(-1, 4)$
e. $(2, 2)$ j. $(3, 5)$

7.7 Classify the points given in Exercise 7.6 with respect to $h_2 = -x + 3$.

7.8 Classify the points given in Exercise 7. 6 with respect to $h_3 = x - y + 2$.

7.9 Classify the points given in Exercise 7.6 with respect to $h_1 \cap h_2 \cap h_3$. Use the halfspaces given in Exercises 7.6–7.8.

7.10 Classify the points given in Exercise 7.6 with respect to $h_4 \cup h_5$, where $h_4 = -(x-4)^2 - (y-3)^2 + 16$ and $h_5 = -x^2 - y^2 + 9$.

7.11 Classify the points given in Exercise 7.6 with respect to $h_5 - h_4$. (The minus sign is the Boolean difference operator.)

7.12 Which of the points given in Exercise 7.6 are outside $h_6 = x^2 + y^2 - 9$?

7.13 Write the equation defining the halfspace whose boundary is the parabola $y = x^2 + 1$, so that the origin is outside.

7.14 Write the equations defining the halfspaces whose combined Boolean intersection produces a unit cube centered at the origin.

7.15 Sketch the plane figure produced by the combined Boolean intersections of the following halfspaces:
$h_1(x, y) = -x + 2$
$h_2(x, y) = x + 2$
$h_3(x, y) = -y + 6$
$h_4(x, y) = y$
$h_5(x, y) = -x - y + 5$
$h_6(x, y) = x - y + 5$
Is this the most efficient combination of halfspaces for this shape? Why?

CHAPTER 8

POINTS

A point is the simplest of the elementary geometric objects: points, lines, and planes. In fact we cannot define a point in terms of anything simpler except as a set of numbers. Points are the basic building blocks for all other geometric objects, and elementary geometry demonstrates how many figures are defined as a locus of points with certain constraining characteristics. For example, in a plane, a circle is the locus of points equidistant from a given point, and a straight line is the locus of points equidistant from two given points. In three-dimensional space, a plane is the locus of points equidistant from two given points. We can also define more complex curves, surfaces, and solids this way, by using equations to define the locus of points. This is a powerful way of describing geometric objects, because it allows us to analyze and quantify their properties and relationships. Most importantly to today's technology, points are indispensable when we create computer graphic displays and geometric models. This chapter discusses the definition of a point as a set of real numbers, point relationships, arrays of points, absolute and relative points, displaying points, pixels and point resolution, and translating and rotating points.

8.1 Definition

A point suggests the idea of place or location. We define a point by a set of one or more real numbers, its coordinates. The coordinates of a point not only locate it in a coordinate system, but also with respect to other points in the system.

A set of n real numbers define a point in n-dimensional space

$$\mathbf{p} = (x_1, x_2, \ldots, x_n) \tag{8.1}$$

where $x_1, x_2, \ldots, x_n$ are the coordinates of $\mathbf{p}$ and n is the number of dimensions of the coordinate system. A boldface, lowercase letter $\mathbf{p}$ will denote a point. This is consistent with the vector notation of Chapter 1. In fact, under certain conditions there is a one-to-one correspondence between the coordinates of a point and the components of its vector representation.

In geometric modeling and computer graphics, most work is in two- or three-dimensional space, so we will not usually see Equation 8.1 in this generalized form. Note, though, that each coordinate has an identifying number, shown as an attached subscript. This is usually not necessary if the geometry is in two or three dimensions, where there are plenty of letters with which to name the coordinates, such as x, y, and z. In fact, a global or world coordinate system in computer graphics is three-dimensional, and points in it are given by

$$\mathbf{p} = (x, y, z) \tag{8.2}$$

However, we often use a subscript to identify a specific point. For example,

$$\mathbf{p}_1 = (x_1, y_1, z_1) \tag{8.3}$$

Points are a key part of the definition of a coordinate system, and we can extend the definition of a point to define a coordinate system. In three dimensions it is the set of all points defined by the triplet of real numbers (x, y, z), where $x, y, z \in (+\infty, -\infty)$. From this we can derive all the other characteristics of the coordinate system. For example, the origin is the point with $x = 0$, $y = 0$, and $z = 0$, or $\mathbf{p} = (0, 0, 0)$. The x axis is the set of points $\mathbf{p} = (x, 0, 0)$, where $x \in (+\infty, -\infty)$. A *rectangular Cartesian coordinate system* is further restricted in that the coordinate axes must be mutually perpendicular and the same distance scale must be used on all the axes.

A point has no geometric or analytic properties other than place. It has no size, orientation, length, area, or volume. It has no inside or outside, nor any other common geometric characteristics. However, things change when we have two or more points, for then we can compute and test many interesting properties describing the relationships of these points to each other. For example, if we are given two points in space $\mathbf{p}_1$ and $\mathbf{p}_2$ (Figure 8.1), we can compute the distance d between them by using the Pythagorean theorem:

$$d = \sqrt{(x_2 - x_1)^2 + (y_2 - y_1)^2 + (z_2 - z_1)^2} \tag{8.4}$$

Note that the distance d is always a positive real number. We can assign a specific unit of measurement to the coordinate system: centimeters, meters, light-years, and so on. The distance is then in terms of this unit of measurement. Although our choice may be arbitrary, we must be consistent throughout an application.

The coordinates of the midpoint, $\mathbf{p}_m$, between two points are

$$\mathbf{p}_m = \left[\frac{(x_1 + x_2)}{2}, \frac{(y_1 + y_2)}{2}, \frac{(z_1 + z_2)}{2} \right] \tag{8.5}$$

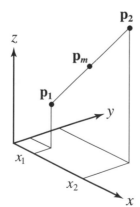

Figure 8.1 *Two points in space and their midpoint.*

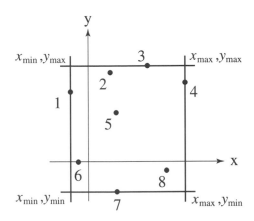

Figure 8.2 *Min-max box enclosing a set of points in the x, y plane.*

For any point in a set of points we can compute which other point is closest and which is farthest. We do this by computing and comparing d for each pair of points. If the absolute distance is not required, then d^2 will do just as well, and we avoid a relatively expensive computation.

Determining the vertices of a minimal rectangular solid, or box, which just contains all the points of a given set, is another useful computation in geometric modeling and computer graphics. This is particularly true when we must test relationships between two or more sets of points or containment of a point in a given volume of space. To compute the coordinates of the eight vertices of the box, we find the maximum and minimum x, y, and z coordinates of the point set. We assume that the edges of the box are parallel to the coordinate axes.

Figure 8.2 is an example of a *min-max* box enclosing a set of points in the x, y plane. The points shown have the following coordinates:

Point 1: $(-2, 7)$. This point has the minimum x value, written as x_{min}, in the set.
Point 2: $(2, 9)$
Point 3: $(6, 10)$. This point has the maximum y value, y_{max}.
Point 4: $(10, 8)$. This point has the maximum x value, x_{max}.
Point 5: $(3, 5)$
Point 6: $(-1, 0)$
Point 7: $(3, -3)$. This point has the minimum y value, y_{min}.
Point 8: $(8, -1)$

8.2 Arrays of Points

When a point is entered into a computer program, its coordinates are stored in an *array*. An array is an ordered arrangement of numbers which also may include an identifying number for each point. Otherwise the position of the coordinates of a point within the array may be all the "identification" needed. Figure 8.3 shows the coordinates of six points arranged in a rectangular array of 18 numbers. This is an efficient

	Column 1	Column 2	Column 3		
	x	y	z		
Row 1	8.1	−2.7	4.0	←	Point 1
Row 2	6.9	3.0	−6.5	←	Point 2
Row 3	5.7	1.1	2.1	←	Point 3
Row 4	0.0	4.2	−7.5	←	Point 4
Row 5	11.2	−0.8	−3.1	←	Point 5
Row 6	−6.6	1.0	−2.9	←	Point 6

Figure 8.3 *Coordinates of six points arranged in an array.*

way to represent and manipulate the coordinates. However, the numbers are not stored in the computer in "rectangular" configurations, as we shall see. We can present this information in a more compact form as

$$A = \begin{bmatrix} 8.1 & -2.7 & 4.0 \\ 6.9 & 3.0 & -6.5 \\ 5.7 & 1.1 & 2.1 \\ 0.0 & 4.2 & -7.5 \\ 11.2 & -0.8 & -3.1 \\ -6.6 & 1.0 & -2.9 \end{bmatrix}$$

We can identify an array with a symbol, say A, and use double subscripts to identify a particular coordinate in the array. For example, if we use A_{ij}, then the subscript i denotes the row and j denotes the column in which the number A_{ij} appears. So, in the array above, the value of A_{43} is −7.5.

Instead of being stored as a rectangular array, coordinates are usually arranged in a linear sequence, or list, of numbers. There are two ways to form a list of coordinate values for n points (Figure 8.4). Point-identifying information is often included in these data arrays, although none is shown here. In Figure 8.4a, the coordinates of each point are grouped together in sequence. In Figure 8.4b, all the x coordinate values are listed first, followed by the y and then z coordinates.

8.3 Absolute and Relative Points

In geometric modeling, or in creating a computer-graphics display, we use two kinds of points: *absolute points* and *relative points*. We make this distinction because of the way we compute the coordinates and the way we plot and display them. We define absolute points directly by their individual coordinates. For example, a set of absolute

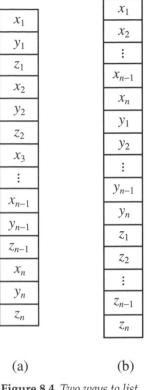

(a)　　　　　　　(b)

Figure 8.4 *Two ways to list point coordinates.*

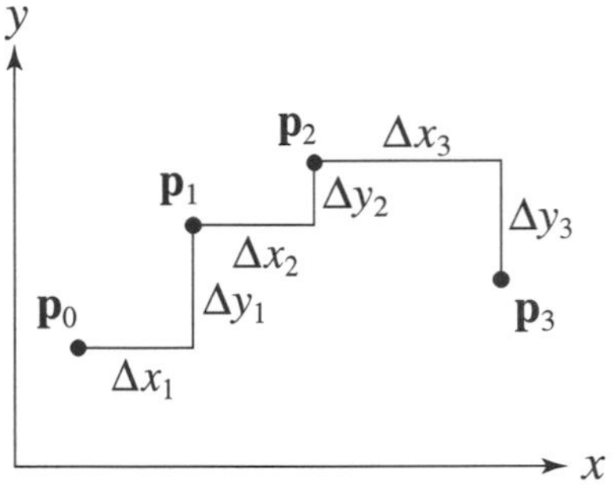

Figure 8.5 *Relative points.*

points is

$$\text{Absolute point } \mathbf{p}_i = (x_i, y_i) \quad \text{for } i = 1, \ldots, n \qquad (8.6)$$

We define the coordinates of each relative point with reference to the coordinates of the point preceding it. This is best demonstrated by the following expression:

$$\text{Relative point } \mathbf{p}_i = \mathbf{p}_{i-1} + \Delta_i \quad \text{for } 1, \ldots, n \qquad (8.7)$$

where $\mathbf{p}_0$ is some initial point and where $\Delta_i = (\Delta x_i, \Delta y_i)$ (Figure 8.5). This sequence may be generated during numerical analysis or when computing points on lines, curves, or surfaces.

8.4 Displaying Points

Now we must digress and discuss another geometry problem of contemporary application, again in computer graphics. A common point-display problem arises when we must define a *window* that just encloses a set of points. To do this, we investigate each point $\mathbf{p}_i$ in a search for the maximum and minimum x and y values, where $W_R = \max x$, $W_L = \min x$, $W_T = \max y$, and $W_B = \min y$, which define the window boundaries.

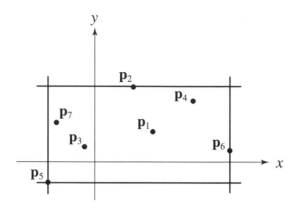

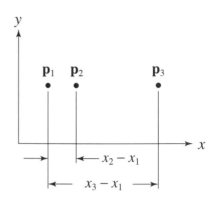

Figure 8.6 *Window boundaries and the min-max box.*   **Figure 8.7** *Limit of resolution of displayed points.*

In Figure 8.6 there are seven points with coordinates $\mathbf{p}_1 = (3.0, 1.5)$, $\mathbf{p}_2 = (2.0, 4.0)$, $\mathbf{p}_2 = (-0.5, 0.7)$, $\mathbf{p}_4 = (5.0, 3.0)$, $\mathbf{p}_5 = (-2.5, -1.0)$, $\mathbf{p}_7 = (-2.0, 2.0)$. It is easy to see that $\max x = x_6$, $\min x = x_5$, $\max y = y_2$, and $\min y = y_5$. This means that $W_R = 7.0$, $W_L = -2.5$, $W_T = 4.0$, and $W_B = -1.0$. We can confirm this visually by inspecting the points in the figure.

An interesting problem may arise when displaying three or more points. It is not always possible to resolve every point in a set of points to be displayed. Let's consider, for example, the three points, $\mathbf{p}_1$, $\mathbf{p}_2$, and $\mathbf{p}_3$, in Figure 8.7. For simplicity, we can arrange them on a common horizontal line so that $y_1 = y_2 = y_3$. This means that only the x values determine their separation. If the number of pixels H between $\mathbf{p}_1$ and $\mathbf{p}_3$ is less than the ratio of the separations of $\mathbf{p}_1$ and $\mathbf{p}_3$ to $\mathbf{p}_1$ and $\mathbf{p}_2$, then $\mathbf{p}_1$ and $\mathbf{p}_2$ will not be resolved. That is, $\mathbf{p}_1$ and $\mathbf{p}_2$ will not be displayed as two separate and distinct points. This relationship is expressed by the inequality $(x_3 - x_1)/(x_2 - x_1) > H_{\text{pixels}}$. If this inequality is true, then $\mathbf{p}_1$ and $\mathbf{p}_2$ cannot be resolved, and they will be assigned the same pixel.

It is also important to know if a given point in the picture plane is inside or outside the window region of a computer-graphics display. If the coordinates of a point are x, y, then the point is inside the window if and only if both of the following inequalities are true:

$$W_L \leq x \leq W_R \quad \text{and} \quad W_B \leq y \leq W_T \tag{8.8}$$

8.5 Translating and Rotating Points

We can move a point from one location to another in two ways: We can *translate* it from its current position to a new one, or we can *rotate* it about some point in the plane or axis in space to a new position. A translation is described by making changes relative to a point's coordinates (Figure 8.8). These changes are denoted as x_T and y_T, and the

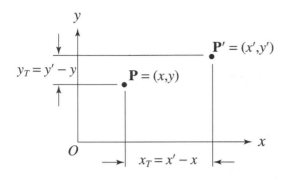

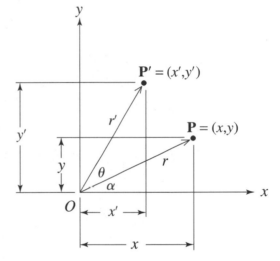

Figure 8.8 *Translating a point.*

Figure 8.9 *Rotating a point about the origin.*

translation equations are

$$x' = x + x_T$$
$$y' = y + y_T$$

(8.9)

where x' and y' are the coordinates of the new location. Generalizing these equations to three or more dimensions is straightforward. Now is a good time to review Chapter 3 on transformations.

The simplest rotation of a point in the coordinate plane is about the origin (Figure 8.9). If we rotate a point **p** about the origin and through an angle θ, then we derive the coordinates of the transformed point **p'** as follows: we express x' and y' in terms of $\alpha + \theta$ and r', thus

$$x' = r' \cos (\alpha + \theta)$$
$$y' = r' \sin (\alpha + \theta)$$

(8.10)

From elementary trigonometry we have

$$\cos (\alpha + \theta) = \cos \alpha \cos \theta - \sin \alpha \sin \theta$$
$$\sin (\alpha + \theta) = \sin \alpha \cos \theta - \cos \alpha \sin \theta$$

(8.11)

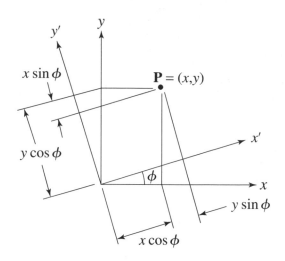

Figure 8.10 *Rotating the coordinate axes.*

and

$$\cos \alpha = \frac{x}{r} \qquad \sin \alpha = \frac{y}{r} \tag{8.12}$$

Because $r = r'$, with appropriate substitutions of the preceding equations into Equation 8.10, we find

$$x' = x \cos \theta - y \sin \theta$$
$$y' = x \sin \theta + y \cos \theta \tag{8.13}$$

This set of equations describes a rotation transformation in the plane about the origin. We can generalize this description to three dimensions if we change from rotation about the origin—a point—to rotation about a straight line, an axis of rotation. In fact, Equation 8.13 describes the rotation of a point about the z axis. We define a positive rotation in the plane as counterclockwise about the origin.

We can just as easily find the coordinates of a point in a new coordinate system which shares the same origin as the original system but which is rotated through some angle, say ϕ, with respect to it. Figure 8.10 illustrates the geometry of this transformation. The coordinates of a point in the new system are

$$x' = x \cos \phi + y \sin \phi$$
$$y' = -x \sin \phi + y \cos \phi \tag{8.14}$$

Exercises

8.1   Compute the distance between each of the following pairs of points:

   a. $(-2.7, 6.5, 0.8)$ and $(5.1, -5.7, 1.9)$ d. $(-3, 0, 0)$ and $(7, 0, 0)$
   b. $(1, 1, 0)$ and $(4, 6, -3)$ e. $(10, 9, -1)$ and $(3, 8, 3)$
   c. $(7, -4, 2)$ and $(0, 2.7, -0.3)$

8.2 Compute the coordinates of the points in Exercise 8.1 relative to a coordinate system centered at (3, 1, 0) in the original system and parallel to it.

8.3 Compute the distance between each of the points found for Exercise 8.2.

8.4 Show that the distance between any pair of points is independent of the coordinate system chosen.

8.5 Compute the midpoint between the pairs of points in Exercise 8.1.

8.6 Given an arbitrary set of points, find the coordinates of the vertices of a rectangular box that just encloses it.

8.7 Find the coordinates of the eight vertices of a rectangular box that just encloses the ten points given in Exercise 8.1.

8.8 Given that Δ_i is a constant for all $\mathbf{p}_i$, that is, $\Delta_i = (\Delta x_i, \Delta y_i) = (\Delta x, \Delta y)$, find $\mathbf{p}_4$ in terms of $\mathbf{p}_0$ and Δ_i.

8.9 Find the set of Δ_is for the vertex points of a square whose sides are three units long and with $\mathbf{p}_0 = (1, 0)$. Assume that the sides of the square are parallel to the x, y coordinate axes, and proceed counterclockwise.

8.10 Repeat Exercise 8.9 for a square whose sides are four units long and with $\mathbf{p}_0 = (-2, -2)$.

8.11 Repeat Exercise 8.10 for $\mathbf{p}_0 = (1, -4)$.

8.12 Given $W_R = 14$, $W_L = -2$, $W_T = 8$, and $W_B = -4$, determine which of the following points are inside this computer-graphics display window:

a. $\mathbf{p}_1 = (15, 4)$ d. $\mathbf{p}_4 = (-2, 8)$
b. $\mathbf{p}_2 = (3, 10)$ e. $\mathbf{p}_5 = (10, 10)$
c. $\mathbf{p}_3 = (14, 2)$

8.13 Find the coordinates of the corners of the window defined in Exercise 8.12.

8.14 State a mathematical test to determine if a point is contained in a rectangular volume in space, whose sides are parallel to the principal planes.

8.15 Derive the equations that describe the resultant transformation of a point that is first translated by x_T, y_T and then rotated by θ.

8.16 Derive the equations that describe the resultant transformations of a point that is first rotated by θ and then translated by x_T, y_T.

8.17 Compare and comment on the results of Exercises 8.15 and 8.16.

CHAPTER 9

LINES

A straight line is the next simplest geometric object after a point. All physical examples of a straight line are finite line segments with well-defined endpoints and length. However, the mathematical or geometric line may be unbounded or infinite, or it may be a semi-infinite half-line or ray. This chapter reviews the mathematical description of lines in two and three dimensions, including linear parametric equations. It describes how to compute points on a line, geometric relationships between points and lines, line intersections, and translating and rotating lines.

9.1 Lines in the Plane

The *slope-intercept form* is algebraically the simplest way to describe a straight line that lies in the x, y plane (Figure 9.1). If we know the slope m of a line and where it intersects the y axis, at $y = b$, then we write the equation of the line as

$$y = mx + b \tag{9.1}$$

The slope m is the ratio of the change in y to the change in x between any two points on the line. If the coordinates of these two points are x_1, y_1 and x_2, y_2, then the slope of the line that passes through them is

$$m = \frac{x_2 - x_1}{y_2 - y_1} \tag{9.2}$$

The *point-slope form* is a variation of the slope intercept form (Figure 9.2). If we know the slope m and a point x_1, y_1 through which the line passes, then

$$y - y_1 = m(x - x_1) \tag{9.3}$$

If we use the point $x_1 = 0$, $y_1 = b$, then Equation 9.3 simplifies to Equation 9.1.

The *two-point form* derives from the proposition that any two distinct points in the plane define a line that passes through both of them (Figure 9.3). Thus, given two

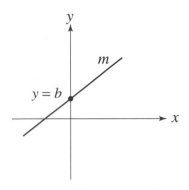

Figure 9.1 *Slope-intercept form.*

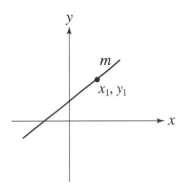

Figure 9.2 *Point-slope form.*

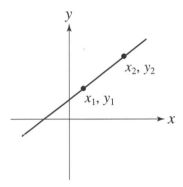

Figure 9.3 *Two-point form.*

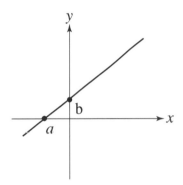

Figure 9.4 *Intercept form.*

points x_1, y_1 and x_2, y_2, we substitute Equation 9.2 into Equation 9.3 for m, and rearrange the terms to obtain

$$\frac{y - y_1}{x - x_1} = \frac{y_2 - y_1}{x_2 - x_1} \tag{9.4}$$

Using Equation 9.4 for the two points $(x_1, y_1) = (a, 0)$ and $(x_2, y_2) = (0, b)$, produces

$$\frac{x}{a} + \frac{y}{b} = 1 \tag{9.5}$$

This is the *intercept form*, aptly named, since it defines a line by its points of intersection with the coordinate axes (Figure 9.4).

Equations 9.1 and 9.3 are *explicit* equations, where x is the *independent variable* and y is the *dependent variable*. That is, we choose arbitrary values for x and compute the corresponding values of y. The value of y depends on the value of x. Equations 9.4 and 9.5 are *implicit equations*, where we can choose either x or y to be the independent variable. However, we can express x and y separately in terms of a third variable, say u. Using this approach we need two equations to define a line in the plane. They are

$$x = au + b$$
$$y = cu + d \tag{9.6}$$

These are *parametric equations*, and u is the *parametric variable* . Ordinarily, we treat u as the independent variable and x and y as dependent variables.

The parametric equations of a line in the plane work like this: Let's say we want to define a line that passes through x_1, y_1 and x_2, y_2 and that $u = 0$ at the first point, $u = 1$ at the second point. This is enough information to write two sets of two simultaneous equations to determine the constant coefficients a, b, c, and d. In this case, for $u = 0$ in Equations 9.6, we have

$$\begin{matrix} x_1 = b \\ y_1 = d \end{matrix} \quad \text{or} \quad \begin{matrix} b = x_1 \\ d = y_1 \end{matrix} \tag{9.7}$$

and for $u = 1$

$$a + b = x_2$$
$$c + d = y_2 \tag{9.8}$$

We find b and d directly, from Equations 9.7, and substitute appropriately into Equations 9.8 to find a and c. Thus

$$\begin{matrix} x_2 = a + x_1 \\ y_2 = c + y_1 \end{matrix} \quad \text{or} \quad \begin{matrix} a = x_2 - x_1 \\ c = y_2 - y_1 \end{matrix} \tag{9.9}$$

Now we know a, b, c, and d for the line passing through the two given points, and we rewrite Equations 9.6 to obtain the parametric equations of this line:

$$x = (x_2 - x_1)u + x_1$$
$$y = (y_2 - y_1)u + y_1 \tag{9.10}$$

To find a set of points on this line, we substitute a set of u values into Equations 9.10. Each u value determines a coordinate pair of x, y values. This way of defining straight lines is often used in computer graphics and geometric modeling, because it allows a sequence of points on the line to be computed and plotted on a display screen. These are *linear parametric equations* . We use higher-order parametric equations to define curves and surfaces. Of course, the extension to three dimensions is obvious.

Because parametric equations allow us to separate the dependent x and y coordinate values of points along a line, the method readily lends itself to vector geometry (see Chapter 1).

We can express the angle θ between two lines in the plane as a function of their two respective slopes, m_1 and m_2 (Figure 9.5):

$$\tan \theta = \frac{m_1 - m_2}{1 + m_1 m_2} \tag{9.11}$$

The derivation of Equation 9.11 is a simple exercise in algebra and trigonometry.

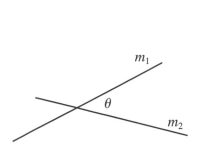

Figure 9.5 *Angle between two lines.*

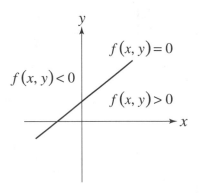

Figure 9.6 *Halfspaces defined by the implicit equation of a line.*

If two lines are parallel, then

$$m_1 = m_2 \tag{9.12}$$

If they are perpendicular, then

$$m_1 m_2 = -1 \tag{9.13}$$

A *halfspace* is defined by the implicit equation of a line (Figure 9.6). For example, given the explicit equation of the line $y = x + 2$, the corresponding implicit equation is $f(x, y) = x - y + 2$. If for any point x, y, $f(x, y) = 0$, then the point lies on the line. If $f(x, y) > 0$ or $f(x, y) < 0$, the point lies on one side or the other of the line. We say that $f(x, y) > 0$ and $f(x, y) < 0$ define two halfspaces, and $f(x, y) = 0$ defines their common boundary (Chapter 7).

9.2 Lines in Space

Three linear parametric equations, one for each coordinate, define a straight line in three-dimensional space:

$$\begin{aligned} x &= a_x u + b_x \\ y &= a_y u + b_y \\ z &= a_z u + b_z \end{aligned} \tag{9.14}$$

where x, y, and z are the dependent variables. Equations 9.14 generate a set of coordinates for each value of the parametric variable u. The coefficients $a_x, a_y, a_z, b_x, b_y,$ and b_z are unique and constant for any given line, their values depending on the endpoint coordinates.

We can think of this set of equations as a point-generating machine. The input is values of u. The machine produces coordinates of points on a line as output (Figure 9.7). It produces a bounded line segment if we limit the range of values we assign to the parametric variable. In computer graphics and geometric modeling, u usually takes

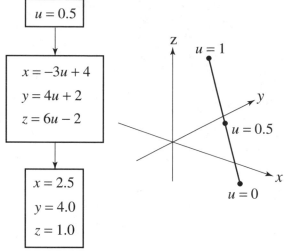

Figure 9.7 *A point-generating machine.*

values in the closed interval from 0 to 1. This is called the *unit interval*, and is expressed as

$$u \in [0, 1] \qquad (9.15)$$

The interval limits determine the nature of the line, making it a line segment, a semi-infinite line (a ray), or an infinite line. If we insert values of $u = 0$ and $u = 1$ into the point-generating machine of Figure 9.7, we obtain the endpoint coordinates $\mathbf{p}_0 = (4, 2, -2)$ and $\mathbf{p}_1 = (1, 6, 4)$. To characterize a line segment by the coordinates of its endpoints, we must modify Equations 9.14, identifying the endpoints of the line segment, as above, by $\mathbf{p}_0$ and $\mathbf{p}_1$ (Figure 9.8).

Substituting $u = 0$ into Equations 9.14 yields

$$\begin{aligned}
b_x &= x_0 \\
b_y &= y_0 \\
b_z &= z_0
\end{aligned} \qquad (9.16)$$

Figure 9.8 *Line segment in space.*

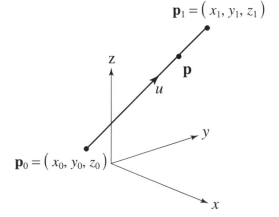

where x_0, y_0, and z_0 are the coordinates at the $u = 0$ endpoint of the line. At $u = 1$,

$$x_1 = a_x + x_0$$
$$y_1 = a_y + y_0 \qquad\qquad (9.17)$$
$$z_1 = a_z + z_0$$

or

$$a_x = x_1 - x_0$$
$$a_y = y_1 - y_0 \qquad\qquad (9.18)$$
$$a_z = z_1 - z_0$$

Substituting the results of Equations 9.16 and 9.18 into Equations 9.14 yields

$$x = (x_1 - x_0)u + x_0$$
$$y = (y_1 - y_0)u + y_0 \qquad u \in [0, 1] \qquad (9.19)$$
$$z = (z_1 - z_0)u + z_0$$

This is a very useful set of equations. It means that if we know the endpoint coordinates of a line segment, we can immediately write a parametric equation for it and find any intermediate points on it. The parametric variable conveniently ranges through the closed interval from zero to one.

To find the length of a line segment we simply apply the Pythagorean theorem to the endpoint coordinate differences:

$$L = \sqrt{(x_1 - x_0)^2 + (y_1 - y_0)^2 + (z_1 - z_0)^2} \qquad (9.20)$$

There are three numbers associated with every line that uniquely describe its angular orientation in space. These numbers are called *direction cosines* (or *direction numbers*). We denote them as d_x, d_y, and d_z. The computations are simple:

$$d_x = \frac{(x_1 - x_0)}{L}$$
$$d_y = \frac{(y_1 - y_0)}{L} \qquad\qquad (9.21)$$
$$d_z = \frac{(z_1 - z_0)}{L}$$

The geometry of the direction cosine computation for d_x is shown in Figure 9.9, and also for d_y and d_z. The edges of the auxiliary rectangular solid are parallel to the coordinate axes, and $\mathbf{p}_0$, A, and $\mathbf{p}_1$ define a right angle at A. From this figure we see that

$$d_x = \cos\theta \qquad\qquad (9.22)$$

The sum of the squares of the direction cosines must equal one. Therefore, any two of them are sufficient to determine the third:

$$d_x^2 + d_y^2 + d_z^2 = 1 \qquad\qquad (9.23)$$

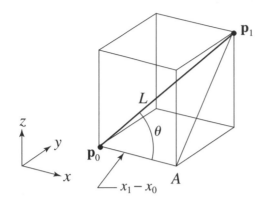

Figure 9.9 *Geometry for the computation of a direction cosine.*

However, direction cosines alone do not completely specify a line, because they tell us nothing about the location of the line. For example, any two parallel lines in space have the same direction cosines. But direction cosines are a good way to test whether they are parallel.

9.3 Computing Points on a Line

The parametric equations of a line are particularly useful for computing points on that line. An important set of points on a line frequently used in geometric modeling and computer graphics is points at equal intervals along the line. There are two ways to compute the coordinates for these points, one considerably faster and more efficient than the other. We will look at both methods.

To find the coordinates of points at n equal intervals on a given line, we use $\mathbf{p}_0$ and $\mathbf{p}_1$ and then compute the remaining $n - 1$ intermediate points. Figure 9.10 shows a line with eight equal intervals.

The first method uses Equation 9.19, for each of the $n - 1$ points. There are $n - 1$ values of the parametric variable, given by

$$u = \frac{1}{n}, \frac{2}{n}, \frac{3}{n}, \ldots, \frac{n-2}{n}, \frac{n-1}{n} \tag{9.24}$$

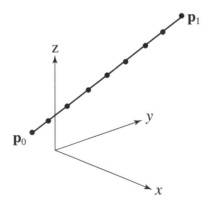

Figure 9.10 *Points at equal intervals along a line segment.*

and requiring $n - 1$ divisions to compute. For the x coordinates there are $n - 1$ multiplications and n additions (including finding $x_1 - x_0$). The computation totals for all coordinates are $3n$ additions, $3(n - 1)$ multiplications, and $n - 1$ divisions.

A second approach uses fewer computations. First, we notice that u always changes by a constant amount Δu, where

$$\Delta u = \frac{1}{n} \tag{9.25}$$

Next, we observe that for any x_i

$$x_i = (x_1 - x_0)u_i + x_0 \tag{9.26}$$

and for x_{i+1}

$$x_{i+1} = (x_1 - x_0)u_{i+1} + x_0 \tag{9.27}$$

However, since $u_{i+1} = u_i + \Delta u$, we can rewrite Equation 9.27 as

$$x_{i+1} = (x_1 - x_0)(u_i + \Delta u) + x_0 \tag{9.28}$$

or

$$x_{i+1} = (x_1 - x_0)u_i + (x_1 - x_0)\Delta u + x_0 \tag{9.29}$$

But $(x_1 - x_0)u_i + x_0 = x_i$, so we can simplify Equation 9.29:

$$x_{i+1} = x_i + (x_1 - x_0)\Delta u \tag{9.30}$$

and since $(x_1 - x_0)\Delta u$ is a constant, we let $\Delta x = (x_1 - x_0)\Delta u$. Therefore,

$$x_{i+1} = x_i + \Delta x \tag{9.31}$$

This tells us that we find each successive x coordinate by adding a constant to the previous value. This derivation applies to the y and z coordinates as well. Now let's count the computations: To compute Δu requires one division. To compute Δx requires one addition and one multiplication. For the x coordinates there are $n - 1$ additions. The totals for all n points are $3n$ additions, 3 multiplications, and one division. This process is called the *forward difference* method. It is commonly used to compute points on curves and surfaces, too.

9.4 Point and Line Relationships

Any point $\mathbf{q}$ is either on or off a given line (Figure 9.11). If it is on the line, it is either between the endpoints, $\mathbf{q}_1$, on the backward extension of the line, $\mathbf{q}_2$, or on the forward extension, $\mathbf{q}_3$.

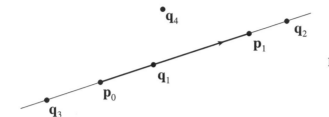

Figure 9.11 *Point and line relationships.*

We can write Equations 9.19 in terms of u to obtain

$$u_x = \frac{x - x_0}{x_1 - x_0}$$

$$u_y = \frac{y - y_0}{y_1 - y_0} \tag{9.32}$$

$$u_z = \frac{z - z_0}{z_1 - z_0}$$

Given the coordinates of any point $\mathbf{q} = (x, y, z)$, we compute u_x, u_y, and u_z. If and only if $u_x = u_y = u_z$ is point $\mathbf{q}$ on the line. Otherwise, it is off the line. We should allow some small but finite deviation. For example, $|u_x - u_y| = \varepsilon$, where $\varepsilon \ll 1$. If point $\mathbf{q}$ is on the line, then the value of u indicates its precise position.

In a plane, we can determine the position of a point relative to a line by solving the parametric equations to obtain the implicit equation

$$f(x, y) = (x - x_0)(y_1 - y_0) - (y - y_0)(x_1 - x_0) \tag{9.33}$$

Then, for a reference point $\mathbf{p}_R$ not on the line, we compute $f(x_R, y_R)$. For an arbitrary test point $\mathbf{p}_T$ we compute $f(x_T, y_T)$. If $f(x_T, y_T) = 0$, then $\mathbf{p}_T$ is on the line. If $f(x_R, y_R)$ and $f(x_T, y_T)$ have the same sign, that is, $f(x_T, y_T) > 0$ and $f(x_R, y_R) > 0$, or $f(x_T, y_T) < 0$ and $f(x_R, y_R) < 0$, then $\mathbf{p}_T$ and $\mathbf{p}_R$ are on the same side of the line. Otherwise they are on opposite sides (Figure 9.12). One way to choose a reference point is to let $\mathbf{p}_R = (x_0 + 1, y_0)$; then $\mathbf{p}_R$ is to the right of the line. If $y_0 = y_1$, the line is horizontal, and we can then let $\mathbf{p}_R = (x_0, y_0 + 1)$, placing $\mathbf{p}_R$ above the line.

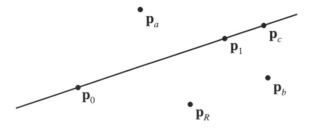

Figure 9.12 *Position of a point relative to a line.*

9.5 Intersection of Lines

If two lines intersect, they have a common point, the *point of intersection*. There are two problems of interest: first, the general problem of determining if two lines in space intersect (Figure 9.13a); second, the special case of finding the intersection of a line in the plane with a second line in the plane that is either horizontal or vertical (Figure 9.13b).

For the general problem, two lines, a and b, intersect if there is a point (the point of intersection) such that

$$x_a = x_b$$
$$y_a = y_b \qquad (9.34)$$
$$z_a = z_b$$

It follows from Equations 9.19 that

$$(x_{1a} - x_{0a})u_a + x_{0a} = (x_{1b} - x_{0b})u_b + x_{0b}$$
$$(y_{1a} - y_{0a})u_a + y_{0a} = (y_{1b} - y_{0b})u_b + y_{0b} \qquad (9.35)$$
$$(z_{1a} - z_{0a})u_a + z_{0a} = (z_{1b} - z_{0b})u_b + z_{0b}$$

where x_{0a} and y_{0a} are the coordinates of the endpoint $u = 0$ of line a, x_{1a} and y_{1a} are the coordinates of the endpoint $u = 1$ of line a, and similarly for line b.

We can use any two of the three equations to solve for u_a and u_b. Then we substitute u_a and u_b into the remaining equation to verify the solution. If the solution is verified, the lines intersect. Finally, both u_a and u_b must be in the interval 0 to 1 for a valid intersection.

For the special problem, given a line in the x, y plane, determine if it intersects with either of the vertical lines W_R, W_L (right or left), or with either of the horizontal lines W_T, W_B (top or bottom) (Figure 9.14). (Note that this nomenclature anticipates the window boundary coordinates of computer graphic displays: Chapter 14.) W_R is a

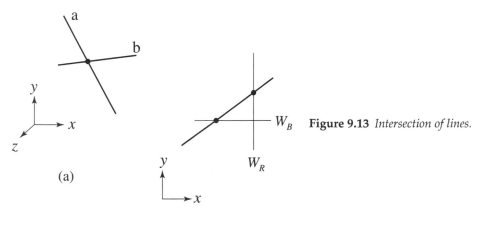

Figure 9.13 *Intersection of lines.*

(a)

(b)

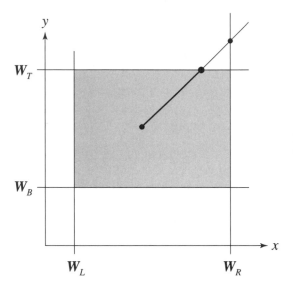

Figure 9.14 *Intersections with horizontal and vertical lines.*

vertical line whose equation is $x = W_R$. W_T is a horizontal line whose equation is $y = W_T$. If $x = W_R$ intersects the arbitrary line, then

$$u = \frac{W_R - x_0}{x_1 - x_0} \qquad (9.36)$$

If Equation 9.36 produces a value of u in the unit interval, we use it to compute the y coordinate of the point of intersection. If the arbitrary line itself is vertical, then it cannot intersect the line $x = W_R$ (for a vertical line $x_0 = x_1$ and, therefore, $x_1 - x_0 = 0$). We use a similar procedure when computing line intersections with W_L, W_T, and W_B.

9.6 Translating and Rotating Lines

We can translate a line by translating its end-point coordinates (Figure 9.15). The end-point translations must be identical, otherwise the transformed line will have a different length or angular orientation, or both.

In the plane, we have

$$\begin{aligned} x_0' &= x_0 + x_T & x_1' &= x_1 + x_T \\ y_0' &= y_0 + y_T & y_1' &= y_1 + t_T \end{aligned} \qquad (9.37)$$

The translated line is always parallel to its original position, and its length does not change. In fact, all points on the line are translated equally. That is why this is called a *rigid body translation*. The generalization to three or more dimensions is straightforward.

The simplest rotation of a line in two dimensions is about the origin (Figure 9.16). To do this we rotate both endpoints $\mathbf{p}_0$ and $\mathbf{p}_1$ through an angle θ about the origin. Then we find the coordinates of the transformed endpoints $\mathbf{p}_0'$ and $\mathbf{p}_1'$ by

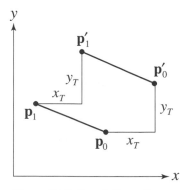

Figure 9.15 *Translating a line.*

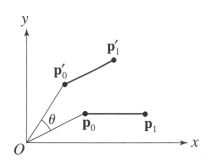

Figure 9.16 *Rotation of a line about a point in the plane.*

applying Equation 8.13:

$$x_0' = x_0 \cos \theta - y_0 \sin \theta \qquad x_1' = x_1 \cos \theta - y_1 \sin \theta$$
$$y_0' = x_0 \sin \theta + y_0 \cos \theta \qquad y_1' = x_1 \sin \theta + y_1 \cos \theta \tag{9.38}$$

This equation produces a rigid-body rotation of the line. Again, a rotation in three dimensions is a direct extension of this (see Section 3.5).

Exercises

9.1 Prove that the sum of the squares of the direction cosines of a line is equal to one.

9.2 Compute the length and direction cosines for each of the following line segments, defined by their endpoints:
   a. $\mathbf{p}_0 = (3.7, 9.1, 0.2)$, $\mathbf{p}_1 = (0.9, -2.6, 2.6)$
   b. $\mathbf{p}_0 = (2.1, -6.4, 0)$, $\mathbf{p}_1 = (3.3, 0.7, -5.1)$
   c. $\mathbf{p}_0 = (10.3, 4.2, 3.7)$, $\mathbf{p}_1 = (6.0, 10.3, 9.2)$
   d. $\mathbf{p}_0 = (5.3, -7.9, 1.4)$, $\mathbf{p}_1 = (0, 4.1, 0.7)$

9.3 Write the parametric equations for each line given in Exercise 9.2.

9.4 How does reversing the order of the end points defining a line segment affect its length and direction cosines?

9.5 Compute the $n-1$ intermediate points on each of the following line segments, defined by their endpoints:
   a. $\mathbf{p}_0 = (7, 3, 9)$, $\mathbf{p}_1 = (7, 3, 0)$, for $n = 3$
   b. $\mathbf{p}_0 = (-4, 6, 0)$, $\mathbf{p}_1 = (2, 11, -7)$, for $n = 4$
   c. $\mathbf{p}_0 = (0, 0, 6)$, $\mathbf{p}_1 = (6, 1, -5)$, for $n = 4$

9.6 Given the line segment defined by its endpoints $\mathbf{p}_0 = (6, 4, 8)$ and $\mathbf{p}_1 = (8, 8, 12)$, write the parametric equations of this line segment and determine for each of the following points if it is on or off the line segment.

a. $\mathbf{q}_1 = (4, 0, 8)$ d. $\mathbf{q}_4 = (10, 4, 2)$

b. $\mathbf{q}_2 = (12, -8, 20)$ e. $\mathbf{q}_5 = (10, 12, 16)$

c. $\mathbf{q}_3 = (7, 6, 10)$

9.7 Write the parametric equations for each of the line segments whose endpoints are given, and find the point of intersection (if any) between each pair of line segments.

a. Line 1: $\mathbf{p}_0 = (2, 4, 6)$, $\mathbf{p}_1 = (4, 6, -4)$
   Line 2: $\mathbf{p}_0 = (0, 0, 1)$, $\mathbf{p}_1 = (6, 8, -6)$

b. Line 1: $\mathbf{p}_0 = (2, 4, 6)$, $\mathbf{p}_1 = (4, 6, -4)$
   Line 2: $\mathbf{p}_0 = (4, 3, 5)$, $\mathbf{p}_1 = (2.5, 4.5, 3.5)$

c. Line 1: $\mathbf{p}_0 = (2, 4, 6)$, $\mathbf{p}_1 = (4, 6, -4)$
   Line 2: $\mathbf{p}_0 = (3, 5, 1)$, $\mathbf{p}_1 = (0, 2, 16)$

d. Line 1: $\mathbf{p}_0 = (10, 8, 0)$, $\mathbf{p}_1 = (-1, -1, 0)$
   Line 2: $\mathbf{p}_0 = (13, 2, 0)$, $\mathbf{p}_1 = (4, 7, 0)$

e. Line 1: $\mathbf{p}_0 = (5, 0, 0)$, $\mathbf{p}_1 = (2, 0, 0)$
   Line 2: $\mathbf{p}_0 = (8, 0, 0)$, $\mathbf{p}_1 = (10, 0, 0)$

CHAPTER 10

PLANES

The mathematics and geometry of planes pervades computer graphics, geometric modeling, computer-aided design and manufacturing, scientific visualization, and many more applications. This chapter looks at various ways to define them, including the normal form and the three-point form, the relationship between a point and plane, and plane intersections.

10.1 Algebraic Definition

Euclidean geometry defines a plane in space as the locus of points equidistant from two fixed points. The resulting plane is the perpendicular bisector of the line joining the two points. We call this definition a *demonstrative* or *constructive* definition. Computer-graphics and geometric-modeling applications require a more quantitative definition, because these planes may be bounded by polygons to form the polyhedral facets representing the surface of some modeled object or the plane of projection of some view of an object, and so on.

The classic algebraic definition of a plane is the first step toward quantification. The implicit Cartesian equation of a plane is

$$Ax + By + Cz + D = 0 \tag{10.1}$$

This is a linear equation in x, y, and z. By assigning numerical values for each of the constant coefficients A, B, C, and D, we define a specific plane in space. It is an arbitrary plane because, depending on the values of the coefficients, we can give it any orientation. Figure 10.1 shows only that part of a plane that is in the positive x, y, and z octant. The three bounding lines are the intersections of the arbitrary plane with the principal planes.

If the coordinates of any point satisfy Equation 10.1, then the point lies on the plane. We can arbitrarily specify any two coordinates x and y and determine the third coordinate z by solving this equation.

We obtain a more restricted version of this general equation by setting one of the coefficients equal to zero. If we set C equal to zero, for example, we obtain

$$Ax + By + D = 0 \tag{10.2}$$

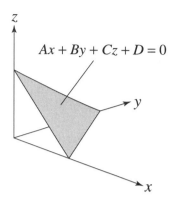

Figure 10.1 *Part of a plane.*

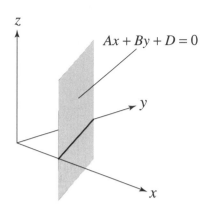

Figure 10.2 *A plane perpendicular to the x, y plane.*

This looks a lot like the equation of a line in the x, y plane. However, in three dimensional space this really means that the z coordinate of any arbitrary point can take on any value. The remaining coordinates are, of course, constrained by the relationship imposed by the equation. All planes defined this way are perpendicular to the principal plane identified by the two restricted coordinate values. In the case of Equation 10.2, the plane is always perpendicular to the x, y plane. Furthermore, the equation defines the line of intersection between the plane and the principal x, y plane (Figure 10.2).

If two of the coefficients equal zero, say $A = B = 0$, we obtain

$$Cz + D = 0 \quad \text{or} \quad z = \frac{D}{C} = k \tag{10.3}$$

where k is a constant determined by D/C. This is a plane perpendicular to the z axis and intersecting it at $z = k$ (Figure 10.3).

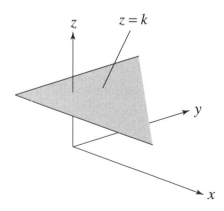

Figure 10.3 *A plane perpendicular to the z axis.*

10.2 Normal Form

A plane can also be defined by giving its *normal* (perpendicular) distance N from the origin and the direction cosines of the line defined by N, namely, d_x, d_y, d_z. These are also the direction cosines, or direction numbers, of the plane. This is all we need to define the intersections a, b, and c of the plane with each coordinate axis. To determine a, b, and c, we observe that $O\mathbf{p}_N\,\mathbf{p}_a$ is a right triangle; because N is perpendicular to the plane, any line in the plane through $\mathbf{p}_N$ is necessarily perpendicular to N (Figure 10.4). Therefore,

$$a = \frac{N}{d_x}, \quad b = \frac{N}{d_y}, \quad c = \frac{N}{d_z} \tag{10.4}$$

This yields the three points $\mathbf{p}_a = (a, 0, 0)$, $\mathbf{p}_b = (0, b, 0)$, and $\mathbf{p}_c = (0, 0, c)$. Now we can write the implicit form of the plane equation in terms of the distance N and its direction cosines as

$$d_x x + d_y y + d_z z - N = 0 \tag{10.5}$$

We can easily verify Equation 10.5 by setting any two of the coordinates equal to zero, say $y = z = 0$, to obtain the third coordinate value, in this case $x = N/d_x = a$. In general, the expressions $d_x = A$, $d_y = B$, $d_z = C$, and $N - D$ are true only if $A^2 + B^2 + C^2 = 1$. However, given $Ax + By + Cz + D = 0$, we find that

$$N = \frac{D}{\sqrt{A^2 + B^2 + C^2}} \tag{10.6}$$

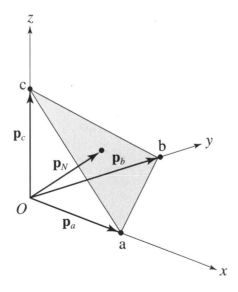

Figure 10.4 *Normal form.*

and

$$
d_x = \frac{A}{\sqrt{A^2 + B^2 + C^2}}
$$

$$
d_y = \frac{B}{\sqrt{A^2 + B^2 + C^2}} \qquad (10.7)
$$

$$
d_z = \frac{C}{\sqrt{A^2 + B^2 + C^2}}
$$

Finally, if two planes have the same direction cosines, then they are parallel.

10.3 Plane Defined by Three Points

Another way to define a plane is by specifying three noncollinear points in space. We demonstrate this as follows: We assume that the coordinates of these three points are $\mathbf{p}_1 = (x_1, y_1, z_1)$, $\mathbf{p}_2 = (x_2, y_2, z_2)$, and $\mathbf{p}_3 = (x_3, y_3, z_3)$. Then we let $A' = A/D$, $B' = B/D$, and $C' = C/D$.

The implicit equation of a plane becomes

$$
A'x + B'y + C'z + 1 = 0 \qquad (10.8)
$$

Using this equation and the coordinates of the three points, we can write

$$
\begin{aligned}
A'x_1 + B'y_1 + C'z_1 + 1 &= 0 \\
A'x_2 + B'y_2 + C'z_2 + 1 &= 0 \\
A'x_3 + B'y_3 + C'z_3 + 1 &= 0
\end{aligned} \qquad (10.9)
$$

Now we have three equations in three unknowns: A', B', and C', which we can readily solve. Two points are not sufficient to determine A', B', and C', and a fourth point would not necessarily fall on the plane.

If the three points are the points of intersection of the plane with the coordinate axes $\mathbf{p}_a = (a, 0, 0)$, $\mathbf{p}_b = (0, b, 0)$, and $\mathbf{p}_c = (0, 0, c)$, then the plane equation is

$$
\frac{x}{a} + \frac{y}{b} + \frac{z}{c} = 1 \qquad (10.10)
$$

10.4 Vector Equation of a Plane

There are at least four ways to define a plane using vectors. We will discuss one of them. Others are thoroughly discussed in Section 1.8. The normal form of the vector equation is easy to express using vectors. All we must do is express the perpendicular distance from the origin to the plane in question as a vector, $\mathbf{n}$. From this the other properties follow. For example, since the components of a unit vector are also its direction cosines, we have

$$
\hat{n}_x = d_x, \quad \hat{n}_y = d_y, \quad \hat{n}_z = d_z \qquad (10.11)
$$

where $\hat{\mathbf{n}}$ is the unit vector corresponding to $\mathbf{n}$. This means we can rewrite Equations 10.4 as follows:

$$a = \frac{|\mathbf{n}|}{\hat{n}_x}, \quad b = \frac{|\mathbf{n}|}{\hat{n}_y}, \quad c = \frac{|\mathbf{n}|}{\hat{n}_z} \tag{10.12}$$

where $|\mathbf{n}|$ is the length of $\mathbf{n}$.

We can also find the vector normal form using the implicit form of Equation 10.1 and Equations 10.6 and 10.7. Thus,

$$|\mathbf{n}| = \frac{D}{\sqrt{A^2 + B^2 + C^2}} \tag{10.13}$$

and

$$\hat{n}_x = \frac{A}{\sqrt{A^2 + B^2 + C^2}}$$
$$\hat{n}_y = \frac{B}{\sqrt{A^2 + B^2 + C^2}} \tag{10.14}$$
$$\hat{n}_z = \frac{C}{\sqrt{A^2 + B^2 + C^2}}$$

10.5 Point and Plane Relationships

Many problems in computer graphics and geometric modeling require us to determine on which side of a plane we find a given point $\mathbf{p}_T$. For example, we may want to know if the point is inside or outside a solid model whose bounding faces are planes. First, we define some convenient reference point $\mathbf{p}_R$ not on the plane (Figure 10.5). Then we use the implicit form of the plane equation, $f(x, y, z) = Ax + By + Cz + D$, and compute $f(x_R, y_R, z_R)$ and $f(x_T, y_T, z_T)$. Where the point lies depends on the

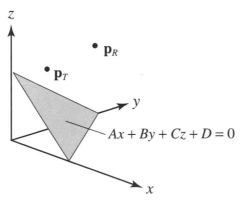

Figure 10.5 *Point and plane relationships.*

following conditions:

1. If $f(x_T, y_T, z_T) = 0$, then $\mathbf{p}_T$ is on the plane.
2. If $f(x_T, y_T, z_T) > 0$ and $f(x_R, y_R, z_R) > 0$, then $\mathbf{p}_T$ is on the same side of the plane as $\mathbf{p}_R$.
3. If $f(x_T, y_T, z_T) < 0$ and $f(x_R, y_R, z_R) < 0$, then $\mathbf{p}_T$ is on the same side of the plane as $\mathbf{p}_R$.
4. If none of the conditions above are true, then $\mathbf{p}_T$ is on the opposite side of the plane relative to $\mathbf{p}_R$.

10.6 Plane Intersections

If a line segment and plane intersect, they intersect at a point $\mathbf{p}_I$ common to both (Figure 10.6). We assume that the line segment is defined by parametric equations whose endpoint coordinates are (x_0, y_0, z_0) and (x_1, y_1, z_1). To find the coordinates of the point of intersection $\mathbf{p}_I$ we must solve the following four equations in four unknowns:

$$
\begin{aligned}
Ax_I + By_I + Cz_I + D &= 0 \\
x_I &= (x_1 - x_0)u_I + x_0 \\
y_I &= (y_1 - y_0)u_I + y_0 \\
z_I &= (z_1 - z_0)u_I + z_0
\end{aligned}
\tag{10.15}
$$

We then solve these equations for u_I:

$$
u_I = \frac{-(Ax_0 + By_0 + Cz_0 + D)}{A(x_1 - x_0) + B(y_1 - y_0) + C(z_1 - z_0)}
\tag{10.16}
$$

If $u_I \in [0, 1]$, then we solve for x_I, y_I, and z_I. If $Ax_0 + By_0 + Cz_0 + D = 0$ and $Ax_1 + By_1 + Cz_1 + D = 0$, then the line lies in the plane. If $u_I = \infty$, that is, if $A(x_1 - x_0) + B(y_1 - y_0) + C(z_1 - z_0) = 0$, then the line is parallel to the plane and does not intersect it.

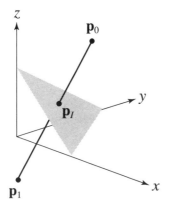

Figure 10.6 *Line and plane intersection.*

A somewhat more difficult problem is that of finding the intersection between two planes P_1 and P_2. We begin with the set of implicit equations of the two planes, solving them as two simultaneous equations in three unknowns:

$$A_1x + B_1y + C_1z + D_1 = 0$$
$$A_2x + B_2y + C_2z + D_2 = 0 \qquad (10.17)$$

Knowing the basic characteristics of possible solutions to a problem helps us design more reliable and efficient algorithms. This system of equations has three possible solutions, and we can state them in terms of the geometry of intersecting planes:

1. The planes do not intersect. (They are parallel.)
2. They intersect in one line.
3. The planes are coincident.

Two unbounded nonparallel planes intersect in a straight line (Figure 10.7), and the extra unknown is more like an extra degree of freedom. Moreover, it indicates that the solution is a line and not a point (which requires a third equation).

Assigning a value to one of the variables, say $z = z_I$, reduces the system to two equations in two unknowns:

$$A_1x + B_1y = -(C_1z_I + D_1)$$
$$A_2x + B_2y = -(C_2z_I + D_2) \qquad (10.18)$$

Solving these for x and y, we obtain one point on the line of intersection: $\mathbf{p}_I = (x_I, y_I, z_I)$.

Next, we assign a new value to z, say $z = z_J$, to obtain

$$A_1x + B_1y = -(C_1z_J + D_1)$$
$$A_2x + B_2y = -(C_2z_J + D_2) \qquad (10.19)$$

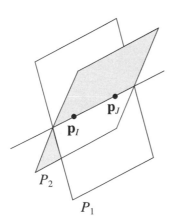

Figure 10.7 *Intersection of two planes.*

Again we solve for x and y to produce another point on the line of intersection: $\mathbf{p}_J = (x_J, y_J, z_J)$. These two points $\mathbf{p}_I$ and $\mathbf{p}_J$ are all we need to define the line of intersection in parametric form:

$$x = (x_J - x_I)u + x_I$$
$$y = (y_J - y_I)u + y_I \qquad (10.20)$$
$$z = (z_J - z_I)u + z_I$$

Exercises

10.1 Given the three points $\mathbf{p}_1 = (a, 0, 0)$, $\mathbf{p}_2 = (0, b, 0)$, $\mathbf{p}_3 = (0, 0, c)$, show that the equation of the plane containing these points can be written as $\frac{x}{a} + \frac{y}{b} + \frac{z}{c} = 1$.

10.2 Find the geometric relationships between the following points and the plane $y = -3$, where the origin is the reference point.
   a. $\mathbf{p}_1 = (4, 2, 5)$ d. $\mathbf{p}_4 = (5, -4, 3)$
   b. $\mathbf{p}_2 = (3, -3, 0)$ e. $\mathbf{p}_5 = (-3, -7, 8)$
   c. $\mathbf{p}_3 = (-2, 4, 6)$

10.3 Find the geometric relationships between the following points and the plane $4y - 3z - 24 = 0$, where the origin is the reference point.
   a. $\mathbf{p}_1 = (6, 2, 4)$ d. $\mathbf{p}_4 = (2, 2, -9)$
   b. $\mathbf{p}_2 = (-3, 6, 0)$ e. $\mathbf{p}_5 = (-1, 7, -1)$
   c. $\mathbf{p}_3 = (8, 0, -8)$

10.4 Find the point of intersection between each of the five lines, whose endpoints are given, and the plane $3x + 4y + z = 24$. First find the parametric equations for each line for $u \in [0, 1]$, and then find the value of the parametric variable corresponding to the intersection point, if it exists.
   a. $\mathbf{p}_0 = (4, 0, 4)$, $\mathbf{p}_1 = (4, 6, 4)$ d. $\mathbf{p}_0 = (0, 0, 24)$, $\mathbf{p}_1 = (8, 0, 0)$
   b. $\mathbf{p}_0 = (10, 0, 2)$, $\mathbf{p}_1 = (10, 2, 2)$ e. $\mathbf{p}_0 = (0, 9, 0)$, $\mathbf{p}_1 = (0, 0, 25)$
   c. $\mathbf{p}_0 = (10, -10, 2)$, $\mathbf{p}_1 = (10, 2, 2)$

CHAPTER 11

POLYGONS

Polygons and polyhedra were among the first forms studied in geometry. Their regularity, symmetry, and orderly lawfulness made them the center of attention of mathematicians, philosophers, artists, architects, and scientists for thousands of years. They are surprisingly simple arrangements of points, lines, and planes. This simplicity, however, harbors subtleties of symmetry and order that deepen our understanding of the physical universe, the atoms of which we are made, and the space-time in which we live.

Polygons are still important today. For example, their use in computer graphics is as common as points and lines, because it is easy to subdivide and approximate the surfaces of solids with planes bounded by polygons. Computer-graphics applications project, fill, and shade these polygonal facets to create realistic images of solids. The study of polygons leads to insights into elementary geometry, algebraic geometry, and number theory.

This chapter discusses two-dimensional polygons whose sides are all straight lines lying in the same plane. Three-dimensional polyhedra whose faces are planar and whose edges are straight lines are the subjects of the next chapter. Polygons and polyhedra also exist in higher-dimension spaces and may have curved edges and faces. Topics covered here include definitions of the various types of polygons, their geometric properties, convex hulls, construction of regular polygons, symmetry, and containment.

11.1 Definitions

A somewhat formal but effective definition goes like this: A *polygon* is a many-sided two-dimensional figure bounded by a closed circuit of straight-line segments joining successive pairs of points. The line segments are *edges* and the points are *vertices*. An edge is a bounded straight-line segment—a finite length of an otherwise unbounded straight line. A polygon, then, encloses a finite, two-dimensional area, bounded by edges and vertices. If the vertices all lie in the same plane, then the polygon is a *plane polygon*; otherwise it is a *skew polygon*. It is possible to construct polygons whose edges are curve segments on curved surfaces. We will study only plane polygons here.

The examples of polygons in Figures 11.1a–11.1d are three-, four-, five-, and six-sided polygons, respectively. There is no limit to the number of edges a polygon can have. And there is an infinite number of each type; that is, there is an infinite number

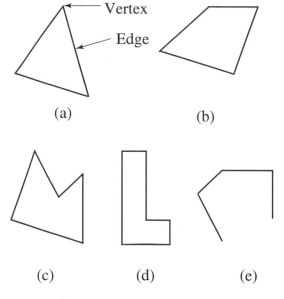

Figure 11.1 *A variety of polygons (a) -(d) and a polyline (e).*

of three-sided or triangular polygons, and the same is true of four-sided, five-sided, or *n*-sided polygons. Figure 11.1e is a polygon because it is not closed.

Every plane polygon is either convex, concave, or stellar. If the straight lines that are prolongations of the bounding edges of a polygon do not penetrate the interior, and if its edges intersect only at vertices, then the polygon is *convex*; otherwise it is *concave*. Any concave polygon can be minimally contained by a convex one. The polygons in Figures 11.1a and 11.1b are convex; those in Figures 11.1c and 11.1d are concave. If the edges of the polygon intersect at points in addition to the vertex points, then it is a *stellar* or *star polygon* (Figure 11.2a). We can construct a special kind of stellar polygon by connecting sequentially a series of points on the circumference of a circle with line segments, say every *m*th point, until a circuit is completed, with the last segment ending back at the starting point. We must not confuse stellar polygons with merely concave polygons that happen to be in the shape of a star (Figure 11.2b).

We are most familiar with the so-called regular polygons. A *regular polygon* lies in a plane, has straight-line edges all of equal length, has equal vertex angles, and

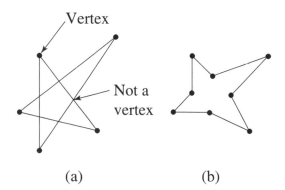

Figure 11.2 *Examples of a star and non-star polygon.*

Figure 11.3 *Regular polygons.*

can be inscribed in a circle with which it shares a common geometric center. Figure 11.3 shows several regular polygons. There is an infinite number of regular polygons, but not regular polyhedra, of which there are only five. Later we will discuss the classical problem that asks which regular polygons can be constructed using only a straightedge and compass.

A polygon is *equilateral* if all of its sides are equal, and *equiangular* if all of its angles are equal. If the number of edges is greater than three, then it can be equilateral without being equiangular (and vice versa). A rhombus is equilateral; a rectangle is equiangular.

In summary, a polygon in a plane is formed by joining line segments at their endpoints. This circuit of line segments and its interior form a polygonal region. The three-dimensional counterpart of a polygon is, of course, a polyhedron.

11.2 Properties

A plane polygon must have at least three edges to enclose a finite area, and in all cases the number of vertices equals the number of edges. In other words,

$$V - E = 0 \qquad (11.1)$$

where V is the number of vertices and E is the number of edges. Curiously, we can create nonplanar polygons that have only two sides. For example, two noncoincidental great-circle arcs connecting antipodal points on a sphere form a *digon* (Figure 11.4).

We can divide any plane polygon into a set of triangles (Figure 11.5) by a procedure called *triangulation*. The minimum number of triangles T, or the minimum triangulation, is simply

$$T = V - 2 \qquad (11.2)$$

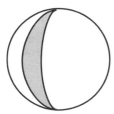

Figure 11.4 *A digon on a sphere.*

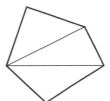
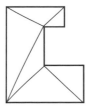

Figure 11.5 *Triangulating a plane polygon.*

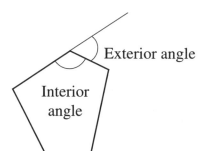

Exterior angle

Figure 11.6 *Exterior and interior angles.*

Interior
angle

The sum of the exterior angles of a plane polygon (not a stellar polygon) is 2π. This means that each exterior angle of a regular polygon is $2\pi/E$ and its interior angle (the supplement) is $(1 - 2/E)\pi$ (Figure 11.6).

The sum of the interior angles of a polygon with E edges is

$$\sum_V \theta = 180(E - 2) \tag{11.3}$$

An easy way to visualize this result is to triangulate the polygon with line segments radiating from an interior point to each of the vertices.

The average angle of a polygon is the sum of the angles divided by the number of angles, which equals E; that is,

$$\theta_{\text{avg}} = \frac{180(E - 2)}{E} \tag{11.4}$$

For regular polygons with E edges, the interior angles are identical (as are the edge lengths), so each interior angle equals the average value.

The perimeter of a regular polygon is

$$\text{Perimeter} = EL \tag{11.5}$$

where E is the number of edges and L is the length of an edge.

The area of a regular polygon is

$$A = \frac{EL^2}{4} \cot\left(\frac{\pi}{E}\right) \tag{11.6}$$

We can describe a polygon by simply listing the coordinates of its vertices: $x_1, y_1, x_2, y_2, \ldots, x_n, y_n$, where n is the number of vertices. We define its edges by successive vertex points, producing the last edge by constructing a line between x_n, y_n and x_1, y_1.

Using this list we can compute the geometric center, gc, of a convex polygon by simply computing the average x and y value of the coordinates of its vertices; thus,

$$x_{gc} = \frac{x_1 + x_2 + \cdots + x_n}{n}$$
$$y_{gc} = \frac{y_1 + y_2 + \cdots + y_n}{n} \tag{11.7}$$

Figure 11.7 *Area of a convex polygon.*

If we are given a convex polygon and its geometric center, we can compute its area as follows: First, we divide the *n*-sided polygon into *n* triangles, using the geometric center as a common vertex point for all the triangles and each edge of the polygon as part of a separate triangle (Figure 11.7). Next, using the Pythagorean theorem, we compute the length of each edge and each radiant from the geometric center to each vertex and then calculate the area of each triangle, given the length of its three sides. Finally, we sum the areas of all the triangles to find the area of the convex polygon.

The perimeter of a convex or concave polygon is the sum of the length of each edge:

$$
\begin{aligned}
L = &\sqrt{(x_1 - x_2)^2 + (y_1 - y_2)^2} \\
&+\sqrt{(x_2 - x_3)^2 + (y_2 - y_3)^2} \\
&+\cdots \\
&+\sqrt{(x_n - x_1)^2 + (y_n - y_1)^2}
\end{aligned}
\tag{11.8}
$$

A stellar polygon is a likely result of connecting vertices with edges out of order. We can test a polygon to make sure we have not accidentally created an unwanted stellar polygon. To detect a stellar polygon, we use the fact that their edges intersect at points in addition to their vertices (Figure 11.8). So we check each edge for the possibility of its intersection with other edges. (Remember that edges of a convex polygon meeting at a vertex do not intersect each other elsewhere.)

Checking for intersections with a given edge is more efficient if we ignore the edges immediately preceding and following it. Edge 2,3 of the polygon in the figure,

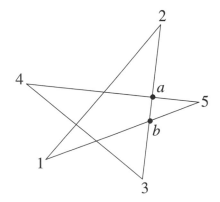

Figure 11.8 *Detecting a stellar polygon.*

for example, can intersect edges 1,2 and 3,4 only at vertices 2 and 3, respectively. Only edges 4,5 and 5,1 remain as candidates for intersection with edge 2,3. Points of intersection are produced at a and b. In a stellar polygon these points must be between the vertex points of the edge in question. It is easy to test to see if this is true; in the example, we compare the x, y coordinates of the point of intersection to the minimum and maximum x, y coordinates of the vertices bounding the edge:

$$x_2 < x_a, \quad x_b < x_3$$
$$y_2 < y_a, \quad y_b < y_3$$

(11.9)

The inequality indicates that points a and b are intersections between vertices 2 and 3 and that the polygon is stellar.

We can use more advanced techniques of vector geometry (Chapter 1) to find the area of a polygon. First, to compute the area of a triangle, we select a reference vertex $\mathbf{p}_1$. The area is one-half the absolute value of a vector product:

$$A = \frac{1}{2}|(\mathbf{p}_2 - \mathbf{p}_1) \times (\mathbf{p}_3 - \mathbf{p}_1)|$$

(11.10)

where the three vertices of the triangle are given by position vectors $\mathbf{p}_i$.

To compute the area of a polygon with four or more edges, we again select a reference vertex $\mathbf{p}_1$. The area of a plane polygon with n vertices, where $n \geq 4$, is

$$A = \frac{1}{2}\left|\sum_{i=1}^{n-2}(\mathbf{p}_{i+1} - \mathbf{p}_1) \times (\mathbf{p}_{i+2} - \mathbf{p}_1)\right|$$

(11.11)

Note that, in general, there are $n - 2$ "signed" (oriented) triangular areas. The vector product nicely accounts for triangle overlaps and, thus, holds for either convex or concave polygons.

11.3 The Convex Hull of a Polygon

The *convex hull* of any polygon is the convex polygon that is formed if we imagine stretching a rubber band over the vertex points (Figure 11.9). The convex hull of a convex polygon has edges corresponding identically to those of the polygon, while

Figure 11.9 *A convex hull.*

the convex hull of a concave polygon always has fewer edges than the polygon, some edges of which are not collinear, and it encloses a larger area. We also associate a convex hull with any arbitrary set of points in the plane. Polyhedra and sets of points in space have three-dimensional convex hulls, analogous to the two-dimensional case. In fact, the convex hull of the points defining a Bézier curve or surface closely approximates the curve or surface, and it is useful in analyzing and modifying its shape.

11.4 Construction of Regular Polygons

Recall that regular polygons have equal edges and interior angles. Their vertices lie on a circle, and another circle is tangent to each of the edges. This means that constructing a regular polygon with n vertices and n edges is equivalent to dividing the circumference of a circle into n equal parts. If we can do this division, then we can construct the corresponding regular polygon.

C. F. Gauss (1777–1855) showed that we can construct a regular convex polygon of n sides with a compass and straight edge if and only if

1. $n = 2^k$, where k is any integer, or
2. $n = 2^k \cdot (2^r + 1) \cdot (2^s + 1) \cdot (2^t + 1) \cdot \ldots$, where $(2^r + 1)$, $(2^s + 1)$, $(2^t + 1)$, $\ldots$ are different Fermat prime numbers.

Gauss's theorem says that we can construct polygons of 3, 4, 5, 6, 8, 10, 12, 15, 16, 17, 20, 24,... sides, but not those with 7, 9, 11, 13, 14, 18, 19, 21, 22, 23, 25,... sides. We cannot construct exactly a regular convex seven-sided polygon (heptagon), so it is impossible to divide the circumference of a circle into seven equal parts. Perhaps this is why the heptagon is rarely used in architecture.

11.5 Symmetry of Polygons Revisited

Regular polygons have three kinds of symmetry: reflection, rotation, and inversion (Chapter 4). A regular polygon with n edges has n rotational and n reflection symmetry transformations. If n is an even number, then it has an inversion symmetry transformation.

An equilateral triangle has six symmetry transformations, consisting of three rotations ($120°$, $240°$, $360°$) and three reflections. The lines of symmetry for reflection are shown in Figure 11.10. Note that inversion of an equilateral triangle produces a different figure. A square has nine symmetry transformations: one inversion, four rotations ($90°$, $180°$, $270°$, $360°$), and four reflections (Figure 11.11). Of course, nonregular polygons can also have symmetry transformations.

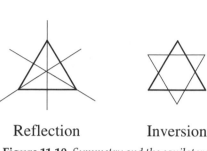

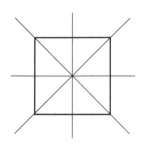

Reflection   Inversion   Reflection

Figure 11.10 *Symmetry and the equilateral triangle.*

Figure 11.11 *Symmetry and the square.*

11.6 Containment

If we are given a test point (x_t, y_t), we can determine if it is inside, outside, or on the boundary of a polygon (Figure 11.12). In the set of vertex points defining the polygon, we find x_{min}, x_{max}, y_{min}, and y_{max}. If the test point is not inside the min-max box, it is not inside the polygon. If it is inside the min-max box, then it still may not be inside the polygon. So we compute the intersections of $y = y_t$ with the edges of the polygon, considering only edges whose endpoints straddle $y = y_t$. We note that there is always an even number of intersections (counting any intersection with a vertex as two). We now pair the x coordinates of the intersections in ascending order; for example, (x_1, x_2), (x_3, x_4), and so on. If x_t falls inside an interval, for example, $x_1 < x_t < x_2$, then it is inside the polygon. If x_t is identically equal to one of the interval limits, then it is on the boundary; otherwise, it is outside the polygon. This procedure works for both convex and concave polygons.

The following procedure applies only to convex polygons (Figure 11.13). It will not work for concave polygons. First, we establish a reference point, (x_{ref}, y_{ref}), that we know to be inside the polygon. Then we write the implicit equation of each edge, $f_{1,2}(x, y)$, $f_{2,3}(x, y)$, $f_{3,4}(x, y)$, and so on. Next, for a given test point (x_t, y_t) and for each and every edge, we compute and compare $f_{i,j}(x_t, y_t)$ and $f_{i,j}(x_{ref}, y_{ref})$. If the evaluated functions, in pairs, have the same sign, then a test point is inside the polygon. If any one

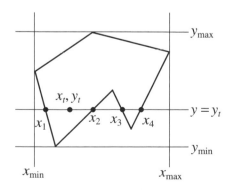

Figure 11.12 *Point-containment test for convex to concave polygons.*

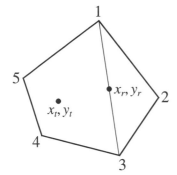

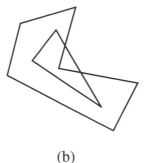

Figure 11.13 *Point-containment test for convex polygons only.*

$f_{i,j}(x_t, y_t) = 0$ and if the preceding conditions obtain, then the point is on the boundary (an edge) of the polygon. If any two sequential f_{ij} both equal zero, then $\mathbf{p}_t$ is on the common vertex; otherwise it is outside the polygon.

Here is a method that applies to both convex and concave polygons: Given a test point (x_t, y_t) we can determine, in two steps, if it is inside, outside, or on the boundary of a polygon.

1. Using the polygon vertex points, we find the min-max box.
   a. If (x_t, y_t) is not inside the min-max box, then it is not inside the polygon.
   b. If (x_t, y_t) is inside, we proceed to the next step.
2. Considering only edges whose endpoints straddle y_t, we compute the intersections of $y = y_t$ with the edges of the polygon (an analogous approach uses $x = x_t$). Again, there is always an even number of intersections, and we count intersections with a vertex as two. We pair the x coordinates of the intersections in ascending order; for example, (x_1, x_2), (x_3, x_4), and so on. Then,
   a. If x_t falls inside an interval, $x_1 \langle x_t \langle x_2$ for example, then $\mathbf{p}_t$ is inside the polygon.
   b. If x_t is identically equal to one of the interval limits, then (x_t, y_t) is on the boundary.
   c. Otherwise, $(x_t, y_t)F$ is outside the polygon.

Given two convex polygons, if all the vertices of one are contained in the other, then the first polygon is inside the second (Figure 11.14a). This test alone does not work

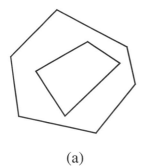

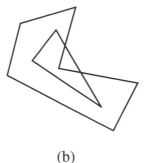

Figure 11.14 *Polygon-polygon containment.*

(a)   (b)

for all cases with concave polygons, where we must compute intersections to eliminate ambiguities (Figure 11.14b).

Exercises

11.1 The following groups of vertex coordinates define polygons. Classify each polygon as concave, convex, or stellar. Assume that the vertices are connected with edges in the order listed and that the first and last vertices are connected with an edge.
   a. $(-2, 8)$, $(1, 13)$, $(0, 7)$, $(-2, 13)$, $(2, 10)$
   b. $(4, 12)$, $(7, 13)$, $(9, 11)$, $(7, 15)$
   c. $(-1, 2)$, $(2, 3)$, $(2, -1)$
   d. $(4, 7)$, $(7, 7)$, $(7, 6)$, $(6, 6)$, $(6, 4)$, $(7, 4)$, $(7, 2)$, $(4, 2)$
   e. $(9, 5)$, $(12, 3)$, $(13, 8)$, $(9, 9)$

11.2 For each polygon given in Exercise 1, list the coordinates of its convex hull.

11.3 Compute the exterior and interior angles for regular polygons with:
   a. Three edges   d. Six edges
   b. Four edges   e. Eight edges
   c. Five edges

11.4 Compute the perimeter and area of each polygon in Exercise 3. (Assume that each edge is 1 cm long.)

11.5 Compute the perimeter for the polygon (4,4), (10,2), (12,8), (6,11), (3,9).

11.6 Compute the coordinates of the geometric center for the polygon in Exercise 5.

11.7 Compute the perimeter and coordinates of the geometric center of each of the following convex polygons:
   a. $(-3, 5)$, $(1, 5)$, $(3, 7)$, $(3, 11)$, $(-3, 7)$
   b. $(5, 3)$, $(7, 6)$, $(5, 10)$
   c. $(9, 4)$, $(11, 4)$, $(14, 6)$, $(14, 9)$, $(12, 9)$, $(9, 7)$
   d. $(-3, 0)$, $(0, -2)$, $(6, 0)$, $(0, 4)$
   e. $(6, -3)$, $(13, -3)$, $(13, -2)$, $(9, 1)$

11.8 Show that any concave polygon can be subdivided into a set of convex polygons.

11.9 List the symmetries of a regular hexagon, and draw one showing its lines of symmetry.

11.10 How many diagonals do each of the following regular polygons have? (A diagonal is a line segment connecting nonconsecutive vertices.)
   a. Equilateral triangle   d. Hexagon
   b. Square   e. Heptagon
   c. Pentagon

11.11 Find the mathematical expression that relates the number of vertices of a regular polygon to the number of its diagonals.

POLYHEDRA

Polyhedra, in their infinite variety, have been objects of study for more than 2500 years. In ancient Greece, contemplation of the regular polyhedra and a few semiregular derivative shapes led to the establishment of the classical world's most famous school, Plato's Academy. We find polyhedral forms expressed concretely in the civil and monumental architecture of antiquity. Their usefulness to us, and our interest in them, continues undiminished today. We now use polyhedra in the mathematics of geometric modeling to represent more complex shapes. We use assemblies of them, like building blocks, to create computer graphics displays. Alone and in Boolean combinations, their properties are relatively easy to compute. And, what is most amazing, all possible polyhedra are defined by only the three simplest of geometric elements—points, lines, and planes.

Regular and semiregular polyhedra are the most interesting. Many are found in nature, as well as in art and architecture. Economy of structure and aesthetically-pleasing symmetries characterize these shapes, and help account for their prevalence in nature and in man's constructions. Salt crystals are cubic polyhedra. Certain copper compounds are octahedral. Buckminster Fuller's geodesic dome, in its simplest form, is an icosahedron whose faces are subdivided into equilateral triangles.

This chapter defines convex, concave, and stellar polyhedra, with particular attention to the five regular polyhedra, or Platonic solids, and the 13 semiregular, or Archimedean, polyhedra and their duals. Using Euler's formula, we will prove that only five regular polyhedra are possible in a space of three dimensions. Other topics include nets, the convex hull of a polyhedron, geometric properties, the connectivity matrix, halfspace representations of polyhedra, model data structure, and so-called "maps".

12.1 Definitions

A *polyhedron* is a multifaceted three-dimensional solid bounded by a finite, connected set of plane polygons such that every edge of each polygon belongs also to just one other polygon. The polygonal faces form a closed surface, dividing space into two regions, the interior of the polyhedron and the exterior. A cube (or hexahedron) is a familiar example of a polyhedron. The number of faces determines the name of a regular polyhedron. For a more complex polyhedron, the name may also indicate the presence of differently shaped faces and special features, such as a truncated or snub

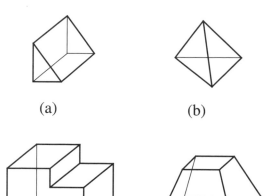

(a) (b)

Figure 12.1 *Examples of polyhedra.*

(c) (d)

polyhedron. Universally applicable naming conventions are still being debated, so they are not discussed here.

All the faces of a polyhedron are plane polygons, and all its edges are straight line segments (Figure 12.1). The simplest possible polyhedron, one with four faces, is the tetrahedron. The polyhedra in Figure 12.1a, 12.1b, and 12.1d are convex, and the one in Figure 12.1c is concave. In every case each polyhedral edge is shared by exactly two polygonal faces.

Three geometric elements define all polyhedra in a three-dimensional space: vertices (V), edges (E), and faces (F). Each vertex is surrounded by an equal number of edges and faces, each edge is bounded by two vertices and two faces, and each face is bounded by a closed loop of coplanar edges that form a polygon. Finally, the angle between faces that intersect at a common edge is the *dihedral angle*. The cube in Figure 12.2 illustrates these elements of a polyhedron.

It is important to understand the geometry that happens at a vertex of a polyhedron. Here is a review of some key concepts and definitions of terms:

Any straight line in a plane divides the plane into two *half-planes*. Two half-planes extending from a common line form a *dihedral angle*.

Figure 12.2 *Elements of a polyhedron.*

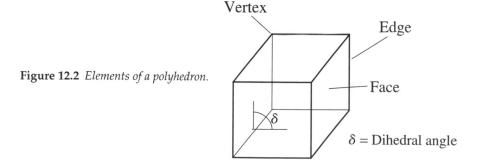

Three or more planes intersecting at a common point form a *polyhedral angle*, and the common point is the *vertex* of this angle. The intersections of the planes are the *edges* of the angle. The parts of the planes lying between the edges are the *faces* of the angle. The angles formed by adjacent edges are the *face angles* of the polyhedral angle. Thus, the vertex, edges, faces, face angles, and dihedral angles formed by the faces are the constituent parts of the polyhedral angle. For any polyhedral angle there is an equal number of edges, faces, face angles, and dihedral angles. A polyhedral angle with three faces is a *trihedral angle* (the polyhedral angle at a vertex of a cube is an example). Polyhedral angles of four, five, six, and seven faces are called tetrahedral, pentahedral, hexahedral, and heptahedral, respectively.

An angle of 360° surrounds a point in the plane. We can compare the sum of the face angles around a vertex of a polyhedron with the 360° angle around a point on the plane. The *angular deficit* at a vertex of a polyhedron is defined as the difference between the sum of the face angles surrounding the vertex and 360°. It is the gap that results if we open the vertex out flat, as in the net diagrams discussed later. The sum of the angular deficits over all the vertices of a polyhedron is its total angular deficit, and the smaller the angular deficit, the more sphere-like the polyhedron.

Any regular polyhedron (and many other polyhedra) is *homeomorphic* to a sphere. Two solid shapes are said to be homeomorphic if their bounding surfaces can be deformed into one another without cutting or gluing. This means that they are *topologically equivalent*. Any polyhedron that is homeomorphic to a sphere is a *simple polyhedron*. For example, a cube is a simple polyhedron because it can be topologically deformed into a sphere. A polyhedron with one or more holes through it is not simple. It is like a donut or torus shape, which cannot be deformed into a sphere without cutting and pasting.

12.2 The Regular Polyhedra

A convex polyhedron is a *regular polyhedron* if the following conditions are true:

1. All face polygons are regular.
2. All face polygons are congruent.
3. All vertices are identical.
4. All dihedral angles are equal.

The cube is a good example. All its faces are identical and all its edges are of equal length. In three-dimensional space we can construct only five regular polyhedra (also called the five Platonic solids): the tetrahedron, hexahedron (cube), octahedron, dodecahedron, and icosahedron (Figure 12.3).

The sum of all face angles at a vertex of a convex polygon is always less than 2π. Otherwise, one of two conditions is present: If the sum of the angles $= 2\pi$, then the edges meeting at the vertex are coplanar; if the sum of the angles $> 2\pi$, then some of the edges at the vertex are reentrant and the polyhedron is concave.

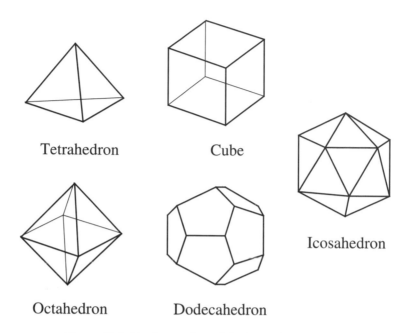

Tetrahedron Cube

Icosahedron

Octahedron Dodecahedron

Figure 12.3 *The five regular polyhedra or Platonic solids.*

Tables 12.1 and 12.2 list characteristic properties of the five regular polyhedra. In Table 12.2, e is the length of the polyhedra's edge, R_I is the radius of the inscribed sphere, R_C is the radius of the circumscribed sphere, and θ is the dihedral angle.

Tables 12.3–12.7 give the vertex coordinates for each of the regular polyhedra. Each has a unit edge length, with the origin at its geometric center and at least one axis aligned with a line of symmetry or edge.

Table 12.1 Properties of regular polyhedra

	Face polygons	vertices	Edges	Faces	Faces at a vertex	Sum face angles at a vertex
Tetrahedron	Triangles	4	6	4	3	180°
Cube	Squares	8	12	6	3	270°
Octahedron	Triangles	6	12	8	4	240°
Dodecahedron	Pentagons	20	30	12	3	324°
Icosahedron	Triangles	12	30	20	5	300°

Table 12.2 More properties of regular polyhedra

	Area	Volume	R_I	R_C	θ
Tetrahedron	$1.7321e^2$	$0.1178\,e^3$	$0.204124e$	$0.612372e$	70°31′44″
Cube	$6e^2$	e^3	$0.500000e$	$0.866025e$	90°
Octahedron	$3.4641e^2$	$0.4714\,e^3$	$0.408248e$	$0.707107e$	109°28′16″
Dodecahedron	$20.6458e^2$	$7.6632\,e^3$	$1.113516e$	$1.401259e$	116°33′54″
Icosahedron	$8.6603e^2$	$2.1817\,e^3$	$0.755761e$	$0.951057e$	138°11′23″

Table 12.3 Tetrahedron

V	x	y	z
1	0.0	0.5774	−0.2041
2	−0.5	−0.2887	−0.2041
3	0.5	−0.2887	−0.2041
4	0.0	0.0	0.6124

Table 12.4 Cube

V	x	y	z
1	0.5	0.5	−0.5
2	−0.5	0.5	−0.5
3	−0.5	−0.5	−0.5
4	0.5	−0.5	−0.5
5	0.5	0.5	0.5
6	−0.5	0.5	0.5
7	−0.5	−0.5	0.5
8	0.5	−0.5	0.5

Table 12.5 Octahedron

V	x	y	z
1	0.5	0.5	0.0
2	−0.5	0.5	0.0
3	−0.5	−0.5	0.0
4	0.5	−0.5	0.0
5	0.0	0.0	0.7071
6	0.0	0.0	−0.7071

Table 12.6 Dodecahedron

V	x	y	z
1	0.0	0.8507	1.1135
2	0.8090	0.2629	1.1135
3	0.5	−0.6882	1.1135
4	−0.5	−0.6882	1.1135
5	−0.8090	0.2629	1.1135
6	0.0	1.3769	0.2065
7	1.3095	0.4255	0.2065
8	0.8093	−1.1139	0.2065
9	−0.8093	−1.1139	0.2065
10	−1.3095	0.4255	0.2065
11	0.8093	1.1139	−0.2065
12	1.3095	−0.4255	−0.2065
13	0.0	−1.3769	−0.2065
14	−1.3095	−0.4255	−0.2065
15	−0.8093	1.1139	−0.2065
16	0.5	0.6882	−1.1135
17	0.8090	−0.2629	−1.1135
18	0.0	0.8507	−1.1135
19	−0.8090	−0.2629	−1.1135
20	−0.5	0.6882	−1.1135

Table 12.7 Icosahedron

V	x	y	z
1	0.0	0.0	0.9512
2	0.0	0.8507	0.4253
3	0.8090	0.2629	0.4253
4	0.5	−0.6882	0.4253
5	−0.5	−0.6882	0.4253
6	−0.0	0.2629	−0.4253
7	0.5	0.6882	−0.4253
8	0.8090	−0.2626	−0.4253
9	0.0	−0.8507	−0.4253
10	−0.8090	−0.2629	−0.4253
11	−0.5	0.6882	0.4253
12	0.0	0.0	−0.9512

12.3 Semiregular Polyhedra

If we relax conditions 2 and 4 above (defining the regular polyhedra) to allow two or more kinds of faces and two or more dihedral angles of differing value, then an infinite number of polyhedra is possible. They include the prisms and antiprisms and the 13 so-called *Archimedean semiregular polyhedra*. Each Archimedean polyhedron has two or three different kinds of faces bounding it. These faces are regular polygons and may be equilateral triangles, squares, pentagons, hexagons, octagons, or decagons.

If we relax conditions 1 and 3 to allow nonregular faces and more than one kind of vertex, then another infinite set of polyhedra is possible. It includes the 13 Archimedean dual or vertically regular polyhedra related to the Archimedean semiregular polyhedra by a correspondence between the faces in one group and the vertices in the other group.

If we appropriately truncate the five regular polyhedra, we can generate all the semiregular polyhedra except two snub forms (not discussed here). Truncating the 13 Archimedean dual polyhedra generates the other polyhedra. For example, the truncated cube is an Archimedean semiregular polyhedron, and the tetrakis cube is an Archimedean dual polyhedron (Table 12.8). A considerable, well-illustrated literature can be found on the subject of the semiregular polyhedra.

Table 12.8 Properties of the semiregular polyhedra

	Vertices	Edges / vertex vertex $E : V$	Edges (F)	Faces and Type	
Truncated tetrahedron	12	3	18	8	4 triangles, 4 hexagons
Cuboctahedron	12	4	24	14	8 triangles, 6 squares
Truncated cube	24	3	36	14	8 triangles, 6 octagons
Snub cube	24	5	60	38	32 triangles, 6 squares
Truncated octahedron	24	3	36	14	6 squares, 8 hexagons
Small rhombicuboctahedron	24	4	48	26	8 triangles, 18 squares
Great rhombicuboctahedron	48	3	72	26	12 squares, 8 hexagons, 6 octagons
Truncated dodecahedron	60	3	90	32	20 triangles, 12 decagons
Snub dodecahedron	60	5	150	92	80 triangles, 12 pentagons
Icosidodecahedron	30	4	60	32	20 triangles, 12 pentagons
Truncated icosahedron	60	3	90	32	12 pentagons, 20 hexagons
Small rhombicosidodecahedron	60	4	120	62	20 triangles, 30 squares, 12 pentagons
Great rhombicosidodecahedron	120	3	180	62	30 squares, 20 hexagons, 12 decagons

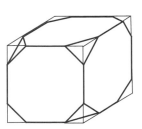

Figure 12.4 *Truncated cube.*

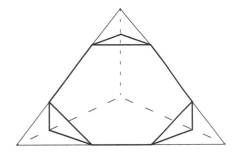

Figure 12.5 *Truncated tetrahedron.*

The truncated tetrahedron and truncated cube are relatively easy to construct and visualize. The other regular and semiregular polyhedra are more difficult. We truncate a cube at each vertex to produce two kinds of regular polygonal faces: eight equilateral triangles and six octagons (Figure 12.4). We truncate each vertex of a tetrahedron to produce four hexagonal faces and four equilateral triangular faces (Figure 12.5—looking perpendicular to the forward, or near, face).

12.4 Dual Polyhedra

The five regular polyhedra are related to each other in a subtle way. The octahedron and cube are reciprocal or dual. This means that if we connect the centers of the faces of one of them with line segments, we obtain the edges of the other. The number of faces of one becomes the number of vertices of the other, and vice versa. The total number of edges does not change. The same is true for the icosahedron and dodecahedron, which form a dual couple. The tetrahedron is self-dual; that is, it reproduces itself by connecting the centers of its faces with line segments, which become the edges of another tetrahedron, albeit smaller than the original. We say then that two polyhedra are dual if the vertices of one can be put into a one-to-one correspondence with the centers of the faces of the other.

Here is another construction we can use to generate the dual of any of the 5 regular and 26 Archimedean polyhedra:

1. Inscribe a sphere in the polyhedron so that the surface of the sphere is tangent to every edge of the polyhedron. The inscribed sphere is tangent to the midpoint of each edge of the regular and Archimedean semiregular polyhedra. The inscribed sphere is tangent to each edge of the Archimedean dual polyhedra, but not necessarily at the midpoint.

2. At each point of tangency add another edge tangent to the sphere and perpendicular to the edge.

3. Extend the added edges until they meet to form the vertices and faces of the dual polyhedron.

Table 12.9 Dual pairs

Tetrahedron ↔ Tetrahedron
Cube ↔ Octahedron
Dodecahedron ↔ Icosahedron
Truncated tetrahedron ↔ Triakis tetrahedron
Cuboctahedron ↔ Rhombic dodecahedron
Truncated cube ↔ Triakis octahedron
Truncated octahedron ↔ Tetrakis cube
Small rhombicuboctahedron ↔ Trapezoidal icositetrahedron
Great rhombicuboctahedron ↔ Hexakis octahedron
Icosidodecahedron ↔ Rhombic triacontahedron
Truncated dodecahedron ↔ Triakis icosahedron
Truncated icosahedron ↔ Pentakis dodecahedron
Small rhombicosidodecahedron ↔ Trapezoidal hexacontahedron
Great rhombicosidodecahedron ↔ Hexakis icosahedron
Snub dodecahedron ↔ Pentagonal hexacontahedron

Table 12.10 Greek roots

on	one	conta	thirty
do	two	triakis	three times
tri	three	tetrakis	four times
tetra	four	pentakis	five times
penta	five	poly	many
hexa	six	hedron	sided
hepta	seven	antiprism	a polyhedron having two regular
			n-gon faces and $2n$ triangular faces
octa	eight		
enna	nine	dextro	right-handed
deca	ten	laevo	left-handed
dodeca	twelve	enantiomorphic	having right- and left-handed forms
icosa	twenty	prism	a polyhedron consisting of two regular
			n-gon faces and n square faces

Table 12.9 lists the 5 regular and 26 Archimedean duals, and Table 12.10 defines Greek roots.

12.5 Star Polyhedra

If we extend the edges of a regular polygon with five or more edges, it will enclose additional regions of the plane and form a star or stellar polygon. If we extend the face planes of an octahedron, dodecahedron, or icosahedron, then new regions are defined, creating star polyhedra. (This does not work for the cube.) For example, the extended face planes of the octahedron enclose eight additional tetrahedra, and the faces of the star polyhedron created are large equilateral triangles of the interpenetrating tetrahedra.

If we extend the face planes of the dodecahedron, we produce three distinct types of cells inside the intersecting planes and three stellated forms. Two of them were discovered by Kepler (1619) and the other by Poinsot (1809). These three star polyhedra and one additional one derived from the icosahedron and also discovered by Poinsot have all the properties of the regular polyhedra. Each face is a regular polygon (or star polygon), and each vertex is surrounded alike. They differ from the regular polyhedra because their faces interpenetrate and they are not topologically simple. Therefore, Euler's formula, unmodified, does not apply.

12.6 Nets

By careful cutting and unfolding, we can open up and flatten out a polyhedron so that it lies in a plane. In doing so, we create a *net* of the polyhedron. There is no single, unique net for a particular polyhedron. A variety of nets for the tetrahedron and cube are shown in Figures 12.6 and 12.7. An easy way to construct a model of a polyhedron is to lay out and cut out one of its nets, folding where appropriate and gluing or taping the matching edges that are otherwise disjoint in the net. The great artist Albrecht Dürer (1471–1528), in a work of his in 1525, showed how to make paper models of the five regular polyhedra.

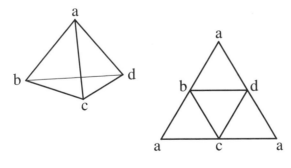

Figure 12.6 *Tetrahedron nets.*

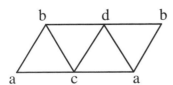

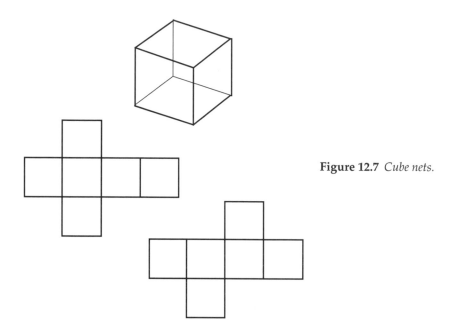

Figure 12.7 *Cube nets.*

12.7 The Convex Hull of a Polyhedron

A polyhedral convex hull is a three-dimensional analog of the convex hull for a polygon. The convex hull of a convex polyhedron is identical to the polyhedron itself. We form the convex hull of a concave polyhedron by wrapping it in a rubber sheet, producing an enveloping convex polyhedron. The concave polyhedron in Figure 12.8a, for example, has the convex hull in Figure 12.8b. The convex hull of a concave polyhedron is the smallest convex polyhedron that will enclose it.

Figure 12.8 *Convex hull of a polyhedron.*

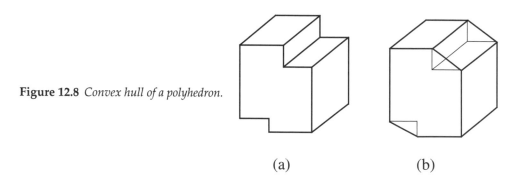

(a)   (b)

12.8 Euler's Formula for Simple Polyhedra

Leonard Euler (1707–1783), one of the greatest mathematicians of any age, rediscovered a mathematical relationship between the number of vertices (V), edges

(E), and faces (F) of a simple polyhedron. (Remember, a simple polyhedron refers to any polyhedron that we can deform continuously into a sphere, assuming that we treat its faces like rubber sheets.) This relationship is now called *Euler's formula* (pronounced "oiler") for polyhedra. It is stated as

$$V - E + F = 2 \qquad (12.1)$$

Applying this formula to a cube yields $8 - 12 + 6 = 2$ and to an octahedron yields $6 - 12 + 8 = 2$. The far-sighted Greek mathematician and engineer Archimedes probably knew of this relationship in 250 BCE, but René Decartes first stated it clearly in 1635.

To apply Euler's formula, other conditions must also be met:

1. All faces must be bounded by a single ring of edges, with no holes in the faces.
2. The polyhedron must have no holes through it.
3. Each edge is shared by exactly two faces and is terminated by a vertex at each end.
4. At least three edges must meet at each vertex.

The polyhedra in Figure 12.9 satisfy the four conditions and, therefore, Euler's formula applies.

If we add vertices, edges, or faces to a polyhedron, we must do so in a way that satisfies Euler's formula and the four conditions. In Figure 12.10a we add an edge, joining vertex 1 to vertex 3 and dividing face 1, 2, 3, 4 into two separate faces. We have added one face and one edge. These additions produce no net change to Euler's formula, so the change to the network of vertices, edges, and faces is legitimate. In Figure 12.10b we add vertices 9 and 10 and join them with an edge. The new vertices divide edges 1, 2 and 3, 4, and the new edge 9, 10 divides face 1, 2, 3, 4. These changes, too, produce no net change to Euler's formula (since $2 - 3 + 1 = 0$). In Figure 12.10c we add one vertex, four edges, and four faces, but we delete the existing Face 2, 6, 7, 3. Again, this action

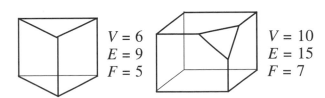

$V = 6$
$E = 9$
$F = 5$

$V = 10$
$E = 15$
$F = 7$

Figure 12.9 *Vertices, edges, and faces satisfying Euler's formula.*

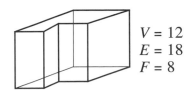

$V = 12$
$E = 18$
$F = 8$

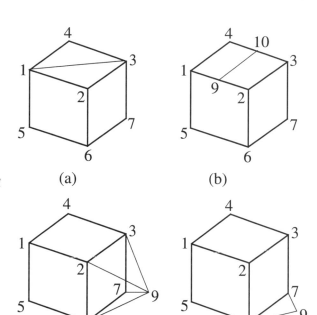

Figure 12.10 *Adding vertices, edges, and faces so that Euler's formula is satisfied.*

produces no net change to Euler's formula (since $1 - 4 + 3 = 0$). In Figure 12.10d, where we attempt to add one vertex, two edges, and one face, the change is not acceptable. Although this change preserves Euler's formula (since $1 - 2 + 1 = 0$), it does not satisfy the conditions requiring each edge to adjoin exactly two faces and at least three edges to meet at each vertex.

Two kinds of changes are illustrated in the figure. In Figures 12.10a and 12.10b the solid shape of the polyhedron (in this case a cube) is preserved, and only the network of vertices, edges, and faces is changed. In Figure 12.10c the solid shape itself is modified by the change in the network defining it.

Euler's formula is only a special case of Ludwig Schläfli's (1814–1895) formula for any number of dimensions. Schläfli generalized Euler's formula using the following set of propositions:

1. An edge, or one-dimensional polytope, has a vertex at each end: $N_0 = 2$.
2. A polygonal face, or two-dimensional polytope, has as many vertices as edges: $N_0 - N_1 = 0$.
3. A polyhedron, or three-dimensional polytope, satisfies Euler's formula: $N_0 - N_1 + N_2 = 2$.
4. A four-dimensional polytope satisfies $N_0 - N_1 + N_2 - N_3 = 0$.
5. Any simply-connected n-dimensional polytope satisfies $N_0 - N_1 + \cdots + (-1)^{n-1} N_{n-1} = 1-(-1)^n$, where *polytope* is the general term of the sequence—point, segment, polygon, polyhedron, and so on. Schläfli also invented the

symbol $\{p, q\}$ for the regular polyhedron whose faces are p-gons, q meeting at each vertex, or the polyhedron with face $\{p\}$ and vertex figure $\{q\}$.

Euler's original formula establishes a relationship between the numbers of vertices, edges, and faces of a simple polyhedron in Euclidean space. Generalizations of this formula and, of course, the original formula are very important in the study of topology (see Section 6.3), since the Euler characteristic obtained by the formula is a topological invariant (unchanged).

On any closed surface without boundary, we can express Euler's formula in the form

$$V - E + F = 2 - 2G \qquad (12.2)$$

where G is the *genus* of the surface and the number $2 - 2G$ is the *Euler characteristic* of the surface. Henri Poincaré (1854–1912), French mathematician extraordinaire, also generalized Euler's formula so that it applied to n-dimensional space. Instead of vertices, edges, and faces, Poincaré defined 0-, 1-, 2-, ..., $(n-1)$-dimensional elements. He denoted these geometric elements of an n-dimensional polytope as N_0, N_1, N_2, ..., N_{n-1}, respectively, and rewrote Euler's formula as

$$N_0 - N_1 + N_2 - \ldots = 1 - (-1)^n \qquad (12.3)$$

Obviously, for $n = 3$ this reduces to Euler's formula.

12.9 Euler's Formula for Nonsimple Polyhedra

If we subdivide a polyhedron into C polyhedral cells, then the vertices, edges, faces, and cells are related by

$$V - E + F - C = 1 \qquad (12.4)$$

A cell is, itself, a closed polyhedron. Adding a vertex to the interior of a cube and joining it with edges to each of the other eight vertices creates a six-cell polyhedron with $V = 9$, $E = 20$, $F = 18$, and $C = 6$ (Figure 12.11). Note that each cell is pyramid shaped, with an external face of the cube as a base.

If a polyhedron has one or more holes (H) in its faces, has passages (P) through it, and/or consists of disjoint bodies (B), then

$$V - E + F - H + 2P = 2B \qquad (12.5)$$

A cube with a passage through it has $V = 16$, $E = 32$, $F = 16$, $H = 0$, $P = 1$, and $B = 1$ (Figure 12.12).

We can define a *connectivity number* N. For a sphere and all topologically equivalent shapes, we find $N = 0$. For torus, or donut-like, shapes $N = 2$. For figure-eight, or pretzel-like, shapes $N = 4$, and so on. Then

$$V - E + F = 2 - N \qquad (12.6)$$

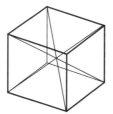

Figure 12.11 *Six-cell polyhedron.*

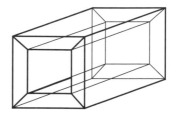

Figure 12.12 *Cube with a passage through it.*

12.10 Five and Only Five Regular Polyhedra: The Proof

Euler's formula provides a direct and simple proof that there are only five regular polyhedra. By using it we can determine all possible regular polyhedra. First, we observe that for a regular polyhedron every face has the same number of edges, say h, every vertex has the same number of edges radiating from it, say k, and every edge has the same length. Because every edge has two vertices and belongs to exactly two faces, it follows that $hF = 2E = kV$. Substituting this into Euler's formula yields

$$\frac{2E}{k} - E + \frac{2E}{h} = 2 \tag{12.7}$$

or

$$\frac{1}{E} = \frac{1}{h} + \frac{1}{k} - \frac{1}{2} \tag{12.8}$$

For a polyhedron we assume that $h, k \geq 3$. If both h and k are larger than 3, then Equation 12.8 implies that

$$0 < \frac{1}{E} = \frac{1}{h} + \frac{1}{k} - \frac{1}{2} \leq \frac{1}{4} + \frac{1}{4} - \frac{1}{2} = 0 \tag{12.9}$$

This is impossible. Either h or k equals 3. If $h = 3$, then

$$0 < \frac{1}{E} = \frac{1}{3} + \frac{1}{k} - \frac{1}{2} \tag{12.10}$$

implies that $3 \leq k \leq 5$. Similarly, if $k = 3$, then $3 \leq h \leq 5$. Therefore, $(h, k, E) = (3, 3, 6), (4, 3, 12), (3, 4, 12), (5, 3, 30)$, and $(3, 5, 30)$ are the only possibilities. They are, of course, the tetrahedron, cube, octahedron, dodecahedron, and icosahedron. It turns out that we did not have to use the fact that all edges of the polyhedron have the same length.

12.11 The Connectivity Matrix

We can efficiently organize the data describing a polyhedron by listing the vertices and their coordinates and by using a connectivity matrix to define its edges. A *connectivity matrix* is a two-dimensional list or table that describes how vertices are connected by edges to form a polyhedron. This matrix is always square, with as many rows and columns as vertices. We denote a particular element or entry in the table as a_{ij}, where the subscript i tells us which row the element is in and j tells us which column. If element $a_{ij} = 1$, then vertices i and j are connected by an edge. If element $a_{ij} = 0$, then vertices i and j are not connected. A connectivity matrix is symmetric about its main diagonal, which is comprised of all zeros. For example, using the general tetrahedron in Figure 12.13, we have

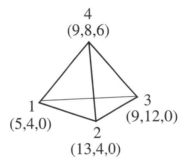

Vertex	Coordinates
1	(5, 4, 0)
2	(13, 4, 0)
3	(9, 12, 0)
4	(9, 8, 6)

Figure 12.13 *Vertex number and coordinates of an example tetrahedron.*

and the tetrahedron connectivity matrix **C**:

$$\mathbf{C} = \begin{bmatrix} 0 & 1 & 1 & 1 \\ 1 & 0 & 1 & 1 \\ 1 & 1 & 0 & 1 \\ 1 & 1 & 1 & 0 \end{bmatrix} \tag{12.11}$$

We number the columns consecutively from left to right and the rows from top to bottom. This method lends itself readily to the design of a database for computer-graphic systems or other computer-aided geometry applications.

Here is another example, this time for a polyhedron that is a rectangular block (Figure 12.14). Instead of a table we list the vertex coordinates in a matrix **V**.

$$\mathbf{V} = \begin{bmatrix} 2 & 2 & 6 \\ 12 & 2 & 6 \\ 12 & 8 & 6 \\ 2 & 8 & 6 \\ 2 & 2 & 2 \\ 12 & 2 & 2 \\ 12 & 8 & 2 \\ 2 & 8 & 2 \end{bmatrix} \tag{12.12}$$

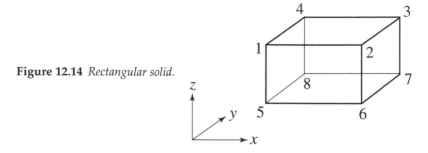

Figure 12.14 *Rectangular solid.*

From the matrix **V**, for example, the coordinates of vertex 3 from the third row are (12, 8, 6). The connectivity matrix for this polyhedron is

$$\mathbf{C} = \begin{bmatrix} 0 & 1 & 0 & 1 & 1 & 0 & 0 & 0 \\ 1 & 0 & 1 & 0 & 0 & 1 & 0 & 0 \\ 0 & 1 & 0 & 1 & 0 & 0 & 1 & 0 \\ 1 & 0 & 1 & 0 & 0 & 0 & 0 & 1 \\ 1 & 0 & 0 & 0 & 0 & 1 & 0 & 1 \\ 0 & 1 & 0 & 0 & 1 & 0 & 1 & 0 \\ 0 & 0 & 1 & 0 & 0 & 1 & 0 & 1 \\ 0 & 0 & 0 & 1 & 1 & 0 & 1 & 0 \end{bmatrix} \tag{12.13}$$

The connectivity matrix **C** has twice as much information as necessary. If we draw a line diagonally from the zero in row 1 and column 1 to the zero in row 8 and column 8, then the triangular array of elements on one side of this diagonal is the mirror image of the array of the other side. It is easy to see why. For example, the entry $a_{73} = 1$ in row 7 and column 3 says that vertex 7 is joined to vertex 3 by an edge; the entry $a_{37} = 1$ in row 3 and column 7 says that vertex 3 is joined to vertex 7 with an edge. These two entries are two identical pieces of information. This redundancy is not only tolerable but useful, because any search of these data for connectivity is more direct. It is important to note that the connectivity matrix reveals little about the actual shape of the object. With appropriate labeling of the vertices, a cube could just as easily have the same connectivity matrix as a rectangular solid or a truncated pyramid.

The information in the **V** and **C** matrices defines a *wireframe* model of the block. But what do we do about the faces? We could devise a procedure to analyze **V** and **C** by covering the wireframe with polygonal faces and searching for planar loops or circuits of edges. But we may, without adequate precautions, find some that penetrate the interior of the polyhedron. There is a better, more direct way. We simply include in the data a sequence of vertices defining closed loops of edges bounding the faces. Using the block previously defined as an example, we form a matrix with each row containing the vertex sequence bounding a face; thus,

$$\mathbf{F} = \begin{bmatrix} 1 & 2 & 3 & 4 \\ 8 & 7 & 6 & 5 \\ 5 & 6 & 2 & 1 \\ 6 & 7 & 3 & 2 \\ 7 & 8 & 4 & 3 \\ 8 & 5 & 1 & 4 \end{bmatrix} \tag{12.14}$$

For example, row 3 gives the vertices surrounding face 3. This means that the vertices are connected by edges in the order given, including an implied closing or loop-completing edge: 5-to-6-to-2-to-1-to-5, producing the correct four edges. The vertex sequences are consistently given in counterclockwise order when we view the face from outside the solid. This convention helps us determine which side of a face is inside and which is outside. If you curl the finger of your right hand in the counterclockwise direction around the edges of a face (that is, in the order in which the vertices surrounding that face are listed), then your thumb points outward from the interior of the polyhedron.

How valid is any set of data defining a polyhedron? One of the first questions we might ask is: Do the specified vertices surrounding each face lie in a plane? This is easy to test. First, using any three vertex points surrounding a face, define the plane containing the face. Then all other vertex points surrounding the face must satisfy this equation within some small specified distance.

12.12 Halfspace Representation of Polyhedra

We can represent a convex polyhedron with n faces by a consistent system of n equations constructed as follows: Express the plane equation of each face as an inequality. For the ith face we write

$$A_i x + B_i y + C_i z + D_i > 0 \tag{12.15}$$

adjusting the sign of the expression on the left, so that arbitrary points on the same side of the plane as the polyhedron vertices satisfy the inequality. Any point that satisfies all n inequalities lies inside the polyhedron; for example,

$$\begin{aligned} x &> 0 \\ -x + 4 &> 0 \\ y &> 0 \\ -y + 4 &> 0 \\ z &> 0 \\ -z + 4 &> 0 \end{aligned} \tag{12.16}$$

This system of six inequalities defines a cube aligned with the coordinate axes, with edges of length 4. We can test a variety of points against these inequalities.

A slight variation on this approach allows us to represent a convex polyhedron with n faces as the intersection of n halfspaces. The plane of each face defines two halfspaces, and an inequality identifies one of them as containing points interior to the polyhedron. The intersection of the set of all such halfspaces defines the polyhedron P.

$$P = \bigcap_{i=1}^{n} f_i(x, y, z) \tag{12.17}$$

where

$$\bigcap_{i=1}^{n} f_i = f_1 \cap f_2 \cap \ldots \cap f_n \tag{12.18}$$

Figure 12.15 is an example of a four-sided polyhedron with vertices V_1, V_2, V_3, and V_4, and with triangular faces F_1, F_2, F_3, and F_4, where

$$P = F_1 \cap F_2 \cap F_3 \cap F_4 \tag{12.19}$$

and

$$
\begin{aligned}
V_1 &= (0, 0, 0) \quad F_1(x, y, z) = x \\
V_2 &= (2, 0, 0) \quad F_2(x, y, z) = y \\
V_3 &= (0, 3, 0) \quad F_3(x, y, z) = z \\
V_4 &= (0, 0, 3) \quad F_4(x, y, z) = -3x - 2y - 2z + 6
\end{aligned}
\tag{12.20}
$$

We adjust the sign of F_4 so that, for any point (x, y, z), if $F_4(x, y, z) > 0$, the point is inside the polyhedron. For example, given a test point $\mathbf{p}_t = (1/3, 1, 1)$, we can classify it with respect to the polyhedron as follows:

$$
F_1 \begin{bmatrix} \dfrac{1}{3} & 1 & 1 \end{bmatrix} = \dfrac{1}{3} \qquad F_2 \begin{bmatrix} \dfrac{1}{3} & 1 & 1 \end{bmatrix} = 1
$$
$$
F_3 \begin{bmatrix} \dfrac{1}{3} & 1 & 1 \end{bmatrix} = 1 \qquad F_4 \begin{bmatrix} \dfrac{1}{3} & 1 & 1 \end{bmatrix} = 1
\tag{12.21}
$$

All $F_i(x, y, z)$ evaluated at $\mathbf{p}_t = (1/3, 1, 1)$ are positive, so $\mathbf{p}_t$ is inside the polyhedron.

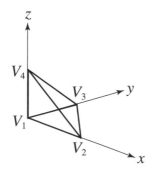

Figure 12.15 *A four-sided polyhedron.*

Table 12.11 Metric data for a cube

Vertex	Coordinates	Face	Face plane coefficients
V_1	$(1, 1, 2)$	F_1	$(1, 0, 0, -1)$
V_2	$(1, 2, 2)$	F_2	$(1, 0, 0, -2)$
V_3	$(2, 2, 2)$	F_3	$(0, 1, 0, -1)$
V_4	$(2, 1, 2)$	F_4	$(0, 1, 0, -2)$
V_5	$(1, 1, 1)$	F_5	$(0, 0, 1, -1)$
V_6	$(1, 2, 1)$	F_6	$(0, 0, 1, -2)$
V_7	$(2, 2, 1)$		
V_8	$(2, 1, 1)$		

Table 12.12 Topology data for a cube

Edge: (Connectivity)	Face: (Vertex circuit)
E_1: (V_1, V_2)	F_1: (V_1, V_2, V_6, V_5)
E_2: (V_2, V_3)	F_2: (V_3, V_4, V_8, V_9)
E_3: (V_3, V_4)	F_3: (V_5, V_8, V_4, V_1)
E_4: (V_4, V_1)	F_4: (V_2, V_3, V_7, V_6)
E_5: (V_5, V_6)	F_5: (V_5, V_6, V_7, V_8)
E_6: (V_6, V_7)	F_6: (V_4, V_3, V_2, V_1)
E_7: (V_7, V_8)	
E_8: (V_8, V_5)	
E_9: (V_1, V_5)	
E_{10}: (V_2, V_6)	
E_{11}: (V_3, V_7)	
E_{12}: (V_4, V_8)	

12.13 Model Data Structure

Any computer program we write for geometry or computer-graphics applications requires a logical method for organizing and storing the data. The above two tables are an example of one form of model data structure for a polyhedron. The polyhedron in this example is a cube.

Table 12.11 contains the metric data for the cube and Table 12.12 contains its topology; that is, how its vertices, edges, and faces are connected. In Table 12.12, the circuit of vertices defining the faces are listed in counterclockwise order as viewed from outside the cube.

12.14 Maps of Polyhedra

A special two-dimensional image of a polyhedron, called a *Schlegel diagram* or *map*, can be constructed by projecting its edges onto a plane from a point directly above the center of one of its faces. A physical analog of this mathematical projection is the

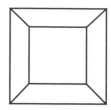

Figure 12.16 *Schlegel diagram of a cube.*

shadow of a polyhedron's edges cast on a plane by a point of light also located above the center of one of the polyhedron's faces. The image constructed in this way is an exaggerated perspective view. One of the faces of the polyhedron, the one closest to and centered on the point of projection, frames all the others, and we must include this face when counting faces.

Although the polyhedron's congruent faces do not map as such on the plane, the Schlegel diagram does enable us to see a realistic two-dimensional representation of a three-dimensional object that preserves such important characteristics of the original as its connectivity of edges and vertices and some of its symmetry. Figure 12.16 is a Schlegel diagram of a cube.

Exercises

12.1   Why is a triangle necessarily a plane figure, while a four-sided polygon in three-dimensional space need not be?

12.2   How many different planes in three-dimensional space are determined by
a. Four points
b. Five points
c. *n* points

12.3   If the center of each of the six faces of a cube is joined by a line segment to the center of each of its adjacent faces, what kind of polyhedron is formed? Sketch the resulting polyhedron, and demonstrate that any two of its edges are equal.

12.4   Same as Exercise 3, except begin with a tetrahedron.

12.5   Same as Exercise 3, except begin with an octahedron.

12.6   A quartz crystal has the form of a hexagonal prism with a pyramid on one of its bases. Show that Euler's formula holds.

12.7   Verify that the five regular polyhedra satisfy Euler's formula.

12.8   Show that a seven-edged polyhedron is impossible.

12.9   Compute the sum of the angles between adjacent edges at a vertex of each of the regular polyhedra.

12.10 Compute the sum of the angles between adjacent edges at vertex 10 of the polyhedron shown in Figure 12.17.

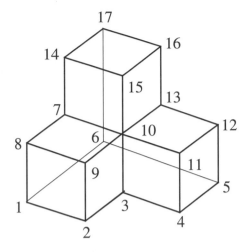

Figure 12.17 *Polyhedron for Exercises 12.10 and 12.11.*

12.11 Compute the sum of the angles between adjacent edges at vertex 3 of the polyhedron shown in Figure 12.17.

12.12 Develop and draw valid nets for the following regular polyhedra:
a. Octahedron
b. Dodecahedron
c. Icosahedron

12.13 Truncate each vertex of a tetrahedron, equally, to form a new polyhedron with four congruent regular hexagonal faces and four congruent equilateral triangle faces. Count the vertices, edges, and faces of this new polyhedron, and verify Euler's formula. Comment on the edge lengths, and sketch the net.

12.14 Truncate each vertex of a cube, equally, to form a new polyhedron with six congruent regular octagonal faces and eight congruent equilateral-triangle faces. Count the vertices, edges, and faces of this new polyhedron, and verify Euler's formula.

12.15 Describe the axes of rotational symmetry of an octahedron.

12.16 Draw the Schlegel diagram for a
a. tetrahedron
b. cube
c. octahedron

CHAPTER 13

CONSTRUCTIVE SOLID GEOMETRY

Traditional geometry, plane and solid—or even analytic—does not tell us how to create even the most simple shapes we see all around us in both the natural world and the made one. The methods of *constructive solid geometry* (CSG) gives us a way to describe complicated solid shapes as combinations of simpler solid shapes. This makes it an important tool for computer graphics and geometric modeling. CSG uses *Boolean operators* to construct a *procedural model* of a complex solid. This model, or rather the data describing the model, is stored in the mathematical form of a *binary tree*, where the leaf nodes are simple shapes, or *primitives*, sized and positioned in space, and each branch node is a Boolean operator: union, difference, or intersection.

This chapter discusses set theory, Boolean algebra, halfspaces, binary trees, solids, the special operators called *regularized Boolean operators*, set- membership classification, the Boolean model and how to construct it using those operators, primitives, and boundary evaluation.

13.1 Set Theory Revisited

Recall that a *set* is a collection of objects, called *elements* or *members* of the set (see Section 7.2). Examples of a set may be mathematical, as in the set of all even numbers; geometric, as in the set of points on a line segment; logical, as in all nonempty sets; or almost any specifiable collection of things, as in all white crows.

A set containing all of the elements of all of the sets in a particular problem is the *universal set*, E. In geometry, E often indicates all of the points in a two- or three-dimensional space.

The *complement* of a set A, written cA, with respect to the universal set E, is the set of all of the elements in E that are not elements of A. The *null set*, or *empty set*, $\emptyset$, is the set having no elements at all.

A set A is a subset of another set B if and only if every member of A is also a member of B. The symbol $\subset$ denotes the relationship between a set and a subset. $A \subset B$ says that A is a subset of B. For example, if A is the set of all numbers between 1 and 10, inclusive, whose names have three letters, and if B is the set of all numbers between 1 and 10, then A is a subset of B, $A \subset C$. A is a *proper subset* of B if every element of A is contained in B and if B has at least one element not in A, which is true of the example just given. Every set is a subset of itself, but not a proper subset.

13.2 Boolean Algebra

George Boole (1815–1864) invented the algebra we use for combining sets. The operators of ordinary algebra are, addition, subtraction, and multiplication. In Boolean algebra the operators are *union, intersection*, and *difference*, and are called *Boolean operators*. We treat sets in much the same way as ordinary algebraic quantities.

The union operator ∪ combines two sets A and B to form a third set C whose members are all of the members of A plus all of the members of B that are not already members of A. We express this as the Boolean algebraic equation

$$C = A \cup B \tag{13.1}$$

For example, if A consists of four members a, b, c, and d, and if B consists of three members c, d, and e, then C consists of five members a, b, c, d, and e. The *Venn diagram* in Figure 13.1 illustrates the union of points defining two rectangular regions in the plane, R and S, such that $T = R \cup S$, where T is the shaded area.

The *intersection* operator ∩ combines two sets A and B to form a third set C, whose members are only those common to both A and B, which we write as

$$C = A \cap B \tag{13.2}$$

Using the example of sets A and B, whose members were described for the union operator, we find that the members of C are c and d. The Venn diagram of Figure 13.2 shows the intersection of R and S such that $T = R \cap S$.

The difference operator combines two sets A and B to form a third set C, whose members are only those of the first set that are not also members of the second. We write this as

$$C = A - B \tag{13.3}$$

Again, using the example of sets A and B whose members we described previously, we find that the members of C are a and b. The set T formed by the difference $R - S$ is the shaded area in Figure 13.3.

The union and intersection operators are commutative, so

$$A \cup B = B \cup A \tag{13.4}$$

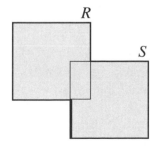

Figure 13.1 *Union of two sets.*

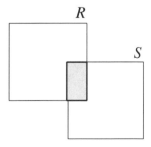

Figure 13.2 *Intersection of two sets.*

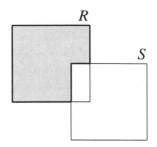

Figure 13.3 *Difference of two sets: R–S.*

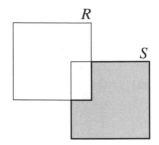

Figure 13.4 *Difference of two sets: S–R.*

and

$$A \cap B = B \cap A \tag{13.5}$$

but the difference operator is not, since $A - B = \{a, b\}$ and $B - A = \{e\}$. Thus,

$$A - B \neq B - A \tag{13.6}$$

Compare Figures 13.3 and 13.4 to see an example of this inequality for the difference of two rectangular regions in the plane: $R - S \neq S - R$.

Set operations obey the rules of Boolean algebra and govern the ways we can combine sets. We can summarize the properties of this algebra in the Table 13.1.

Table 13.1 Properties of operations on sets

Union Properties

1. $A \cup B$ is a set	Closure
2. $A \cup B = B \cup A$	Commutativity
3. $(A \cup B) \cup C = A \cup (B \cup C)$	Associativity
4. $A \cup \emptyset = A$	Identity
5. $A \cup A = A$	Idempotency
6. $A \cup cA = E$	Complementation

Intersection Properties

1. $A \cap B$ is a set	Closure
2. $A \cap B = B \cap A$	Commutativity
3. $(A \cap B) \cap C = A \cap (B \cap C)$	Associativity
4. $A \cap E = A$	Identity
5. $A \cap A = A$	Idempotency
6. $A \cap cA = \emptyset$	Complementation

Distributive Properties

1. $A \cup (B \cap C) = (A \cup B) \cap (A \cup C)$
2. $A \cap (B \cup C) = (A \cap B) \cup (A \cap C)$

Complementation Properties

1. $cE = \emptyset$
2. $c\emptyset = E$
3. $c(cA) = A$
4. $c(A \cup B) = cA \cap cB$
5. $c(A \cap B) = cA \cup cB$

13.3 Halfspaces Revisited

This section revisits the ideas of halfspaces presented in Chapter 7, but now in the context of constructive solid geometry. Recall that an unbounded straight line or open curve divides the plane in which it lies into two regions, called halfspaces. A closed curve, such as a circle or ellipse, also divides the plane into halfspaces, one finite and one infinite. A plane or an arbitrary surface similarly divides three-dimensional space into two halfspaces. An n-dimensional space is divided into halfspaces by a hyperplane, or surface of $n - 1$ dimensions.

A halfspace is given by an implicit function $f(x, y)$ in the plane and $f(x, y, z)$ in three dimensions. If, for a particular pair or triplet of coordinates representing a point in the plane or in three-dimensional space, respectively, the value of the function is zero, then the point is on the curve or surface defined by the function, where the curve or surface is the boundary between the two regions. In three dimensions, this boundary surface is sometimes called a *directed surface*. If the value of the function is positive, then we say the point is in the *inside* halfspace. If the value is negative, then the point is in the *outside* halfspace. The coordinates of all points on one side of the boundary produce values of the function with the same algebraic sign. Points on the other side produce values with the opposite sign. The following conventions are used to classify a point with respect to a halfspace.

1. If the coordinates of a point yield $h(x, y, z) = 0$, then the point is on the boundary of the halfspace.
2. If the coordinates of a point yield $h(x, y, z) > 0$, then the point is in the *inside* halfspace.
3. If the coordinates of a point yield $h(x, y, z) < 0$, then the point is in the *outside* halfspace; it is in the complement of $h(x, y, z)$.

These conventions establish the semi-infinite region, all of whose points yield positive values for $h(x, y)$ as the halfspace of interest. Changing the sign of $h(x, y)$ reverses the inside/outside point classification. Thus,

$$-h(x, y) = ch(x, y) \tag{13.7}$$

where c denotes the complement. Figure 13.5 shows the division of the plane into two regions by $h(x, y)$, where the region to the right of the boundary, $h(x, y) = 0$, produces positive values of $h(x, y)$.

We can combine n halfspaces using the Boolean intersection operator to produce a closed region. Mathematically, we express this as the Boolean summation of intersections. In three dimensions this is

$$g = \bigcap_{i=1}^{n} h_i(x, y, z) \tag{13.8}$$

or

$$g = h_1(x, y, z) \cap h_2(x, y, z) \cap \ldots \cap h_n(x, y, z) \tag{13.9}$$

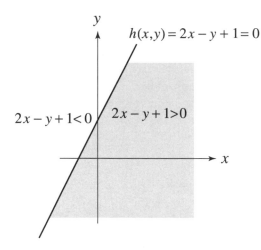

Figure 13.5 *Example of a halfspace.*

This is one way to create the initial simple shapes, or *primitives*, of constructive solid geometry. We can then combine these primitives themselves using Boolean operators to form more complex shapes.

13.4 Binary Trees

A *graph-based model* records topological structure, where data pointers link the faces, edges, and vertices of a model. For example, we can represent a solid object as a list of its faces and their respective surface equations. We represent the edges of each face as curve equations, with pointers to their endpoints (vertices) and adjoining faces. We represent the vertices as lists of coordinates, with pointers to the edges meeting at each vertex. There are two kinds of information here—pointers defining the connectivity or topology between vertices, edges, and faces, and numeric data defining curve and surface equations and vertex coordinates. The model of a polyhedron is a more specific example (Chapter 12). Its graph lists the polyhedron vertices, with their coordinates in one array and their connectivity in another (using a connectivity matrix).

A graph is a set of *nodes* (or points) connected by *branches* (lines). Figure 13.6 shows two examples of graphs. If a branch has a direction associated with it, then it is called a *directed graph* (Figure 13.6a). A *path* from one node to another is the sequence of branches traversed. If the start and end nodes of a path are the same, then the path is a *circuit* (Figure 13.6b). If there is a path between any pair of nodes, then the graph is *connected*.

A *tree* is a connected graph without circuits (Figure 13.7) and has the following additional properties:

1. There is one and only one node, called the *root*, which no branches enter. (In other words, it has no *parent node*.)
2. Every node except the root node has one and only one *entering branch*.
3. There is a unique path from the root node to each other node.

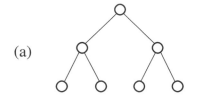

(a)

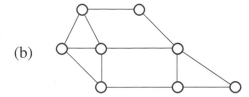

(b)

Figure 13.6 *Two different graphs.*

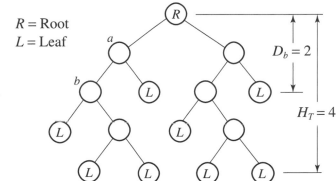

R = Root
L = Leaf

Figure 13.7 *Elements of a tree.*

The terms *entering branch, exiting branch, parent node,* and *descendant node* are all relative to the root node. For example, we assume that a path begins at the root node and proceeds to a leaf node. Then the branch (line) connecting nodes a and b in the figure is an *exiting branch* of node a and an *entering branch* of node b. Furthermore, node a is the *parent* of b and, conversely, b is a descendant of a. If the descendants of each node are in order, say from left to right, then the tree is *ordered*.

The *height* H_T of a tree is the number of branches that must be traversed to reach the farthest leaf from the root. The height of the tree in Figure 13.7 is $H_T = 4$. The *depth, D,* of a node in a tree is the number of branches that must be traversed to reach it. In the figure, the depth of node b is $D_b = 2$.

Constructive solid geometry uses a special tree called a *binary tree*. A binary tree is an ordered tree where each node, except the leaf nodes, has two descendants, a left descendant and a right descendant. Figure 13.7 is an example of a binary tree. A binary tree is *complete* if for some integer k every node of depth less than k has a left and right descendant, and every node of depth k is a leaf (Figure 13.8). The total number of nodes, N_B, in a complete binary tree of height $H_T = k$ is

$$N_B = 2^{k+1} - 1 \tag{13.10}$$

Thus, the total number of nodes for a complete binary tree with $k = 3$ is 15.

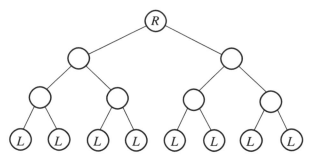

Figure 13.8 *Example of a* binary tree.

To *traverse* a binary tree means to "visit" or query each node to get information about it or the path on which it lies. We must devise an algorithm to do this systematically and efficiently. We will look at three ways of doing this: *preorder traversal, postorder traversal,* and *inorder traversal.* For any given node n, it and all of its descendants form a subtree t_n of a tree T. The node n is the root of t_n, and d_L and d_R are its immediate descendants. In the traversals to follow, we assign a number, in sequence, to each node in the order it is visited.

We recursively define the preorder traversal of T as follows (Figure 13.9):

1. Visit the root, r.
2. Visit the subtrees with roots d_L and d_R, in that order.

The postorder traversal of T (Figure 13.10) is:

1. Visit in postorder the subtrees with roots d_L and d_R, in that order.
2. Visit the root of that subtree.

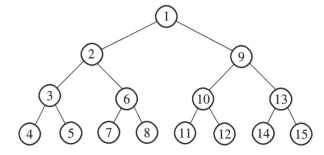

Figure 13.9 *A preorder traversal of a binary tree.*

Figure 13.10 *A postorder traversal of a binary tree.*

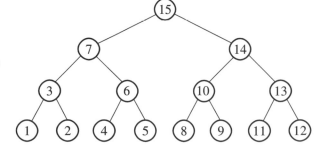

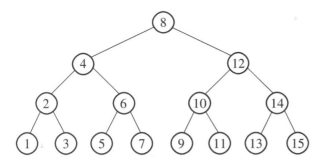

Figure 13.11 *An* inorder *traversal of a binary tree.*

The inorder traversal (Figure 13.11) is:

1. Visit in inorder the left subtree of root r (if it exists).
2. Visit r.
3. Visit in inorder the right subtree of r (if it exists).

It is difficult to describe these traversals in 25 words or less. However, the node numbering of each in Figures 13.9–13.11 is more revealing of the properties of the traversals. For example, in Figure 13.11 we see that the inorder traversal occurs such that each node in the left subtree of a root node has a number less than n_R, while each node in the right subtree has a number greater than n_R.

13.5 Solids

What is a solid? To answer this we will first define what we mean by the boundary of an object. We denote an n-dimensional Cartesian space by the symbol E^n. Thus, E^3 is the familiar three-dimensional space, and E^2 is the two-dimensional plane. Two real numbers, or coordinates, define points in E^2. Three real numbers define points in E^3.

A *region* R^n is a finite, bounded portion of any space E^n. The points belonging to a given region lie either entirely inside it or on its boundary. Thus, the total set of points R in a region lie in one of two subsets: Those points inside R are members of the subset R_I, and those on its boundary are members of the subset R_B, so that

$$R = R_I \cup R_B \tag{13.11}$$

Every physical object occupies a three-dimensional region R^3. Its boundary R_B^3 is a closed surface. It is self-evident that, given any point in E^3, it is either outside the region of the object, in the boundary set R_B^3, or in the interior set R_I^3.

Every curve segment is a one-dimensional region R^1. It has only two points in the set R_B^1, and all of the rest are in R_I^1.

A surface is a two-dimensional region R^2. An open surface is bounded by one or more nonintersecting closed curves (Figure 13.12). If the surface has holes in it, then

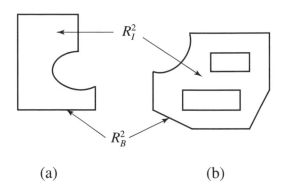

Figure 13.12 *Bounded two-dimensional regions.*

(a)   (b)

a closed curve is required for each hole (Figure 13.12b). All of the points on all of the closed curves comprise the set R_B^2. All of the other points of the surface are in R_I^2.

Let's generalize this notation scheme to make it more powerful. To do this we let $R^{m,n}$ denote a region of E^n, where m indicates the dimensionality of R and n indicates the dimensionality of the space E in which R is located and where it follows that $m \le n$. Thus,

$$R^{m,n} = B^{m-1,n} \cup I^{m,n} \tag{13.12}$$

where $B^{m-1,n}$ is the set of all points on the boundary of $R^{m,n}$ and $I^{m,n}$ is the set of all of its interior points. Figure 13.13 shows a two-dimensional region with an outer boundary and two inner boundaries defining holes.

An arbitrary point has one and only one of the following three properties with respect to any region $R^{m,n}$:

1. It is inside the region (that is, it is a member of the set $I^{m,n}$).
2. It is on the boundary of the region (that is, it is a member of the set $B^{m-1,n}$).
3. It is outside (not a member of) the set $R^{m,n}$.

For a region without holes—a homogeneous region—if $m = n$, then $I^{m,n}$ can be implied by an explicit formulation of $B^{m-1,n}$. Given a solid $R^{3,3}$ in E^3 without holes, the

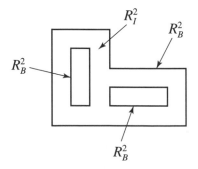

Figure 13.13 *Multiple boundaries.*

explicit definition of $B^{2,3}$ is sufficient to represent it unambiguously. $B^{2,3}$ is the bounding surface of the solid. Points on the inside, that is, points in $I^{3,3}$, are implied by $B^{2,3}$.

Table 13.2 defines allowable regions in E^3.

Table 13.2 Allowable regions in E^3

$R^{m,n}$	Type	$B^{m-1,n}$	$I^{m,n}$
$R^{0,3}$	Point	The point itself.	No interior points.
$R^{1,3}$	Curve	Two endpoints.	All points on the curve, excluding the two boundary points in $B^{0,3}$.
$R^{2,3}$	Surface	Closed boundary curve	All points on the surface, excluding those on the bounding curves in $B^{1,3}$.
$R^{3,3}$	Solid	Closed boundary surface	All points in the solid, excluding those on the bounding surface in $B^{2,3}$.

13.6 Boolean Operators

In CSG we work with three-dimensional solid objects defined by closed sets of points. There is a boundary subset of points and an interior subset of points. We can study examples of closed and open sets of points using an interval on the real line. When we write the expression for an *open interval*, $a < x < b$, we mean any value of x between, but not including, a and b. There are no definite boundary points to the interval that are also in the set x, so we say that it is an open set. On the other hand, when we write the expression for a *closed interval*, $a \leq x \leq b$, we mean any value of x between a and b, including a and b. There are definite boundary points, namely a and b, so we call it a closed set. (See also Section 5.9.)

CSG uses special Boolean operators similar to those that produce set union, intersection, and difference. The computer algorithms that perform these operations must generate objects that are also closed sets of points, having boundary and interior subsets that preserve the dimensions of the initial objects. This means that in any operation, such as $A \cap B = C$, all three objects must have the same spatial dimension.

It is possible to generate an unacceptable result when doing an ordinary Boolean intersection. For example, in Figure 13.14 sets A and B are two well-defined two-dimensional objects, yet their intersection is a one-dimensional line segment. Sets A and B each possess a boundary set, bA and bB, and an interior set, iA and iB. The resulting intersection, $A \cap B = C$, is mathematically correct according to set theory, but it is geometrically unacceptable because C has no interior. So the ordinary intersection operator apparently does not preserve the dimensionality of the initial two sets A and B. For such cases we would prefer an operator that recognizes dangling edges or disconnected parts of a lower dimension. To this end, CSG uses *regularized Boolean operators*. We will see how important they are in the following discussion.

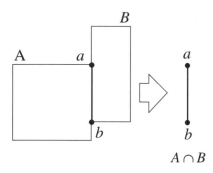

Figure 13.14 *An intersection with no interior points.*

Regularized Set Operators

Let's see what must happen in order to combine two simple polygonal shapes *A* and *B* to create a third shape *C* (Figure 13.15). Both *A* and *B* are defined by their boundaries, each of which consists of a closed string of line segments corresponding to the polygon's edges. The arrows indicate the order in which the segments are joined and traversed, as well as the direction of their parameterization (Figure 13.15a). Note that each straight segment line can be represented mathematically by parametric equations,

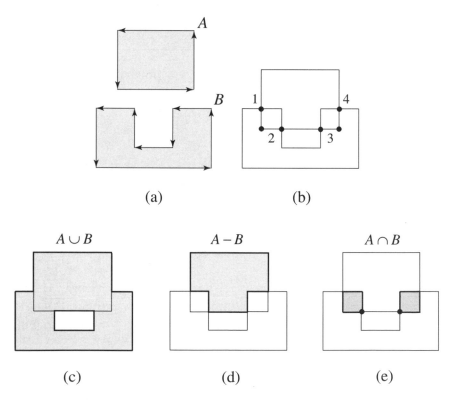

Figure 13.15 *Three different ways to combine two simple shapes.*

where the parametric variable u takes values in the closed unit interval $0 \le u \le 1$. For example, the equations $x = 12u + 2$ and $y = 6$ represent a line connecting the points $(2, 6)$ and $(14, 6)$. Note also that the boundaries of both A and B are traversed in the same direction—counterclockwise. Thus, if we imagine walking around the boundary of A or B in a counterclockwise direction, then the inside of the polygon is always on the left. This is the usual convention, and it will help us later to decide if any given point is inside or outside a polygon.

Figure 13.15b shows A and B translated into the position we want them in before applying a Boolean operation. The four points of intersection of the boundaries are 1, 2, 3, and 4. Here is an algorithm for finding the union of A and B, where $C = A \cup B$ (Figure 13.15c):

1. Find all the intersection points between the boundaries of A and B (in this case, points 1, 2, 3, and 4).
2. Define new segments for each edge of A and B that contains a point of intersection. Two points of intersection lie on edge bc of A, so it is replaced by three line segments.
3. Construct a revised list of vertices in sequence for both A and B.
4. Use the midpoint (or any other internal point) of each new segment (A and B) and a point-containment test to determine if the segment is inside or outside the other polygon. Segments of one polygon that are inside the other cannot be on the boundary of C.
5. Trace and sort all outside segments to find the boundary of C. Start tracing from a point outside B. (The intersection points mark the transition of the boundary of segments of C from those belonging to A to those belonging to B, and vice versa.)
6. Trace and sort the inside loops. These are the boundaries of holes in C. (It turns out that loops are automatically traversed in a clockwise direction. However, the inside of C is still on the left when we "walk" in this direction.)

The boundary segments of A and B that contribute to the outer boundary of C and loops defining the boundaries of holes in C are called *active segments*; the others are *inactive segments*.

The difference operation, $C = A - B$, is the same as union except that the direction of parameterization and traversal is reversed to clockwise (Figure 13.15d). This, in effect, turns the region B into a hole and subtracts material from A. The intersection operation is also similar to union, except that tracing should start from some point on the boundary of A that is inside B (Figure 13.15e).

Although the previous example is a relatively simple problem, it does illustrate what is required of algorithms for Boolean combinations of two geometric objects: computing intersections of lines, testing for point containment, and tracing and sorting line segments to define new boundaries. But what if the combination does not produce a result that is dimensionally homogeneous?

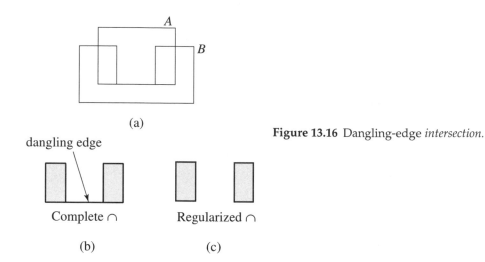

Figure 13.16 Dangling-edge *intersection.*

Look at what happens when we translate A or B to produce the result illustrated in Figure 13.16a. The complete intersection of A and B, $C = A \cap B$, is not dimensionally homogeneous (Figure 13.16b). We have combined two homogeneous two-dimensional polygons and produced two separate two- dimensional polygons joined by a line, called a *dangling edge*. We have a mix of one- and two-dimensional shapes. Although this is the correct and complete intersection, it is not an acceptable answer in CSG. Instead, we would like the intersection operator to produce the shapes shown in Figure 13.16c. Here the dangling edge does not appear, and the shapes are dimensionally homogeneous. The Boolean operators that produce dimensionally homogeneous shapes are *regularized operators*.

So, how do we compute the regularized intersection of A and B? We begin with the ordinary set intersection operation

$$C = A \cap B \tag{13.13}$$

and rewrite it as

$$C = (bA \cup iA) \cap (bB, \cup iB) \tag{13.14}$$

Using the associative property, this expands to

$$C = (bA \cap bB) \cup (iA \cap bB) \cup (bA \cap iB) \cup (iA \cap iB) \tag{13.15}$$

Figure 13.17 shows a geometric interpretation of each of these four terms. We know that

$$C = bC \cup iC \tag{13.16}$$

and that we must now find subsets of bC and iC that form a closed dimensionally homogeneous object, say $\hat{C}$, using the terms of Equation 13.15. In the example, we see that $\hat{C}$ consists of two disjoint two-dimensional regions. Figure 13.17d shows the two-

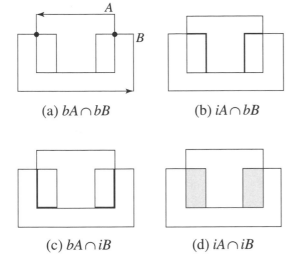

(a) $bA \cap bB$ (b) $iA \cap bB$

Figure 13.17 *Pieces of an intersection.*

(c) $bA \cap iB$ (d) $iA \cap iB$

dimensional interior of these parts of $\hat{C}$, and we conclude that

$$i\hat{C} = iC = iA \cap iB \tag{13.17}$$

Only the boundary $b\hat{C}$ remains to be determined. We observe that the boundaries of the new geometric object(s) always consist of bits and pieces of boundary segments of the initial combining objects. In other words, initial boundary points can become interior points, whereas initial interior points cannot become boundary points. For a regularized intersection, it is always true that

$$iA \cap bB \subset b\hat{C} \tag{13.18}$$

and

$$bA \cap iB \subset b\hat{C} \tag{13.19}$$

The boundary and interior elements shown in Figures 13.17b–13.17d are valid subsets of $\hat{C}$. But what about the elements shown in Figure 13.17a? We must analyze them to determine which elements are valid subsets of $\hat{C}$. The two isolated points are valid members of $b\hat{C}$ because they are members of both $iA \cap bB$ and $bA \cap iB$. We are left with overlapping segments of boundaries from A and B, neither of which are interior to A or B. Now we must find a way to show that they are not part of the regularized intersection.

The simplest test of overlapping segments is to compare their directions of traversal in the overlap. This will give a correct solution only if a consistent direction of parameterization or traversal holds for both A and B. If the directions of the A and B segments are the same in the overlap, then the overlap is part of a valid boundary of $\hat{C}$, because both interiors are on the same side of the overlap. If they are in the opposite direction, then they are not a valid part of $b\hat{C}$, because the interiors of A and B are on opposite sides of the overlap. This test can be done to compare vector representations of the overlapping segments. Other tests are also available.

We express the process for finding the regularized intersection of two bounded sets as

$$\hat{C} = b\hat{C} \cup i\hat{C} \tag{13.20}$$

and expand this to

$$\hat{C} = \text{Valid}_b\,(bA \cap bB) \cup (iA \cap bB) \cup (bA \cap iB) \cup (iA \cap iB) \tag{13.21}$$

We can separate this last expression into interior and exterior parts as follows:

$$b\hat{C} = \text{Valid}_b\,(bA \cap bB) \cup (iA \cap bB) \cup (bA \cap iB) \tag{13.22}$$

and

$$i\hat{C} = iA \cap iB \tag{13.23}$$

Nothing in Equations 13.22 and 13.23 indicates dimensionality, because they are equally applicable to one-, two-, three-, or n-dimensional objects. Similar procedures apply to the union and difference operators.

Set-Membership Classification

One of the geometric tools we can use to regularize sets produced by a Boolean combination of other sets is to determine whether a given point is inside, outside, or on the boundary of a set. Any regularized set A has three important subsets associated with it: the set of all of its interior points, denoted iA, the set of all points on its boundary, denoted bA, and all points outside of it, denoted cA. Determining which set a given point belongs to is called *set-membership classification*.

Set-membership classification helps us solve several important problems in geometric modeling and computer graphics. Here are four of them:

1. The *point-inclusion* problem: Given some arbitrary geometric solid and some point, is the point inside, outside, or on the boundary of the solid?
2. The *clipping* problem: Given a polygon and a line segment, what parts of the line are inside, outside, or on the boundary of the polygon?
3. The *polygon-intersection* problem: Given two polygons A and B, what, if any, is the intersection polygon $A \cap B$?
4. The *solid-interference* problem: Given two solids A and B, do they interfere (that is, intersect unintentionally)?

13.7 Boolean Models

A *Boolean model* is a complex solid formed by the Boolean combination of two or more simpler solids. If A and B are regularized geometric solids, that is, solids that are dimensionally homogeneous, and if $C = A < OP > B$, where $< OP >$ is any regularized Boolean operator (union, difference, or intersection), then $A < OP > B$ is a

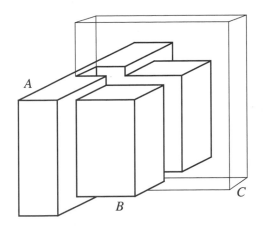

Figure 13.18 *A Boolean model.*

Boolean model of C. We will assume in what follows that the symbols ∪, ∩, and − denote regularized operators.

A, B, and C are simple rectangular solids whose size and position we have arranged as shown in Figure 13.18. We define a Boolean model D as

$$D = (A \cup B) - C \tag{13.24}$$

Equation 13.24 says nothing quantitative about the new solid it creates. It only represents a certain combination of *primitive* solids. It does not directly give us the vertex coordinates of D or any information about its edges and faces. All it tells us is how to construct the model, how to fit A, B, and C together to create D. For this reason, a Boolean model is a *procedural* or *unevaluated* model. If we want to know more about D we must evaluate its Boolean model. This means that we must compute intersections to determine its vertices and edges. Then we must analyze the connectivity of these new elements to reveal the topological characteristics of the model.

The binary tree for the model defined by Equation 13.24 is shown in Figure 13.19. The leaf nodes are the primitive solids, with the appropriate Boolean operator and its product shown at each interval node and the root. CSG models are ordered binary trees whose leaf nodes are either primitive solids or rigid-body transformations. Each internal node is a regularized Boolean operator that combines the products or contents of the two nodes immediately below it in the tree and, if necessary, transforms the result in readiness for the next operation. Each subtree of a node represents a solid produced by the Boolean operations and primitives arrayed in its leaf and internal nodes. The root represents the complete model.

Primitives

Most CSG systems use a few basic primitive shapes. We define their size, shape, position, and orientation by specifying the values for a small set of variables. For example, rectangular solids, or blocks, are almost always available. We can produce a particular *instance* of a block by specifying its length, width, height, and location within the model's coordinate system.

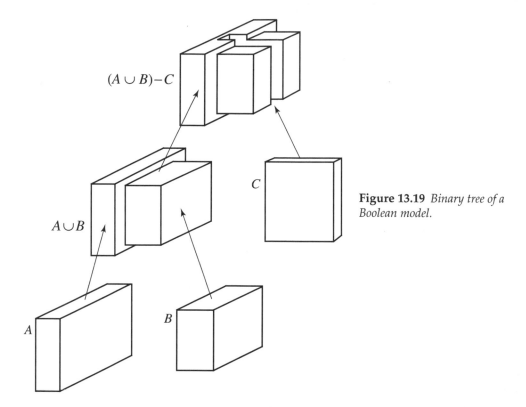

$(A \cup B) - C$

$A \cup B$

A

B

C

Figure 13.19 *Binary tree of a Boolean model.*

Figure 13.20 shows several common CSG primitives. Note that some of them can be constructed by a Boolean combination of others. For example, using a properly sized block and cylinder, we can produce an inside fillet primitive. Furthermore, the block and cylinder primitives alone have the same descriptive power as a set of primitives consisting of a block, cylinder, wedge, inside fillet, cylindrical segment, and tetrahedron.

Each primitive, in turn, is usually represented in a CSG system as the intersection of halfspaces. For example, the regularized intersection of six planar halfspaces represents a block-type primitive, and the regularized intersection of two planar halfspaces and a cylindrical halfspace represents a cylindrical primitive.

Boundary Evaluation

Usually we are most interested in the boundary of a complex Boolean, or CSG, model. To construct a boundary representation, computer graphics or modeling programs use a set of algorithms called the *boundary evaluator*, which determines where faces are truncated and new edges and vertices are created or deleted. For example, where boundary elements of combined solids overlap or coincide, the evaluator merges them into a single element to maintain a consistent and nonredundant data structure.

The boundary evaluator creates new edges where the surfaces of two combined solids intersect. The boundary evaluator algorithms find these intersections and determine what parts of the intersections are valid edges of the new solid. New edges are terminated by new vertices, so vertices, too, must be classified.

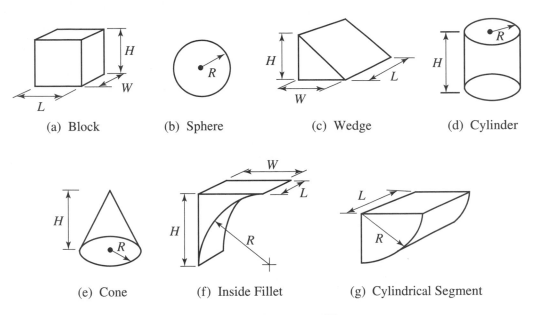

Figure 13.20 Primitive *solids.*

Exercises

13.1 Given sets $A = \{a, b, c, d, e\}$ and $B = \{c, d, f, g, h\}$, find $A \cup B$.

13.2 Using sets A and B defined in Exercise 13.1, find $A \cap B$.

13.3 Using sets A and B defined in Exercise 13.1, find $A - B$.

13.4 Given sets $A = \{a, b, c\}$ and $B = \{d, e\}$, find $A \cap B$.

13.5 Using sets A and B defined in Exercise 13.4, and $E = \{a, b, c, d, e, f, g\}$,
    a. Find cA b. Find cB c. Find $c\,(A \cup B)$

13.6 Show that $A - (A - B) = A \cap B$.

13.7 Find the two-dimensional halfspaces that define a rectangle two units wide and five units high whose lower left corner is at the origin of the coordinate system.

13.8 If each node, except leaf nodes, of a binary tree has two descendants, how do you interpret the terms *quadtree* and *octree*?

13.9 How many nodes are there in a complete quadtree?

13.10 How many nodes are there in a complete octree?

13.11 Derive expressions for the valid regularized union of two bounded and dimensionally homogeneous objects A and B. (The expressions will be similar to those for the intersection of two such objects, as given by Equations 13.21, 13.22, and 13.23.)

13.12 Derive expressions for the valid regularized difference of two bounded and dimensionally homogeneous objects A and B (similar to Exercise 13.11).

CHAPTER 14

CURVES

We all have a strong intuitive sense of what a curve is. Although we never see a curve floating around free of any object, we can readily identify the curved edges and silhouettes of objects and easily imagine the curve that describes the path of a moving object. This chapter explores the mathematical definition of a curve in a form that is very useful to geometric modeling and other computer graphics applications: that definition consists of a set of *parametric equations*. The mathematics of parametric equations is the basis for Bézier, NURBS, and Hermite curves. The curves discussed in this chapter may be placed in the Hermite family of curves. Bézier curves are the subject of the next chapter, and NURBS are best left for more advanced texts on geometric modeling. Both plane curves and space curves are introduced here, followed by discussions of the tangent vector, blending functions, conic curves, reparameterization, and continuity and composite curves.

14.1 Parametric Equations of a Curve

A parametric curve is one whose defining equations are given in terms of a single, common, independent variable called the *parametric variable*. We have already encountered parametric variables in earlier discussions of vectors, lines, and planes.

Imagine a curve in three-dimensional space. Each point on the curve has a unique set of coordinates: a specific x value, y value, and z value. Each coordinate is controlled by a separate parametric equation, whose general form looks like

$$x = x(u), \quad y = y(u), \quad z = z(u) \tag{14.1}$$

where $x(u)$ stands for some as yet unspecified function in which u is the independent variable; for example, $x(u) = au^2 + bu + c$, and similarly for $y(u)$ and $z(u)$. It is important to understand that each of these is an independent expression. This will become clear as we discuss specific examples later.

The dependent variables are the x, y, and z coordinates themselves, because their values depend on the value of the parametric variable u. Engineers and programmers who do geometric modeling usually prefer these kinds of expressions because the coordinates x, y, and z are independent of each other, and each is defined by its own parametric equation.

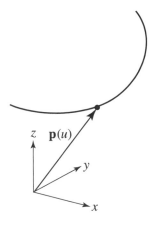

Figure 14.1 *Point on a curve defined by a vector.*

Each point on a curve is defined by a vector **p** (Figure 14.1). The components of this vector are $x(u)$, $y(u)$, and $z(u)$. We express this as

$$\mathbf{p} = \mathbf{p}(u) \tag{14.2}$$

which says that the vector **p** is a function of the parametric variable u.

There is a lot of information in Equation 14.2. When we expand it into component form, it becomes

$$\mathbf{p}(u) = [x(u) \quad y(u) \quad z(u)] \tag{14.3}$$

The specific functions that define the vector components of **p** determine the shape of the curve. In fact, this is one way to define a curve—by simply choosing or designing these mathematical functions. There are only a few simple rules that we must follow: 1) Define each component by a single, common parametric variable, and 2) make sure that each point on the curve corresponds to a unique value of the parametric variable. The last rule can be put the another way: Each value of the parametric variable must correspond to a unique point on the curve.

14.2 Plane Curves

To define plane curves, we use parametric functions that are second degree polynomials:

$$\begin{aligned}
x(u) &= a_x u^2 + b_x u + c_x \\
y(u) &= a_y u^2 + b_y u + c_y \\
z(u) &= a_z u^2 + b_z u + c_z
\end{aligned} \tag{14.4}$$

where the a, b, and c terms are constant coefficients.

We can combine $x(u)$, $y(u)$, $z(u)$, and their respective coefficients into an equivalent, more concise, vector equation:

$$\mathbf{p}(u) = \mathbf{a}u^2 + \mathbf{b}u + \mathbf{c} \tag{14.5}$$

We allow the parametric variable to take on values only in the interval $0 \leq u \leq 1$. This ensures that the equation produces a bounded line segment. The coefficients **a**, **b**, **c**, in this equation are vectors, and each has three components; for example, $\mathbf{a} = [a_x \quad a_y \quad a_z]$.

This curve has serious limitations. Although it can generate all the conic curves, or a close approximation to them, it cannot generate a curve with an inflection point, like an S-shaped curve, no matter what values we select for the coefficients **a**, **b**, **c**. To do this requires a cubic polynomial (Section 14.3).

How do we define a specific plane curve, one that we can display, with definite end points, and a precise orientation in space? First, note in Equation 14.4 or 14.5 that there are nine coefficients that we must determine: $a_x, b_x, \ldots, c_z$. If we know the two end points and an intermediate point on the curve, then we know nine quantities that we can express in terms of these coefficients (3 points × 3 coordinates each = 9 known quantities), and we can use these three points to define a unique curve (Figure 14.2). By applying some simple algebra to these relationships, we can rewrite Equation 14.5 in terms of the three points. To one of the two end points we assign $u = 0$, and to the other $u = 1$. To the intermediate point, we arbitrarily assign $u = 0.5$. We can write this points as

$$\mathbf{p}_0 = [x_0 \quad y_0 \quad z_0]$$
$$\mathbf{p}_{0.5} = [x_{0.5} \quad y_{0.5} \quad z_{0.5}] \tag{14.6}$$
$$\mathbf{p}_1 = [x_1 \quad y_1 \quad z_1]$$

where the subscripts indicate the value of the parametric variable at each point.

Now we solve Equations 14.4 for the $a_x, b_x, \ldots, c_z$ coefficients in terms of these points. Thus, for x at $u = 0$, $u = 0.5$, and $u = 1$, we have

$$x_0 = c_x$$
$$x_{0.5} = 0.25a_x + 0.5b_x + c_x \tag{14.7}$$
$$x_1 = a_x + b_x + c_x$$

with similar equations for y, and z.

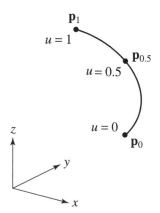

Figure 14.2 *A plane curve defined by three points.*

Next we solve these three equations in three unknowns for $a_x, b_x,$ and $c_x,$ finding

$$a_x = 2x_0 - 4x_{0.5} + 2x_1$$
$$b_x = -3x_0 + 4x_{0.5} - x_1 \qquad (14.8)$$
$$c_x = x_0$$

Substituting this result into Equation 14.4 yields

$$x(u) = (2x_0 - 4x_{0.5} + 2x_1)u^2 + (-3x_0 + 4x_{0.5} - x_1)u + x_0 \qquad (14.9)$$

Again, there are equivalent expressions for $y(u)$ and $z(u)$.

We rewrite Equation 14.9 as follows:

$$x(u) = (2u^2 - 3u + 1)x_0 + (-4u^2 + 4u)x_{0.5} + (2u^2 - u)x_1 \qquad (14.10)$$

Using this result and equivalent expressions for $y(u)$ and $z(u)$, we combine them into a single vector equation:

$$\mathbf{p}(u) = (2u^2 - 3u + 1)\mathbf{p}_0 + (-4u^2 + 4u)\mathbf{p}_{0.5} + (2u^2 - u)\mathbf{p}_1 \qquad (14.11)$$

Equation 14.11 produces the same curve as Equation 14.5. The curve will always lie in a plane no matter what three points we choose. Furthermore, it is interesting to note that the point $\mathbf{p}_{0.5}$ which is on the curve at $u = 0.5$, is not necessarily half way along the length of the curve between $\mathbf{p}_0$ and $\mathbf{p}_1$. We can show this quite convincingly by choosing three points to define a curve such that two of them are relatively close together (Figure 14.3). In fact, if we assign a different value to the parametric variable for the intermediate point, then we obtain different values for the coefficients in Equations 14.8. This, in turn, means that a different curve is produced, although it passes through the same three points.

Equation 14.5 is the *algebraic form* and Equation 14.11 is the *geometric form*. Each of these equations can be written more compactly with matrices. Compactness is not the only advantage to matrix notation. Once a curve is defined in matrix form, we can use the full power of matrix algebra to solve many geometry problems. So now we rewrite

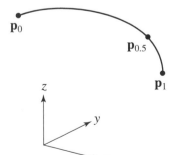

Figure 14.3 *Curve defined by three nonuniformly spaced points.*

Equation 14.5 using the following substitutions:

$$[u^2 \quad u \quad 1] \begin{bmatrix} a \\ b \\ c \end{bmatrix} = au^2 + bu + c \tag{14.12}$$

$$U = [u^2 \quad u \quad 1] \tag{14.13}$$

$$A = [a \quad b \quad c]^T \tag{14.14}$$

and finally, we obtain

$$p(u) = UA \tag{14.15}$$

Remember that **A is** really a matrix of vectors, so that

$$A = \begin{bmatrix} a \\ b \\ c \end{bmatrix} = \begin{bmatrix} a_x & a_y & a_z \\ b_x & b_y & b_z \\ c_x & c_y & c_z \end{bmatrix} \tag{14.16}$$

The nine terms on the right are called the *algebraic coefficients*.

Next, we convert Equation 14.11 into matrix form. The right-hand side looks like the product of two matrices: $[(2u^2 - 3u + 1) \quad (-4u^2 + 4u) \quad (2u^2 - u)]$ and $[p_0 \quad p_{0.5} \quad p_1]$. This means that

$$p(u) = [(2u^2 - 3u + 1) \quad (-4u^2 + 4u) \quad (2u^2 - u)] \begin{bmatrix} p_0 \\ p_{0.5} \\ p_1 \end{bmatrix} \tag{14.17}$$

Using the following substitutions:

$$F = [(2u^2 - 3u + 1) \quad (-4u^2 + 4u) \quad (2u^2 - u)] \tag{14.18}$$

and

$$P = \begin{bmatrix} p_0 \\ p_{0.5} \\ p_1 \end{bmatrix} = \begin{bmatrix} x_0 & y_0 & z_0 \\ x_{0.5} & y_{0.5} & z_{0.5} \\ x_1 & y_1 & z_1 \end{bmatrix} \tag{14.19}$$

where **P** is the *control point matrix* and the nine terms on the right are its elements or the *geometric coefficients*, we can now write

$$p(u) = FP \tag{14.20}$$

This is the matrix version of the geometric form.

Because it is the same curve in algebraic form, $p(u) = UA$, or geometric form, $p(u) = FP$, we can write

$$FP = UA \tag{14.21}$$

The **F** matrix is itself the product of two other matrices:

$$F = [u^2 \quad u \quad 1] \begin{bmatrix} 2 & -4 & 2 \\ -3 & 4 & -1 \\ 1 & 0 & 0 \end{bmatrix} \tag{14.22}$$

The matrix on the left we recognize as **U**, and we can denote the other matrix as

$$\mathbf{M} = \begin{bmatrix} 2 & -4 & 2 \\ -3 & 4 & -1 \\ 1 & 0 & 0 \end{bmatrix} \tag{14.23}$$

This means that

$$\mathbf{F} = \mathbf{UM} \tag{14.24}$$

Using this we substitute appropriately to find

$$\mathbf{UMP} = \mathbf{UA} \tag{14.25}$$

Premultiplying each side of this equation by $\mathbf{U}^{-1}$ yields

$$\mathbf{MP} = \mathbf{A} \tag{14.26}$$

This expresses a simple relationship between the algebraic and geometric coefficients

$$\mathbf{A} = \mathbf{MP} \tag{14.27}$$

or

$$\mathbf{P} = \mathbf{M}^{-1}\mathbf{A} \tag{14.28}$$

The matrix **M** is called a *basis transformation matrix*, and **F** is called a *blending function matrix*. There are other basis transformation matrices and blending function matrices, as we shall see in the following sections.

14.3 Space Curves

A space curve is not confined to a plane. It is free to twist through space. To define a space curve we must use parametric functions that are cubic polynomials. For $x(u)$ we write

$$x(u) = a_x u^3 + b_x u^2 + c_x u + d_x \tag{14.29}$$

with similar expressions for $y(u)$ and $z(u)$. Again, the a, b, c, and d terms are constant coefficients. As we did with Equation 14.5 for a plane curve, we combine the $x(u)$, $y(u)$, and $z(u)$ expressions into a single vector equation:

$$\mathbf{p}(u) = \mathbf{a}u^3 + \mathbf{b}u^2 + \mathbf{c}u + \mathbf{d} \tag{14.30}$$

If $\mathbf{a} = 0$, then this equation is identical to Equation 14.5.

To define a specific curve in space, we use the same approach as we did for a plane curve. This time, though, there are 12 coefficients to be determined. We specify four points through which we want the curve to pass, which provides all the information we need to determine $\mathbf{a}$, $\mathbf{b}$, $\mathbf{c}$, and $\mathbf{d}$. But which four points? Two are obvious: $\mathbf{p}(0)$ and $\mathbf{p}(1)$, the end points at $u = 0$ and $u = 1$. For various reasons beyond the scope of this text, it turns out to be advantageous to use two intermediate points that we assign parametric values of $u = \frac{1}{3}$ and $u = \frac{2}{3}$, or $\mathbf{p}(\frac{1}{3})$ and $\mathbf{p}(\frac{2}{3})$. So we now have the four points we need:

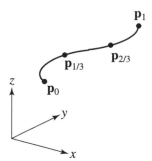

Figure 14.4 *Four points define a cubic space curve.*

$p(0)$, $p(\frac{1}{3})$, $p(\frac{2}{3})$, and $p(1)$, which we can rewrite as the more convenient p_1, p_2, p_3, and p_4 (Figure 14.4).

Substituting each of the values of the parametric variable ($u = 0, \frac{1}{3}, \frac{2}{3}, 1$) into Equation 14.29, we obtain the following four equations in four unknowns:

$$x_1 = d_x$$
$$x_2 = \frac{1}{27}a_x + \frac{1}{9}b_x + \frac{1}{3}c_x + d_x$$
$$x_3 = \frac{8}{27}a_x + \frac{4}{9}b_x + \frac{2}{3}c_x + d_x$$
$$x_4 = a_x + b_x + c_x + d_x$$

(14.31)

Now we can express a_x, b_x, c_x, and d_x in terms of x_1, x_2, x_3, and x_4. After doing the necessary algebra, we obtain

$$a_x = -\frac{9}{2}x_1 + \frac{27}{2}x_2 - \frac{27}{2}x_3 + \frac{9}{2}x_4$$
$$b_x = 9x_1 - \frac{45}{2}x_2 + 18x_3 - \frac{9}{2}x_4$$
$$c_x = -\frac{11}{2}x_1 + 9x_2 - \frac{9}{2}x_3 + x_4$$
$$d_x = x_1$$

(14.32)

We substitute these results into Equation 14.29, producing

$$x(u) = \left(-\frac{9}{2}x_1 + \frac{27}{2}x_2 - \frac{27}{2}x_3 + \frac{9}{2}x_4\right)u^3$$
$$+ \left(9x_1 - \frac{45}{2}x_2 + 18x_3 - \frac{9}{2}x_4\right)u^2$$
$$+ \left(-\frac{11}{2}x_1 + 9x_2 - \frac{9}{2}x_3 + x_4\right)u$$
$$+ x_1$$

(14.33)

All this looks a bit messy right now, but we can put it into a neater, much more compact form. We begin by rewriting Equation 14.33 as follows:

$$x(u) = \left(-\frac{9}{2}u^3 + 9u^2 - \frac{11}{2}u + 1\right) x_1$$
$$+ \left(\frac{27}{2}u^3 - \frac{45}{2}u^2 + 9u\right) x_2$$
$$+ \left(-\frac{27}{2}u^3 + 18u^2 - 9u\right) x_3 \qquad (14.34)$$
$$+ \left(\frac{9}{2}u^3 - \frac{9}{2}u^2 + u\right) x_4$$

Using equivalent expressions for $y(u)$ and $z(u)$, we can summarize them with a single vector equation:

$$\mathbf{p}(u) = \left(-\frac{9}{2}u^3 + 9u^2 - \frac{11}{2}u + 1\right) \mathbf{p}_1$$
$$+ \left(\frac{27}{2}u^3 - \frac{45}{2}u^2 + 9u\right) \mathbf{p}_2$$
$$+ \left(-\frac{27}{2}u^3 + 18u^2 - \frac{9}{2}u\right) \mathbf{p}_3 \qquad (14.35)$$
$$+ \left(\frac{9}{2}u^3 - \frac{9}{2}u^2 + u\right) \mathbf{p}_4$$

This means that, given four points assigned successive values of u (in this case at $u = 0, \frac{1}{3}, \frac{2}{3}, 1$), Equation 14.35 produces a curve that starts at $\mathbf{p}_1$, passes through $\mathbf{p}_2$ and $\mathbf{p}_3$, and ends at $\mathbf{p}_4$.

Now let's take one more step toward a more compact notation. Using the four parametric functions appearing in Equation 14.35, we define a new matrix, $\mathbf{G} = [G_1 \quad G_2 \quad G_3 \quad G_4]$, where

$$G_1 = \left(-\frac{9}{2}u^3 + 9u^2 - \frac{11}{2}u + 1\right)$$
$$G_2 = \left(\frac{27}{2}u^3 - \frac{45}{2}u^2 + 9u\right)$$
$$G_3 = \left(-\frac{27}{2}u^3 + 18u^2 - \frac{9}{2}u\right) \qquad (14.36)$$
$$G_4 = \left(\frac{9}{2}u^3 - \frac{9}{2}u^2 + u\right)$$

and then define a matrix $\mathbf{P}$ containing the control points, $\mathbf{P} = \begin{bmatrix} \mathbf{p}_1 & \mathbf{p}_2 & \mathbf{p}_3 & \mathbf{p}_4 \end{bmatrix}^T$, so that

$$\mathbf{p}(u) = \mathbf{GP} \tag{14.37}$$

The matrix $\mathbf{G}$ is the product of two other matrices, $\mathbf{U}$ and $\mathbf{N}$:

$$\mathbf{G} = \mathbf{UN} \tag{14.38}$$

where $\mathbf{U} = \begin{bmatrix} u^3 & u^2 & u & 1 \end{bmatrix}$ and

$$\mathbf{N} = \begin{bmatrix} -\dfrac{9}{2} & \dfrac{27}{2} & -\dfrac{27}{2} & \dfrac{9}{2} \\[2mm] 9 & -\dfrac{45}{2} & -\dfrac{9}{2} & 1 \\[2mm] -\dfrac{11}{2} & 9 & -\dfrac{9}{2} & 1 \\[2mm] 1 & 0 & 0 & 0 \end{bmatrix} \tag{14.39}$$

(Note that $\mathbf{N}$ is another example of a basis transformation matrix.)

Now we let

$$\mathbf{A} = \begin{bmatrix} \mathbf{a} \\ \mathbf{b} \\ \mathbf{c} \\ \mathbf{d} \end{bmatrix} = \begin{bmatrix} a_x & a_y & a_z \\ b_x & b_y & b_z \\ c_x & c_y & c_z \\ d_x & d_y & d_z \end{bmatrix} \tag{14.40}$$

Using matrices, Equation 14.30 becomes

$$\mathbf{p}(u) = \mathbf{UA} \tag{14.41}$$

which looks a lot like Equation 14.15 for a plane curve, except that we have defined new $\mathbf{U}$ and $\mathbf{A}$ matrices. In fact, Equation 24.15 is a special case of the formulation for a space curve.

To convert the information in the $\mathbf{A}$ matrix into that required for the $\mathbf{P}$ matrix, we do some simple matrix algebra, using Equations 14.37, 14.38, and 14.41. First we have

$$\mathbf{GP} = \mathbf{UNP} \tag{14.42}$$

and then

$$\mathbf{UA} = \mathbf{UNP} \tag{14.43}$$

or more simply

$$\mathbf{A} = \mathbf{NP} \tag{14.44}$$

14.4 The Tangent Vector

Another way to define a space curve does not use intermediate points. It uses the tangents at each end of a curve, instead. Every point on a curve has a straight line associated with it called the tangent line, which is related to the first derivative of the parametric functions $x(u)$, $y(u)$, and $z(u)$, such as those given by Equation 14.30. Thus

$$\frac{d}{du}x(u), \quad \frac{d}{du}y(u), \quad \text{and} \quad \frac{d}{du}z(u) \tag{14.45}$$

From elementary calculus, we can compute, for example,

$$\frac{dy}{dx} = \frac{dy(u)/du}{dx(u)/du} \tag{14.46}$$

We can treat $dx(u)/du$, $dy(u)/du$, and $dz(u)/du$ as components of a vector along the tangent line to the curve. We call this the *tangent vector*, and define it as

$$\mathbf{p}^u(u) = \left[\frac{d}{du}x(u)\mathbf{i} \quad \frac{d}{du}y(u)\mathbf{j} \quad \frac{d}{du}z(u)\mathbf{k}\right] \tag{14.47}$$

or more simply as

$$\mathbf{p}^u = [x^u \quad y^u \quad z^u] \tag{14.48}$$

(Here the superscript u indicates the first derivative operation with respect to the independent variable u.) This is a very powerful idea, and we will now see how to use it to define a curve.

In the last section, we discussed how to define a curve by specifying four points. Now we have another way to define a curve. We will still use the two end points, but instead of two intermediate points, we will use the tangent vectors at each end to supply the information we need to define a curve (Figure 14.5). By manipulating these tangent vectors, we can control the slope at each end. The set of vectors $\mathbf{p}_0$, $\mathbf{p}_1$, $\mathbf{p}_0^u$, and $\mathbf{p}_1^u$ are called the *boundary conditions*. This method itself is called the *cubic Hermite interpolation*, after C. Hermite (1822–1901) the French mathematician who made significant contributions to our understanding of cubic and quintic polynomials.

We differentiate Equation 14.29 to obtain the x component of the tangent vector:

$$\frac{d}{du}dx(u) = x^u = 3a_x u^2 + 2b_x u + c_x \tag{14.49}$$

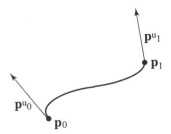

Figure 14.5 *Defining a curve using end points and tangent vectors.*

Evaluating Equations 14.29 and 14.49 at $u = 0$ $u = 1$, yields

$$
\begin{aligned}
x_0 &= d_x \\
x_1 &= a_x + b_x + c_x + d_x \\
x_0^u &= c_x \\
x_1^u &= 3a_x + 2b_x + c_x
\end{aligned}
\tag{14.50}
$$

Using these four equations in four unknowns, we solve for a_x, b_x, c_x, and d_x in terms of the boundary conditions

$$
\begin{aligned}
a_x &= 2(x_0 - x_1) + x_0^u + x_1^u \\
b_x &= 3(-x_0 + x_1) - 2x_0^u - x_1^u \\
c_x &= x_0^u \\
d_x &= x_0
\end{aligned}
\tag{14.51}
$$

Substituting the result into Equation 14.29, yields

$$
x(u) = \left(2x_0 - 2x_1 + x_0^u + x_1^u\right)u^3 + \left(-3x_0 + 3x_1 - 2x_0^u - x_1^u\right)u^2 + x_0^u u + x_0
\tag{14.52}
$$

Rearranging terms we can rewrite this as

$$
\begin{aligned}
x(u) = {}&(2u^3 - 3u^2 + 1)x_0 + (-2u^3 + 3u^2)x_1 \\
&+ (u^3 - 2u^2 + u)x_0^u + (u^3 - u^2)x_1^u
\end{aligned}
\tag{14.53}
$$

Because $y(u)$ and $z(u)$ have equivalent forms, we can include them by rewriting Equation 14.53 in vector form:

$$
\begin{aligned}
\mathbf{p}(u) = {}&(2u^3 - 3u^2 + 1)\mathbf{p}_0 + (-2u^3 + 3u^2)\mathbf{p}_1 \\
&+ (u^3 - 2u^2 + u)\mathbf{p}_0^u + (u^3 - u^2)\mathbf{p}_1^u
\end{aligned}
\tag{14.54}
$$

To express Equation 14.54 in matrix notation, we first define a blending function matrix $\mathbf{F} = [F_1 \quad F_2 \quad F_3 \quad F_4]$, where

$$
\begin{aligned}
F_1 &= 2u^3 - 3u^2 + 1 \\
F_2 &= -2u^3 + 3u^2 \\
F_3 &= u^3 - 2u^2 + u \\
F_4 &= u^3 - u^2
\end{aligned}
\tag{14.55}
$$

These matrix elements are the polynomial coefficients of the vectors in Equation 14.54, which we rewrite as

$$
\mathbf{p}(u) = F_1\mathbf{p}_0 + F_2\mathbf{p}_1 + F_3\mathbf{p}_0^u + F_4\mathbf{p}_1^u
\tag{14.56}
$$

If we assemble the vectors representing the boundary conditions into a matrix $\mathbf{B}$,

$$
\mathbf{B} = \begin{bmatrix} \mathbf{p}_0 & \mathbf{p}_1 & \mathbf{p}_0^u & \mathbf{p}_1^u \end{bmatrix}^{\mathrm{T}}
\tag{14.57}
$$

then

$$
\mathbf{p}(u) = \mathbf{FB}
\tag{14.58}
$$

Here, again, we write the matrix **F** as the product of two matrices, **U** and **M**, so that

$$\mathbf{F} = \mathbf{UM} \tag{14.59}$$

where

$$\mathbf{U} = \begin{bmatrix} u^3 & u^2 & u & 1 \end{bmatrix} \tag{14.60}$$

and

$$\mathbf{M} = \begin{bmatrix} 2 & -2 & 1 & 1 \\ -3 & 3 & -2 & -1 \\ 0 & 0 & 1 & 0 \\ 1 & 0 & 0 & 0 \end{bmatrix} \tag{14.61}$$

Rewriting Equation 14.58 using these substitutions, we obtain

$$\mathbf{p}(u) = \mathbf{UMB} \tag{14.62}$$

It is easy to show that the relationship between the algebraic and geometric coefficients for a space curve is the same form as Equation 14.27 for a plane curve. Since

$$\mathbf{p}(u) = \mathbf{UA} \tag{14.63}$$

the relationship between **A** and **B** is, again,

$$\mathbf{A} = \mathbf{MB} \tag{14.64}$$

Consider the four vectors that make up the boundary condition matrix. There is nothing extraordinary about the vectors defining the end points, but what about the two tangent vectors? A tangent vector certainly defines the slope at one end of the curve, but a vector has characteristics of both direction and magnitude. All we need to specify the slope is a unit tangent vector at each end, say $\mathbf{t}_0$ and $\mathbf{t}_1$. But $\mathbf{p}_0$, $\mathbf{p}_1$, $\mathbf{t}_0$, and $\mathbf{t}_1$ supply only 10 of the 12 pieces of information needed to completely determine the curve. So the magnitude of the tangent vector is also necessary and contributes to the shape of the curve. In fact, we can write $\mathbf{p}_0^u$ and $\mathbf{p}_1^u$ as

$$\mathbf{p}_0^u = m_0 \mathbf{t}_0 \tag{14.65}$$

and

$$\mathbf{p}_1^u = m_1 \mathbf{t}_1 \tag{14.66}$$

Clearly, m_0 and m_1 are the magnitudes of $\mathbf{p}_0^u$ and $\mathbf{p}_1^u$.

Using these relationships, we modify Equation 14.54 as follows:

$$\begin{aligned} \mathbf{p}(u) = {} & (2u^3 - 3u^2 + 1)\mathbf{p}_0 + (-2u^3 + 3u^2)\mathbf{p}_1 \\ & + (u^3 - 2u^2 + u)m_0 \mathbf{t}_0 + (u^3 - u^2)m_1 \mathbf{t}_1 \end{aligned} \tag{14.67}$$

Now we can experiment with a curve (Figure 14.6). Let's hold $\mathbf{p}_0$, $\mathbf{p}_1$, $\mathbf{t}_0$, and $\mathbf{t}_1$ constant and see what happens to the shape of the curve as we vary m_0 and m_1. For simplicity we will consider a curve in the x, y plane. This means that z_0, z_1, z_0^u, and z_1^u

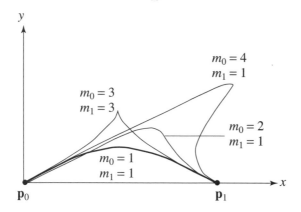

Figure 14.6 *The effect of tangent vector magnitude on curve shape.*

are all equal to zero. The **B** matrix for the curve drawn with the bold line (and with $m_0 = m_1 = 1$) is

$$
\mathbf{B} = \begin{bmatrix} \mathbf{p}_0 \\ \mathbf{p}_1 \\ m_0\mathbf{t}_0 \\ m_1\mathbf{t}_1 \end{bmatrix} = \begin{bmatrix} 0 & 0 & 0 \\ 1 & 0 & 0 \\ 0.707 & 0.707 & 0 \\ 0.707 & -0.707 & 0 \end{bmatrix}
\tag{14.68}
$$

Carefully consider this array of 12 elements; they uniquely define the curve. By changing either m_0 or m_1, or both, we can change the shape of the curve. But it is a restricted kind of change, because not only do the end points remain fixed, but the end slopes are also unchanged!

The three curves drawn with light lines in Figure 14.6 show the effects of varying m_0 and m_1. This is a very powerful tool for designing curves, making it possible to join up end to end many curves in a smooth way and still exert some control over the interior shape of each individual curve. For example, as we increase the value of m_0 while holding m_1 fixed, the curve seems to be pushed toward $\mathbf{p}_1$. Keeping m_0 and m_1 equal but increasing their value increases the maximum deflection of the curve from the x axis and increases the curvature at the maximum. (Under some conditions, not necessarily desirable, we can force a loop to form.)

14.5 Blending Functions

The elements of the blending function matrix **F** in Equation 14.55 apply to all parametric cubic curves defined by ends points and tangent vectors at $u = 0$ and $u = 1$. We discussed other blending functions that apply to parametric cubic curves defined by four points. These are the elements of the matrix **G** (Equation 14.36). In fact we can design just about any kind of blending functions, although they may not have many desirable properties.

What blending functions do is "blend" the effects of given geometric constraints, or boundary conditions. Thus, **F** blends the contributions of $\mathbf{p}_0$, $\mathbf{p}_1$, $\mathbf{p}_0^u$, and $\mathbf{p}_1^u$ to create each point on a curve, and **G** blends the contributions of $\mathbf{p}_1$, $\mathbf{p}_2$, $\mathbf{p}_3$, and $\mathbf{p}_4$. The

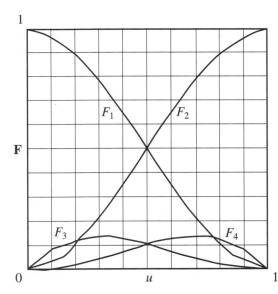

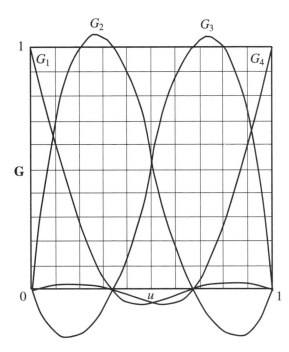

Figure 14.7 F *blending functions.*

graphs of F_1, F_2, F_3, and F_4 (Figure 14.7) reveal the mirror-image symmetry between F_1 and F_2, about the line $u = 0.5$. This is also true for F_3 and F_4. We expect this, because there is nothing intrinsically unique about F_1 with respect to F_2, nor about F_3 with respect to F_4. The end point $\mathbf{p}_0$ dominates the shape of the curve for low values of u, through the effect of F_1, while point $\mathbf{p}_1$ acting through F_2 has the greatest influence for values of u near 1.

Next, we consider the graphs of G_1, G_2, G_3, and G_4 (Figure 14.8). Clearly, G_1 and G_4 are symmetrical, as are G_2 and G_3. Note that at $u = 0$, $G_1 = 1$ and G_2, G_3, and

Figure 14.8 G *blending functions.*

G_4 equal zero; at $u = \frac{1}{3}$, $G_2 = 1$ and G_1, G_3, and G_4 equal zero; at $u = \frac{2}{3}$, $G_3 = 1$ and G_1, G_2, and G_4 equal zero, and finally at $u = 1$, $G_4 = 1$ and G_1, G_2, and G_3 equal zero.

In each case, the blending functions must have certain properties. These properties are determined primarily by the type of boundary conditions we use to define a curve, and how we may want to alter and control the shape of the curve.

14.6 Approximating a Conic Curve

It is usually possible to substitute a cubic Hermite curve for many other kinds of curves. For example, let us try the conic curves: hyperbola, parabola, ellipse, and circle. Given three points, $\mathbf{p}_0$, $\mathbf{p}_1$, and $\mathbf{p}_2$, there is a conic curve whose tangents at $\mathbf{p}_0$ and $\mathbf{p}_1$ lie along $\mathbf{p}_2 - \mathbf{p}_0$ and $\mathbf{p}_1 - \mathbf{p}_2$, respectively (Figure 14.9). The conic is also tangent to a line parallel to $\mathbf{p}_1 - \mathbf{p}_0$ and offset a distance ρH from that same line. The value of ρ determines the type of conic curve, where

$$\text{Hyperbola: } 0.5 < \rho \le 1$$
$$\text{Parabola: } \rho = 0.5 \tag{14.69}$$
$$\text{Ellipse: } 0 \le \rho < 0.5$$

The complete development and proof of this can be found in advanced textbooks.

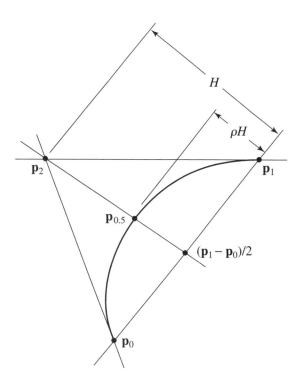

Figure 14.9 *Approximating conic curves.*

The three points $\mathbf{p}_0$, $\mathbf{p}_1$, $\mathbf{p}_2$ and ρ define a cubic Hermite curve that is tangent to the lines mentioned earlier and its equation is

$$\mathbf{p}(u) = \mathbf{F}\begin{bmatrix} \mathbf{p}_0 & \mathbf{p}_1 & 4\rho(\mathbf{p}_2 - \mathbf{p}_0) & 4\rho(\mathbf{p}_1 - \mathbf{p}_2) \end{bmatrix}^{\mathrm{T}} \tag{14.70}$$

It turns out that this equation exactly fits a parabola and produces good approximations to the hyperbola and ellipse. It is interesting to note that the line connecting the points $\mathbf{p}_2$ and $(\mathbf{p}_1 - \mathbf{p}_0)/2$ intersects the curve at exactly $\mathbf{p}_{0.5}$, and that the tangent vector $\mathbf{p}_{0.5}^u$ is tangent to $\mathbf{p}_1 - \mathbf{p}_0$.

14.7 Reparameterization

We can change the parametric interval in such a way that neither the shape nor the position of the curve is changed. A linear function $v = f(u)$ describes this change. For example, sometimes it is useful to reverse the direction of parameterization of a curve. This is the simplest form of reparameterization. It is quite easy to do. In this example, $v = -u$, where v is the new parametric variable.

Figure 14.10 shows two identically shaped cubic Hermite curves. Their only difference is that they have opposite directions of parameterization. This means that

$$\begin{aligned} \mathbf{q}_0 &= \mathbf{p}_1 & \mathbf{q}_0^v &= -\mathbf{p}_1^u \\ \mathbf{q}_1 &= \mathbf{p}_0 & \mathbf{q}_1^v &= -\mathbf{p}_0^u \end{aligned} \tag{14.71}$$

which means we merely interchange $\mathbf{p}_0$ and $\mathbf{p}_1$ and reverse the directions of the tangent vectors.

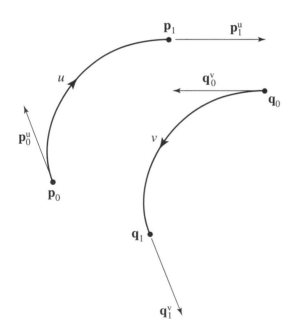

Figure 14.10 *Reversing the direction of parameterization.*

Here is a more general form of reparameterization for cubic Hermite curves. We have a curve that is initially parameterized from u_i to u_j and we must change this so that the parametric variable ranges from v_i to v_j. The initial coefficients are $\mathbf{p}_i$, $\mathbf{p}_j$, $\mathbf{p}_i^u$, and $\mathbf{p}_j^u$, and after reparameterization they are $\mathbf{q}_i$, $\mathbf{q}_j$, $\mathbf{q}_i^v$, and $\mathbf{q}_j^v$.

There is a simple relationship between these sets of coefficients. The end points are related like this: $\mathbf{q}_i = \mathbf{p}_i$ and $\mathbf{q}_j = \mathbf{p}_j$. The tangent vectors require more thought and adjustment. Because they are defined by the first derivative of the parametric basis functions, they are sensitive to the relationship between u and v. A linear relationship is required to preserve the degree of the parametric equations and the directions of the tangent vectors. This means that

$$v = au + b \tag{14.72}$$

Differentiating Equation 14.72, we obtain $dv = a\,du$. Furthermore, we know that $v_i = au_i + b$ and $v_j = au_j + b$, and we can easily solve for a. Then, since

$$\frac{dx}{du} = a\frac{dx}{dv} \tag{14.73}$$

we find that

$$\mathbf{q}^v = \frac{u_j - u_i}{v_j - v_i}\mathbf{p}^u \tag{14.74}$$

Now we are ready to state the complete relationship between the two sets of geometric coefficients:

$$\mathbf{q}_i = \mathbf{p}_i \qquad \mathbf{q}_i^v = \frac{u_j - u_i}{v_j - v_i}\mathbf{p}_i^u$$

$$\mathbf{q}_j = \mathbf{p}_j \qquad \mathbf{q}_j^v = \frac{u_j - u_i}{v_j - v_i}\mathbf{p}_j^u \tag{14.75}$$

This tells us that the tangent vector magnitudes must change to accommodate a change in the range of the parametric variable. The magnitudes are scaled by the ratio of the ranges of the parametric variables. The directions of the tangent vectors and the shape and position of the curve are preserved.

14.8 Continuity and Composite Curves

There are many situations in which a single curve is not versatile enough to model a complex shape, and we must join two or more curves together end to end to achieve a design objective. In most cases, but certainly not all, a smooth transition from one curve to the next is a desirable property. We can do this by making the tangent vectors of adjoining curves collinear. However, it is not necessary that their magnitudes are equal, just their direction.

Figure 14.11 shows two curves, $\mathbf{p}(u)$ and $\mathbf{q}(v)$, with tangent continuity. This imposes certain constraints on the geometric coefficients. First, since $\mathbf{p}_1$ and $\mathbf{q}_0$ must

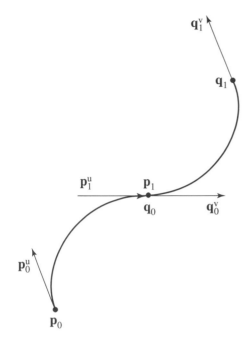

Figure 14.11 *Two curves joined with tangent continuity.*

coincide, we have $\mathbf{q}_0 = \mathbf{p}_1$. Second, the tangent vectors $\mathbf{p}_1^u$ and $\mathbf{q}_0^v$ must be in the same direction, although their magnitudes may differ. This means that $\mathbf{q}_0^v = k\mathbf{p}_1^u$, and the geometric coefficients of $\mathbf{q}(v)$ satisfying these constraints are

$$\mathbf{B}_q = \begin{bmatrix} \mathbf{p}_1 & \mathbf{q}_1 & k\mathbf{p}_1^u & \mathbf{q}_1^v \end{bmatrix}^{\mathrm{T}} \tag{14.76}$$

A composite curve like this has a total of 19 degrees of freedom (compared with 24 for two disjoint curves).

There are various degrees of *parametric continuity* denoted C^n, where n is the degree. C^0 is the minimum degree of continuity between two curves, and indicates that the curves are joined without regard for tangent continuity (i.e., the tangent line is discontinuous at their common point). C^1 indicates first derivative or tangent continuity (discussed earlier), which, of course, presupposes C^0. C^2 indicates second derivative continuity, and is necessary when continuity of curvature at the joint is required. Higher-degree continuity across a joint between two curves is seldom used. There is a related kind of continuity called *geometric continuity*, denoted as G^n, which is not discussed here, but is accessible in more advanced texts.

The notation used in Figure 14.11 and Equation 14.76 is inadequate if more than two or three curves must be used to define a complex curve. A more practical system is suggested here. If n piecewise cubic Hermite curves are joined to form a composite curve of C^1 continuity, we may proceed as follows (Figure 14.12):

1. Label the points consecutively; $\mathbf{p}_1, \mathbf{p}_2, \ldots, \mathbf{p}_i, \ldots, \mathbf{p}_{n-1}, \mathbf{p}_n$.
2. Define unit tangent vectors, $\mathbf{t}_1, \mathbf{t}_2, \ldots, \mathbf{t}_i, \ldots, \mathbf{t}_{n-1}, \mathbf{t}_n$.
3. Define tangent vector magnitudes, $m_{1,0}, m_{1,1}, \ldots, m_{i,0}, m_{i,1}, \ldots, m_{n,0}, m_{n,1}$.

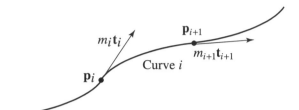

Figure 14.12 *General notation for composite curves.*

Using this notation, the geometric coefficients for curve i are

$$\mathbf{B} = \begin{bmatrix} \mathbf{p}_i & \mathbf{p}_{i+1} & m_{i,0}\mathbf{t}_i & m_{i,1}\mathbf{t}_{i+1} \end{bmatrix}^{\mathrm{T}} \tag{14.77}$$

Exercises

14.1 Find $\mathbf{a}$, $\mathbf{b}$, and $\mathbf{c}$ for each of the curves defined by the following sets of points:
   a. $\mathbf{p}_0 = \begin{bmatrix} 0 & 2 & 2 \end{bmatrix}$, $\mathbf{p}_{0.5} = \begin{bmatrix} 1 & 4 & 0 \end{bmatrix}$, $\mathbf{p}_1 = \begin{bmatrix} 3 & 1 & 6 \end{bmatrix}$
   b. $\mathbf{p}_0 = \begin{bmatrix} -1 & 0 & 4 \end{bmatrix}$, $\mathbf{p}_{0.5} = \begin{bmatrix} 0 & 0 & 0 \end{bmatrix}$, $\mathbf{p}_1 = \begin{bmatrix} 0 & -2 & -2 \end{bmatrix}$
   c. $\mathbf{p}_0 = \begin{bmatrix} -3 & 7 & 1 \end{bmatrix}$, $\mathbf{p}_{0.5} = \begin{bmatrix} 5 & 1 & 4 \end{bmatrix}$, $\mathbf{p}_1 = \begin{bmatrix} 6 & 0 & 0 \end{bmatrix}$
   d. $\mathbf{p}_0 = \begin{bmatrix} 7 & 7 & 8 \end{bmatrix}$, $\mathbf{p}_{0.5} = \begin{bmatrix} 2 & 0 & 3 \end{bmatrix}$, $\mathbf{p}_1 = \begin{bmatrix} 2 & -4 & 1 \end{bmatrix}$
   e. $\mathbf{p}_0 = \begin{bmatrix} 0 & -1 & 2 \end{bmatrix}$, $\mathbf{p}_{0.5} = \begin{bmatrix} -1 & -3 & 7 \end{bmatrix}$, $\mathbf{p}_1 = \begin{bmatrix} 0 & 5 & 2 \end{bmatrix}$

14.2 Find $\mathbf{p}_0$, $\mathbf{p}_{0.5}$, and $\mathbf{p}_1$ for each of the curves defined by the following sets of algebraic vectors:
   a. $\mathbf{a} = \begin{bmatrix} 1 & 0 & 0 \end{bmatrix}$, $\mathbf{b} = \begin{bmatrix} -3 & -3 & 0 \end{bmatrix}$, $\mathbf{c} = \begin{bmatrix} 3 & 0 & 0 \end{bmatrix}$
   b. $\mathbf{a} = \begin{bmatrix} 6 & 9 & 8 \end{bmatrix}$, $\mathbf{b} = \begin{bmatrix} -8 & -2 & 4 \end{bmatrix}$, $\mathbf{c} = \begin{bmatrix} -4 & 6 & 1 \end{bmatrix}$
   c. $\mathbf{a} = \begin{bmatrix} 8 & 1 & -1 \end{bmatrix}$, $\mathbf{b} = \begin{bmatrix} 5 & 4 & -5 \end{bmatrix}$, $\mathbf{c} = \begin{bmatrix} -10 & 4 & -3 \end{bmatrix}$
   d. $\mathbf{a} = \begin{bmatrix} 10 & 6 & 6 \end{bmatrix}$, $\mathbf{b} = \begin{bmatrix} -15 & -17 & -13 \end{bmatrix}$, $\mathbf{c} = \begin{bmatrix} 7 & 7 & 8 \end{bmatrix}$
   e. $\mathbf{a} = \begin{bmatrix} -2 & -4 & 4 \end{bmatrix}$, $\mathbf{b} = \begin{bmatrix} 3 & 2 & -10 \end{bmatrix}$, $\mathbf{c} = \begin{bmatrix} -1 & 0 & 4 \end{bmatrix}$

14.3 What are the dimensions of the matrices in Equation 14.25? Verify the dimensions of the product.

14.4 Find $\mathbf{A}$ when $\mathbf{P} = \begin{bmatrix} 0 & 1 & 1 \\ 3 & -2 & 0 \\ 2 & 5 & -4 \end{bmatrix}$

14.5 Find $\mathbf{M}^{-1}$

14.6 Describe the curve that results if $\mathbf{p}_0$, $\mathbf{p}_{0.5}$, and $\mathbf{p}_1$ are collinear.

14.7 Give the general geometric coefficients of a curve that lies in the x, y plane.

14.8 Give the general geometric coefficients of a curve that lies in the $y = -3$ plane.

14.9 Compute G_1, G_2, G_3, and G_4 at $u = 0$.

14.10 Compute G_1, G_2, G_3, and G_4 at $u = 1$.

14.11 Compute G_1, G_2, G_3, and G_4 at $u = \frac{1}{3}$.

14.12 Compute G_1, G_2, G_3, and G_4 at $u = \frac{2}{3}$.

14.13 What general conditions must be imposed on the four control points p_1, p_2, p_3, and p_4 to produce a curve that lies in the x, y plane?

14.14 Compute dy/dx for the following functions:
   a. $y = 4x^2$
   b. $y = x + 3$
   c. $y = x^3 - 3x + 1$
   d. $y = x^2 + 2x + 1$
   e. $y = 2x^4 + x^3 + 3$

14.15 Find the coordinates of the point of zero slope for each of the curves defined in Exercise 14.14.

14.16 Compute m_0 and t_0 for the following tangent vectors:
   a. $p_0^u = \begin{bmatrix} 3 & -1 & 6 \end{bmatrix}$
   b. $p_0^u = \begin{bmatrix} 0 & 2 & 0 \end{bmatrix}$
   c. $p_0^u = \begin{bmatrix} 1 & 5 & -1 \end{bmatrix}$
   d. $p_0^u = \begin{bmatrix} 7 & 2 & 0 \end{bmatrix}$
   e. $p_0^u = \begin{bmatrix} 4 & 4 & -3 \end{bmatrix}$

14.17 Find d^2y/dx^2 for the functions given in Exercise 14.14.

14.18 Given two disjoint (unconnected) curves, $p(u)$ and $q(u)$, join p_1 to q_0 with a curve $r(u)$ such that there is C^1 continuity across the two joints. Write the geometric coefficients B_r in terms of the coefficients of $p(u)$ and $q(u)$.

14.19 How many degrees of freedom (unique coefficients) are required to define the system of three curves created in Exercise 14.18?

14.20 Use the results of Exercise 14.18 to construct a closed composite curve with C^1 continuity by joining q_1 to p_0 with a curve $s(u)$. Write the geometric coefficients B_s in terms of the coefficients of $p(u)$ and $q(u)$.

14.21 How many degrees of freedom does the closed composite curve of Exercise 14.20 have?

CHAPTER 15

THE BÉZIER CURVE

The Bézier curve is an important part of almost every computer-graphics illustration program and computer-aided design system in use today. It is used in many ways, from designing the curves and surfaces of automobiles to defining the shape of letters in type fonts. And because it is numerically the most stable of all the polynomial-based curves used in these applications, the Bézier curve is the ideal standard for representing the more complex piecewise polynomial curves.

In the early 1960s, Peter Bézier (pronounced *bay-zee-aye*) began looking for a better way to define curves and surfaces, one that would be useful to a design engineer. He was familiar with the work of Ferguson and Coons and their parametric cubic curves and bicubic surfaces. However, these did not offer an intuitive way to alter and control shape. The results of Bézier's research led to the curves and surfaces that bear his name and became part of the UNISURF system. The French automobile manufacturer, Renault, used UNISURF to design the sculptured surfaces of many of its products.

This chapter begins by describing a surprisingly simple geometric construction of a Bézier curve, followed by a derivation of its algebraic definition, basis functions, control points, degree elevation, and truncation. It concludes by showing how to join two curves end-to-end to form a single composite curve.

15.1 A Geometric Construction

We can draw a Bézier curve using a simple recursive geometric construction. Let's begin by constructing a second-degree curve (Figure 15.1). We select three points A, B, C, so that line AB is tangent to the curve at A, and line BC is tangent at C. The curve begins at A and ends at C. For any ratio u_i, where $0 \leq u_i \leq 1$, we construct points D and E so that

$$\frac{AD}{AB} = \frac{BE}{BC} = u_i \tag{15.1}$$

On DE we construct F so that $DF/DE = u_i$. Point F is on the curve.

Repeating this process for other values of u_i, we produce a series of points on a Bézier curve. Note that we must be consistent in the order in which we sub-divide AB and BC. For example, $AD/AB \neq EC/BC$.

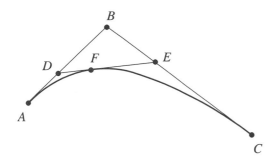

Figure 15.1 *Geometric construction of a second-degree Bézier curve.*

To define this curve in a coordinate system, let point $A = x_A, y_A$, $B = x_B, y_B$, and $C = x_C, y_C$. Then coordinates of points D and E for some value of u_i are

$$x_D = x_A + u_i(x_B - x_A)$$
$$y_D = y_A + u_i(y_B - y_A)$$

(15.2)

and

$$x_E = x_B + u_i(x_C - x_B)$$
$$y_E = y_B + u_i(y_C - y_B)$$

(15.3)

The coordinates of point F for some value of u_i are

$$x_F = x_D + u_i(x_E - x_D)$$
$$y_F = y_D + u_i(y_E - y_D)$$

(15.4)

To obtain x_F and y_F in terms of the coordinates of points A, B, and C, for any value of u_i in the unit interval, we substitute appropriately from Equations 15.2 and 15.3 into Equations 15.4. After rearranging terms to simplify, we find

$$x_F = (1 - u_i)^2 x_A + 2u_i(1 - u_i)x_B + u_i^2 x_C$$
$$y_F = (1 - u_i)^2 y_A + 2u_i(1 - u_i)y_B + u_i^2 y_C$$

(15.5)

We generalize this set of equations for any point on the curve using the following substitutions:

$$x(u) = x_F$$
$$y(u) = y_F$$

(15.6)

and we let

$$x_0 = x_A \quad x_1 = x_B \quad x_2 = x_C$$
$$y_0 = y_A \quad y_1 = y_B \quad y_2 = y_C$$

(15.7)

Now we can rewrite Equation 15.5 as

$$x(u) = (1 - u)^2 x_0 + 2u(1 - u)x_1 + u^2 x_2$$
$$y(u) = (1 - u)^2 y_0 + 2u(1 - u)y_1 + u^2 y_2$$

(15.8)

This is the set of second-degree equations for the coordinates of points on a Bézier curve, based on our construction.

We express this construction process and Equations 15.8 in terms of vectors with the following substitutions: Let the vector $\mathbf{p}_0$ represent point A, $\mathbf{p}_1$ point B, and $\mathbf{p}_2$ point C. From vector geometry we have $D \equiv \mathbf{p}_0 + u(\mathbf{p}_1 - \mathbf{p}_0)$, and $E \equiv \mathbf{p}_1 + u(\mathbf{p}_2 - \mathbf{p}_1)$. If we let $F \equiv \mathbf{p}(u)$, we see that

$$\mathbf{p}(u) = \mathbf{p}_0 + u(\mathbf{p}_1 - \mathbf{p}_0) + u[\mathbf{p}_1 + u(\mathbf{p}_2 - \mathbf{p}_1) - \mathbf{p}_0 + u(\mathbf{p}_1 - \mathbf{p}_0)] \tag{15.9}$$

We rearrange terms to obtain a more compact vector equation of a second-degree Bézier curve:

$$\mathbf{p}(u) = (1 - u)^2\mathbf{p}_0 + 2u(1 - u)\mathbf{p}_1 + u^2\mathbf{p}_2 \tag{15.10}$$

where the points $\mathbf{p}_0$, $\mathbf{p}_1$, and $\mathbf{p}_2$ are called *control points*, and $0 \le u \le 1$. The ratio u is the *parametric variable*. Later, we will see that this equation is an example of a *Bernstein polynomial*. Note that the curve will always lie in the plane containing the three control points, but the points do not necessarily lie in the xy plane.

Similar constructions apply to Bézier curves of any degree. In fact the degree of a Bézier curve is equal to $n - 1$, where n is the number of control points.

Figure 15.2 shows the construction of a point on a cubic Bézier curve, which requires four control points A, B, C, and D to define it. The curve begins at point A tangent to line AB, and ends at D and tangent to CD. We construct points E, F, and G so that

$$\frac{AE}{AB} = \frac{BF}{BC} = \frac{CG}{CD} = u_i \tag{15.11}$$

On EF and FG we locate H and I, respectively, so that

$$\frac{EH}{EF} = \frac{FI}{FG} = u_i \tag{15.12}$$

Finally, on HI we locate J so that

$$\frac{HJ}{HI} = u_i \tag{15.13}$$

We can make no more subdivisions, which means that point J is on the curve. If we continue this process for a sequences of points, then their locus defines the curve.

If points A, B, C, and D are represented by the vectors $\mathbf{p}_0$, $\mathbf{p}_1$, $\mathbf{p}_2$, and $\mathbf{p}_3$, respectively, then expressing the construction of the intermediate points E, F, G, H, and

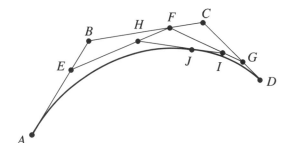

Figure 15.2 *Geometric construction of a cubic Bézier curve.*

I in terms of these vectors to produce point *J*, or $\mathbf{p}(u)$, yields

$$\mathbf{p}(u) = \mathbf{p}_0 + u(\mathbf{p}_1 - \mathbf{p}_0) + u[\mathbf{p}_1 + u(\mathbf{p}_2 - \mathbf{p}_1) - \mathbf{p}_0 - u(\mathbf{p}_1 - \mathbf{p}_0)]$$
$$+ u\{\mathbf{p}_1 + u(\mathbf{p}_2 - \mathbf{p}_1) + u[\mathbf{p}_2 + u(\mathbf{p}_3 - \mathbf{p}_2) - \mathbf{p}_1 - u(\mathbf{p}_2 - \mathbf{p}_0)] - \mathbf{p}_0 \quad (15.14)$$
$$- u(\mathbf{p}_1 - \mathbf{p}_0) - u[\mathbf{p}_1 + u(\mathbf{p}_2 - \mathbf{p}_1) - \mathbf{p}_0 - u(\mathbf{p}_1 - \mathbf{p}_0)]\}$$

This awkward expression simplifies nicely to

$$\mathbf{p}(u) = (1 - u)^3\mathbf{p}_0 + 3u(1 - u)^2\mathbf{p}_1 + 3u^2(1 - u)\mathbf{p}_2 + u^3\mathbf{p}_3 \quad (15.15)$$

Of course, this construction of a cubic curve with its four control points is done in the plane of the paper. However, the cubic polynomial allows a curve that is nonplanar; that is, it can represent a curve that twists in space.

The geometric construction of a Bézier curve shows how the control points influence its shape. The curve begins on the first point and ends on the last point. It is tangent to the lines connecting the first two points and the last two points. The curve is always contained within the *convex hull* of the control points.

No one spends time constructing and plotting the points of a Bézier curve by hand, of course. A computer does a much faster and more accurate job. However, it is worth doing several curves this way for insight into the characteristics of Bézier curves.

15.2 An Algebraic Definition

Bézier began with the idea that any point $\mathbf{p}(u)$ on a curve segment should be given by an equation such as the following:

$$\mathbf{p}(u) = \sum_{i=0}^{n} \mathbf{p}_i f_i(u) \quad (15.16)$$

where $0 \leq u \leq 1$, and the vectors $\mathbf{p}_i$ are the control points (Figure 15.3).

Equation 15.16 is a compact way to express the sum of several similar terms, because what it says is this:

$$\mathbf{p}(u) = \mathbf{p}_0 f_0(u) + \mathbf{p}_1 f_1(u) + \cdots + \mathbf{p}_n f_n(u) \quad (15.17)$$

of which Equation 15.10 and 15.15 are specific examples, for $n = 2$ and $n = 3$, respectively.

The $n + 1$ functions, that is the $f_i(u)$, must produce a curve that has certain well-defined characteristics. Here are some of the most important ones:

1. The curve must start on the first control point, $\mathbf{p}_0$, and end on the last, $\mathbf{p}_n$. Mathematically, we say that the functions must interpolate these two points.
2. The curve must be tangent to the line given by $\mathbf{p}_1 - \mathbf{p}_0$ at $\mathbf{p}_0$ and to $\mathbf{p}_n - \mathbf{p}_{n-1}$ at $\mathbf{p}_n$.
3. The functions $f_i(u)$ must be symmetric with respect to u and $(1 - u)$. This lets us reverse the sequence of control points without changing the shape of the curve.

Other characteristics can be found in more advanced works on this subject.

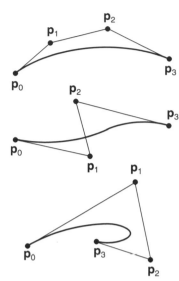

Figure 15.3 *Bézier curves and their control points.*

A family of functions called *Bernstein polynomials* satisfies these requirements. They are the *basis functions* of the Bézier curve. (Other curves, such as the *NURBS curves*, use different, but related, basis functions.) We rewrite Equation 15.16 using them, so that

$$\mathbf{p}(u) = \sum_{i=0}^{n} \mathbf{p}_i B_{i,n}(u) \tag{15.18}$$

where the basis functions are

$$B_{i,n}(u) = \binom{n}{i} u^i (1-u)^{n-i} \tag{15.19}$$

The term $\binom{n}{i}$ is the binomial coefficient function from probability theory and statistics, defined as

$$\binom{n}{i} = \frac{n!}{i!(n-1)!} \tag{15.20}$$

The symbol ! is the factorial operator. For example, $3! = 3 \times 2 \times 1, 5! = 5 \times 4 \times 3 \times 2 \times 1$, and so forth. We use the following conventions when evaluating Equation 15.20: If i and u equal zero, then $u^i = 1$ and $0! = 1$. We see that for $n+1$ control points, the basis functions produce an nth-degree polynomial.

Expanding Equation 15.18 for a second-degree Bézier curve (where $n = 2$ and there are three control points) produces

$$\mathbf{p}(u) = \mathbf{p}_0 B_{0,2}(u) + \mathbf{p}_1 B_{1,2}(u) + \mathbf{p}_2 B_{2,2}(u) \tag{15.21}$$

From Equation 15.20 we find

$$B_{0,2}(u) = (1 - u)^2 \tag{15.22}$$

$$B_{1,2}(u) = 2u(1 - u) \tag{15.23}$$

$$B_{2,2}(u) = u^2 \tag{15.24}$$

These are the basis functions for a second-degree Bézier curve. Substituting them into Equation 15.21 and rearranging terms, we find

$$\mathbf{p}(u) = (1 - u)^2 \mathbf{p}_0 + 2u(1 - u)\mathbf{p}_1 + u^2 \mathbf{p}_2 \tag{15.25}$$

This is the same expression we found from the geometric construction, Equation 15.10. The variable u is now called the *parametric variable*.

Now, let's expand Equation 15.18 for a cubic Bézier curve, where $n = 3$:

$$\mathbf{p}(u) = \mathbf{p}_0 B_{0,3}(u) + \mathbf{p}_1 B_{1,3}(u) + \mathbf{p}_2 B_{2,3}(u) + \mathbf{p}_3 B_{3,3}(u) \tag{15.26}$$

and from Equation 15.20 we find

$$B_{0,3}(u) = (1 - u)^3 \tag{15.27}$$

$$B_{1,3}(u) = 3u(1 - u)^2 \tag{15.28}$$

$$B_{2,3}(u) = 3u^2(1 - u) \tag{15.29}$$

$$B_{3,3}(u) = u^3 \tag{15.30}$$

Substituting these into Equation 15.26 and rearranging terms produces

$$\mathbf{p}(u) = (1 - u)^3 \mathbf{p}_0 + 3u(1 - u)^2 \mathbf{p}_1 + 3u^2(1 - u)\mathbf{p}_2 + u^3 \mathbf{p}_3 \tag{15.31}$$

Bézier curve equations are well suited for expression in matrix form. We can expand the cubic parametric functions and rewrite Equation 15.31 as

$$\mathbf{p}(u) = \begin{bmatrix} (1 - 3u + 3u^2 - u^3) \\ (3u - 6u^2 + 3u^3) \\ (3u^2 - 3u^3) \\ u^3 \end{bmatrix}^T \begin{bmatrix} \mathbf{p}_0 \\ \mathbf{p}_1 \\ \mathbf{p}_2 \\ \mathbf{p}_3 \end{bmatrix} \tag{15.32}$$

or as

$$\mathbf{p}(u) = \begin{bmatrix} u^3 & u^2 & u & 1 \end{bmatrix} \begin{bmatrix} -1 & 3 & -3 & 1 \\ 3 & -6 & 3 & 0 \\ -3 & 3 & 0 & 0 \\ 1 & 0 & 0 & 0 \end{bmatrix} \begin{bmatrix} \mathbf{p}_0 \\ \mathbf{p}_1 \\ \mathbf{p}_2 \\ \mathbf{p}_3 \end{bmatrix} \tag{15.33}$$

If we let

$$\mathbf{U} = \begin{bmatrix} u^3 & u^2 & u & 1 \end{bmatrix} \tag{15.34}$$

$$\mathbf{P} = \begin{bmatrix} \mathbf{p}_0 & \mathbf{p}_1 & \mathbf{p}_2 & \mathbf{p}_3 \end{bmatrix}^T \tag{15.35}$$

and

$$\mathbf{M} = \begin{bmatrix} -1 & 3 & -3 & 1 \\ 3 & -6 & 3 & 0 \\ -3 & 3 & 0 & 0 \\ 1 & 0 & 0 & 0 \end{bmatrix} \tag{15.36}$$

then we can write Equation 15.33 even more compactly as

$$\mathbf{p}(u) = \mathbf{UMP} \tag{15.37}$$

Note that the composition of the matrices **U**, **M**, and **P** varies according to the number of control points (that is, the degree of the Bernstein polynomial basis functions).

15.3 Control Points

In Equation 15.18, we see that the control points are coefficients of the Bernstein polynomial basis functions. Connecting the control points in sequence with straight lines yields the Bézier *control polygon*, and the curve lies entirely within its convex hull. The control polygon establishes the initial shape of a curve, and also crudely approximates this shape. It also gives us a way to change a curve's shape. Again, notice that $n + 1$ control points produce an nth-degree curve.

Figure 15.4 shows how different sequences of three identical sets of points affect the shape of the curve. The order in Figure 15.4a is reversed in Figure 15.4b and curve shape was not affected. Reversing the control point sequence reverses the direction of parameterization. In Figure 15.4c, the sequence is more dramatically altered, creating an altogether different curve. Note, though, that the curve is always tangent to the first and last edges of the control polygon.

Studying the basis functions helps us to understand curve behavior. Figure 15.5 shows basis function plots over the unit interval for cubic Bézier curves. The contribution of the first control point, $\mathbf{p}_0$, is propagated throughout the curve by $B_{0,3}$, and it is most influential at $u = 0$. The other control points do not contribute to $\mathbf{p}(u)$ at $u = 0$, because

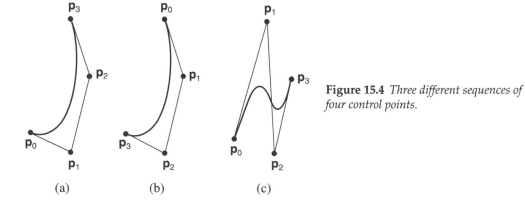

(a)   (b)   (c)

Figure 15.4 *Three different sequences of four control points.*

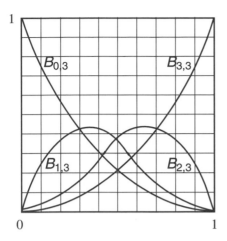

Figure 15.5 *Bézier curve basis functions.*

$B_{0,3}(u) = 1$ and $B_{1,3}(0) = B_{2,3}(0) = B_{3,3}(0) = 0$. Control point $\mathbf{p}_1$ is most influential at $u = 1/3$, and $\mathbf{p}_2$ at $u = 2/3$. At $u = 1$, only $\mathbf{p}_3$ affects $\mathbf{p}(u)$. Note the symmetry of $B_{0,1}(u)$ and $B_{3,3}(u)$, as well as that of $B_{1,3}(u)$ and $B_{2,3}(u)$.

We see that the effect of any control point is weighted by its associated basis function. This means that if we change the position of a control point $\mathbf{p}_i$, the greatest effect on the curve shape is at or near the parameter value $u = i/n$.

Figure 15.6 shows two examples of how we can modify the shape of a Bézier curve. In Figure 15.6a, moving $\mathbf{p}_2$ to $\mathbf{p}_2'$ pulls the curve toward that point. In Figure 15.6b, we add one or two extra control points at $\mathbf{p}_1$. The multiply-coincident points pull the curve closer and closer to that vertex. In this case, each additional control point raises the degree of the basis function polynomials.

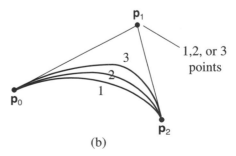

(a)

Figure 15.6 *Modifying the shape of a Bézier curve.*

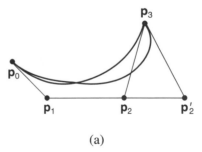

(b)

We asserted previously that we could traverse a set of control points in either direction without affecting shape. They could be ordered $\mathbf{p}_0, \mathbf{p}_1, \ldots, \mathbf{p}_n$ or $\mathbf{p}_n, \mathbf{p}_{n-1}, \ldots, \mathbf{p}_0$. The curve is the same, and only the direction of parameterization is reversed. We express this equivalence as

$$\sum_{i=0}^{n} \mathbf{p}_i\, B_{i,n}(u) = \sum_{i=0}^{n} \mathbf{p}_{n-i}\, B_{i,n}(1-u) \tag{15.38}$$

which follows from the basis function identity

$$B_{i,n}(u) = B_{n-i,n}(1-u) \tag{15.39}$$

15.4 Degree Elevation

Each control point we add to the definition of a Bézier curve raises its degree by one. We might choose to do this if we are not satisfied with a curve's original shape or the possible shapes available to us by moving any of the original control points. Usually it is advisable to add another point in a way that does not initially change the shape of the curve (Figure 15.7). After we have added a point, we can move it or any of the other control points to change the shape of the curve. If a new set of control points $^1\mathbf{p}_i$ generates the same curve as the original set $\mathbf{p}_i$, then it follows that

$$\sum_{i=0}^{n+1} {}^1\mathbf{p}_i\, B_{i,n+1}(u) = \sum_{i=0}^{n} \mathbf{p}_i\, B_{i,n}(u) \tag{15.40}$$

or

$$\sum_{i=0}^{n+1} {}^1\mathbf{p}_i \binom{n+1}{i} u^i (1-u)^{n+1-i} = \sum_{i=0}^{n} \mathbf{p}_i \binom{n}{i} u^i (1-u)^{n-i} \tag{15.41}$$

Equation 15.41 is the result of substituting Equation 15.19 into Equation 15.18. If we multiply the right side of Equation 15.41 by $u + (1-u)$, we obtain

$$\sum_{i=0}^{n+1} {}^1\mathbf{p}_i \binom{n+1}{i} u^i (1-u)^{n+1-i} = \sum_{i=0}^{n} \mathbf{p}_i \binom{n}{i} [u^i (1-u)^{n+1-i} + u^{i+1}(1-u)^{n-i}] \tag{15.42}$$

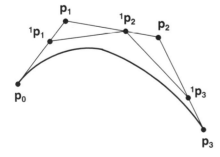

Figure 15.7 *Degree elevation: adding a control point that does not initially change the shape of the curve.*

Summing the right side of this equation produces $2(n + 1)$ terms, and summing the left side produces $n + 2$ terms. By rearranging and grouping terms on the right side so that we can compare and equate the coefficients of $u^i (1 - u)^{n+1-i}$ on both sides, we can write

$$^1\mathbf{p}_i \binom{n+1}{i} = \mathbf{p}_i \binom{n}{i} + \mathbf{p}_{i-1} \binom{n}{i-1} \tag{15.43}$$

Next, we expand the binomial coefficient terms and simplify, to obtain

$$^1\mathbf{p}_i = \left(\frac{i}{n+1}\right) \mathbf{p}_{i-1} + \left(1 - \frac{i}{n+1}\right) \mathbf{p}_i \tag{15.44}$$

for $i = 0, 1, \ldots, n + 1$.

Equation 15.44 says that we can compute a new set of control points $^1\mathbf{p}_i$ from the original points. Figure 15.7 shows what happens when we add a point to a cubic Bézier curve. Note that the new interior points fall on the sides of the original control polygon. We can repeat this process until we have added enough control points to satisfactorily control the curve's shape.

15.5 Truncation

If we want to retain only a part of a Bézier curve, for example, the curve segment between u_i and u_j, then we must truncate the segments from $u = 0$ to $u = u_i$ and from $u = u_j$ to $u = 1$ (Figure 15.8). We must find the set of control points that defines the segment that remains. To do this we must change the parametric variable so that it again varies over the unit interval, instead of over $u_i \leq u \leq u_j$, as in the original curve. We introduce a new parameter v, where

$$v = au + b \tag{15.45}$$

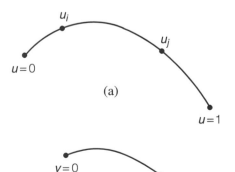

(a)

(b)

Figure 15.8 *Truncating a Bézier curve.*

This transformation applies to parametric polynomial equations of any degree. The linear relationship between u and v (Equation 15.45) preserves the degree of the polynomial.

If $v = 0$ at $u = -u_i$, and $v = 1$ at $u = u_j$, then

$$u = (u_j - u_i)v + u_i \tag{15.46}$$

or

$$u = \Delta u_i v + u_i \tag{15.47}$$

where

$$\Delta u_i = u_j - u_i \tag{15.48}$$

The general transformation equation for the parametric variable (not derived here) is

$$u^n = \sum_{k=0}^{n} \frac{n!}{k!(n-k)!} u_i^k (\Delta u_i v)^{n-k} \tag{15.49}$$

This equation looks more formidable than it is. From it we extract a transformation matrix **T** such that

$$[u^n \quad u^{n-1} \quad \cdots \quad u \quad 1] = [v^n \quad v^{n-1} \quad \cdots \quad v \quad 1]\mathbf{T} \tag{15.50}$$

or

$$\mathbf{U} = \mathbf{VT} \tag{15.51}$$

Here is an example: Given the three control points that define a second-degree Bézier curve, we can find three new control points that define the segment of the curve from u_i to u_j in terms of a new parameter v, which spans the unit interval, $0 \le v \le 1$. The control points that define the original curve are (in matrix form)

$$\mathbf{P} = \begin{bmatrix} \mathbf{p}_0 \\ \mathbf{p}_1 \\ \mathbf{p}_2 \end{bmatrix} \tag{15.52}$$

and the control points that define the truncated segment are

$$\mathbf{P}' = \begin{bmatrix} \mathbf{p}'_0 \\ \mathbf{p}'_1 \\ \mathbf{p}'_2 \end{bmatrix} \tag{15.53}$$

So for a second-degree Bézier curve, we have (similar to the matrix Equation 15.37 for a cubic curve)

$$\mathbf{p}(u) = \mathbf{UMP} \tag{15.54}$$

where $\mathbf{U} = [u^2 \quad u \quad 1]$ and

$$\mathbf{M} = \begin{bmatrix} 1 & -2 & 1 \\ -2 & 2 & 0 \\ 1 & 0 & 0 \end{bmatrix} \tag{15.55}$$

Because $\mathbf{U} = \mathbf{VT}$, from Equation 15.50, we can write

$$\mathbf{p}(v) = \mathbf{VTMP} = \mathbf{VMP'} \tag{15.56}$$

This means that

$$\mathbf{MP'} = \mathbf{TMP} \tag{15.57}$$

or

$$\mathbf{P'} = \mathbf{M}^{-1}\mathbf{TMP} \tag{15.58}$$

where $\mathbf{V} = [v^2 \quad v \quad 1]$, and from Equation 15.49 for $n = 2$

$$\mathbf{T} = \begin{bmatrix} \Delta u_i^2 & 0 & 0 \\ 2u_i \Delta u_i & \Delta u_i & 0 \\ u_i^2 & u_i & 1 \end{bmatrix} \tag{15.59}$$

where the inverse of $\mathbf{M}$ is

$$\mathbf{M}^{-1} = \begin{bmatrix} 0 & 0 & 1 \\ 0 & \frac{1}{2} & 1 \\ 1 & 1 & 1 \end{bmatrix} \tag{15.60}$$

Substituting appropriately into Equation 15.58 yields the new control points

$$\mathbf{p}_0' = (1 - u_i)^2 \mathbf{p}_0 + 2u_i(1 - u_i)\mathbf{p}_1 + u_i^2 \mathbf{p}_2 \tag{15.61}$$

$$\mathbf{p}_1' = (1 - u_i)(1 - u_j)\mathbf{p}_0 + (-2u_iu_j + u_j + u_i)\mathbf{p}_1 + u_iu_j\mathbf{p}_2 \tag{15.62}$$

$$\mathbf{p}_2' = (1 - u_j)^2 \mathbf{p}_0 + 2u_j(1 - u_j)\mathbf{p}_1 + u_j^2 \mathbf{p}_2 \tag{15.63}$$

15.6 Composite Bézier Curves

We can join two or more Bézier curves together, end-to-end, to create longer, more complex curves. These curves are called *composite curves*. We usually want a smooth transition from one curve to the next. One way to do this is to make sure that the tangent lines of the two curves meeting at a point are collinear. In Figure 15.9 a second-degree curve defined by control points $\mathbf{p}_0$, $\mathbf{p}_1$, and $\mathbf{p}_2$ smoothly blends with a cubic curve defined by control points $\mathbf{q}_0$, $\mathbf{q}_1$, $\mathbf{q}_2$, and $\mathbf{q}_3$, because the control points $\mathbf{p}_1$, $\mathbf{p}_2$, $\mathbf{q}_0$, and $\mathbf{q}_1$ are collinear, and $\mathbf{p}_2 = \mathbf{q}_0$.

If we differentiate Equation 15.15 with respect to u and rearrange terms, we find

$$\frac{d\mathbf{p}(u)}{du} = (-3 + 6u - 3u^2)\mathbf{p}_0 + (3 - 12u + 9u^2)\mathbf{p}_1 + (6u - 9u^2)\mathbf{p}_2 + 3u^2\mathbf{p}_3 \tag{15.64}$$

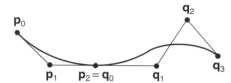

Figure 15.9 *Joining two Bézier curves.*

Evaluating this equation at $u = 0$ and $u = 1$ produces

$$\frac{d\mathbf{p}(0)}{du} = 3(\mathbf{p}_1 - \mathbf{p}_0) \tag{15.65}$$

and

$$\frac{d\mathbf{p}(1)}{du} = 3(\mathbf{p}_3 - \mathbf{p}_2) \tag{15.66}$$

This tells us that the tangent line at $u = 0$ is indeed determined by the vector $\mathbf{p}_1 - \mathbf{p}_0$, and at $u = 1$ by $\mathbf{p}_3 - \mathbf{p}_2$. These are, of course, the first and last edges of the control polygon for the cubic Bézier curve. We can obtain similar results for other Bézier curves.

Exercises

15.1 Construct enough points on the Bézier curve, whose control points are $\mathbf{p}_0 = (4, 2)$, $\mathbf{p}_1 = (8, 8)$, and $\mathbf{p}_2 = (16, 4)$, to draw an accurate sketch.
  a. What degree is the curve?
  b. What are the coordinates at $u = 0.5$?

15.2 The partition-of-unity property of Bernstein basis functions states that $\sum_i B_{i,n}(u) = 1$. Show that this is true for $n = 2$.

15.3 At what value of u is $B_{i,3}(u)$ a maximum?

15.4 Find $\mathbf{M}^{-1}\mathbf{TM}$ (Equation 15.58) for a second-degree curve.

15.5 Find $\mathbf{T}$ for $n = 3$ (Equation 15.49).

CHAPTER 16

SURFACES

Just as with curves, we all have an intuitive sense of what a surface is. A surface is perhaps a little less abstract a concept than a curve, if only because we can actually touch the surfaces of the ordinary objects that fill our environment. It is light reflected off an object's surfaces that makes it visible. It didn't take long for mathematicians and computer graphics applications developers to realize that to create a convincing display of an object they had to thoroughly understand the mathematics of surfaces.

The parametric functions that describe a surface are somewhat different from those used for curves. Two independent parametric variables are required for surfaces, so the equations' general form looks like

$$x = x(u, w), y = y(u, w), z = z(u, w) \tag{16.1}$$

Both parametric variables are usually limited to the unit interval $0 \leq u, w \leq 1$, which is often expressed as $u, w \in [0, 1]$. This defines the simplest element of a surface, called a *surface patch*, or just *patch*. We model very complex surfaces by joining together several patches.

16.1 Planes

The simplest patch is a plane. The following equations define a plane patch in the x, y plane:

$$\begin{aligned} x &= (c - a)u + a \\ y &= (d - b)w + b \\ z &= 0 \end{aligned} \tag{16.2}$$

where $u, w \in [0, 1]$ and $a, b, c,$ and d are constant coefficients. Figure 16.1 illustrates this patch, which has some rather easy-to-identify characteristics. Its four corner points

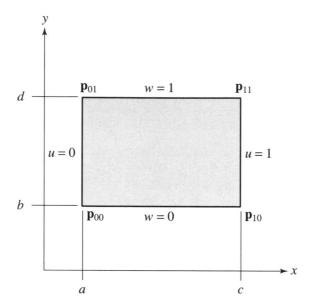

Figure 16.1 *A bounded plane.*

correspond to vectors whose components are defined by

$$\text{For } u = 0 \text{ and } w = 0, \mathbf{p}_{00} = [a \quad b \quad 0]$$
$$\text{For } u = 0 \text{ and } w = 1, \mathbf{p}_{00} = [a \quad d \quad 0]$$
$$\text{For } u = 1 \text{ and } w = 0, \mathbf{p}_{00} = [c \quad b \quad 0]$$
$$\text{For } u = 1 \text{ and } w = 1, \mathbf{p}_{00} = [c \quad d \quad 0]$$

(16.3)

The boundaries are, of course, straight lines derived when the following conditions are imposed on Equations 16.2:

$$
\begin{array}{llll}
u = 0, & x = a, & y = (d-b)w + b, & z = 0 \\
u = 1, & x = c, & y = (d-b)w + b, & z = 0 \\
w = 0, & x = (c-a)u + a, & y = b, & z = 0 \\
w = 1, & x = (c-a)u + a, & y = d, & z = 0
\end{array}
$$

(16.4)

Equations 16.2 obviously produce a very limited and not too exciting variety of plane patches. For example, they generate only patches that lie in the x, y plane, whose boundaries are straight lines.

16.2 Cylindrical Surfaces

A cylindrical surface is generated by a straight line as it moves parallel to itself along some curve (Figure 16.2). This surface is easy to define with parametric functions and vector equations. The following equation produces points on a cylindrical surface:

$$\mathbf{p}(u, w) = \mathbf{p}(u) + w\mathbf{r}$$

(16.5)

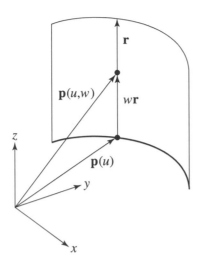

Figure 16.2 *A cylindrical surface.*

Just how does this equation generate a cylindrical surface? First, $\mathbf{p}(u)$ generates a curve, which could be a plane curve, a space curve, or a cubic Hermite, Bézier, or other type of curve. Next, the term $w\mathbf{r}$ generates points along $\mathbf{r}$ from $w = 0$ through $w = 1$ as $\mathbf{r}$ sweeps along $\mathbf{p}(u)$ for values of u from $u = 0$ through $u = 1$.

The four boundary points of this surface are

$$
\begin{aligned}
u = 0, \quad & w = 0, \quad \mathbf{p}_{00} = \mathbf{p}_0 \\
u = 0, \quad & w = 1, \quad \mathbf{p}_{01} = \mathbf{p}_0 + \mathbf{r} \\
u = 1, \quad & w = 0, \quad \mathbf{p}_{10} = \mathbf{p}_1 \\
u = 1, \quad & w = 1, \quad \mathbf{p}_{11} = \mathbf{p}_1 + \mathbf{r}
\end{aligned}
\tag{16.6}
$$

The four boundary curves are

$$
\begin{aligned}
u = 0, \quad & \mathbf{p}_0 + w\mathbf{r} \\
u = 1 \quad & \mathbf{p}_1 + w\mathbf{r} \\
w = 0 \quad & \mathbf{p}(u) \\
w = 1 \quad & \mathbf{p}(u) + \mathbf{r}
\end{aligned}
\tag{16.7}
$$

16.3 The Bicubic Surface

A more general kind of surface is the bicubic surface patch, given by the equation

$$
\mathbf{p}(u, w) = \sum_{i=0}^{3} \sum_{j=0}^{3} \mathbf{a}_{ij} u^i w^j
\tag{16.8}
$$

The $\mathbf{a}_{ij}$ vectors are the algebraic coefficients. The equation is *bicubic* since two parametric variables, u and w, appear as cubic terms. There are 16 $\mathbf{a}_{ij}$ vectors, each with three components, which means that there are 48 degrees of freedom or coefficients we must specify to define a unique surface. The double subscripts are necessary because there are two parametric variables.

Expanding Equation 16.8 we see just what we have to work with:

$$\begin{aligned}
\mathbf{p}(u, w) = {} & \mathbf{a}_{33}u^3w^3 + \mathbf{a}_{32}u^3w^2 + \mathbf{a}_{31}u^3w + \mathbf{a}_{30}u^3 \\
& + \mathbf{a}_{23}u^2w^3 + \mathbf{a}_{22}u^2w^2 + \mathbf{a}_{21}u^2w + \mathbf{a}_{20}u^2 \\
& + \mathbf{a}_{13}uw^3 + \mathbf{a}_{12}uw^2 + \mathbf{a}_{11}uw + \mathbf{a}_{10}u \\
& + \mathbf{a}_{03}w^3 + \mathbf{a}_{02}w^2 + \mathbf{a}_{01}w + \mathbf{a}_{00}
\end{aligned} \tag{16.9}$$

This sixteen-term polynomial in u and w defines the set of all points lying on the surface. It is the algebraic form of the bicubic patch. Using matrix notation, we can rewrite Equation 16.9 as

$$\mathbf{p}(u, w) = \mathbf{UAW}^{\mathrm{T}} \tag{16.10}$$

where $\mathbf{U} = \begin{bmatrix} u^3 & u^2 & u & 1 \end{bmatrix}$, $\mathbf{W} = \begin{bmatrix} w^3 & w^2 & w & 1 \end{bmatrix}$, and

$$\mathbf{A} = \begin{bmatrix} \mathbf{a}_{33} & \mathbf{a}_{32} & \mathbf{a}_{31} & \mathbf{a}_{30} \\ \mathbf{a}_{23} & \mathbf{a}_{22} & \mathbf{a}_{21} & \mathbf{a}_{20} \\ \mathbf{a}_{13} & \mathbf{a}_{12} & \mathbf{a}_{11} & \mathbf{a}_{10} \\ \mathbf{a}_{03} & \mathbf{a}_{02} & \mathbf{a}_{01} & \mathbf{a}_{00} \end{bmatrix} \tag{16.11}$$

Note that the subscripts of the vector elements in the **A** matrix correspond to those in Equation 16.9. As written here they have no direct relationship to the normal indexing convention for matrices. Furthermore, the **a** elements are three-component vectors, so the **A** matrix is really a $4 \times 4 \times 3$ array.

As with curves, the algebraic coefficients of a patch determine its shape and position in space. However, patches of the same size and shape have a different set of coefficients if they occupy different positions in space. Change any one of the 48 coefficients, and a completely different patch results. Maybe its shape has changed, or maybe its position. More analysis and insight is required to resolve this problem.

Equation 16.9 generates a point on the patch each time we insert a specific pair of u, w values into it. But because the range of the dependent variables x, y, and z is not restricted, the range of the algebraic coefficients is also not restricted.

A bicubic patch is bounded by four curves (Figure 16.3), and each one is a parametric cubic curve. It is a simple exercise to demonstrate this. For example, if $w = 0$, then all terms containing w vanish, and Equation 16.9 becomes

$$\mathbf{p}(u) = \mathbf{a}_{30}u^3 + \mathbf{a}_{20}u^2 + \mathbf{a}_{10}u + \mathbf{a}_{00} \tag{16.12}$$

This is the equation for a parametric cubic curve. Similar curves result when we set $u = 0$, $u = 1$, or $w = 1$. In fact, setting either u or w equal to some constant produces a curve on the surface. Curves of constant u or w are called *isoparametric* curves on the surface.

Using algebraic coefficients to define and control the shape of a patch are not convenient or user-friendly, to say the least. They do not contribute much to our understanding of surface behavior. So we will develop a geometric form, defining a patch in

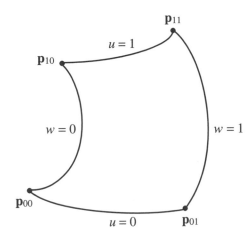

$\mathbf{p}_{11}$

$u = 1$

$\mathbf{p}_{10}$

$w = 0$

$w = 1$

$\mathbf{p}_{00}$

$u = 0$

$\mathbf{p}_{01}$

Figure 16.3 *Some elements of a bicubic patch.*

terms of its boundary geometry. This geometry is given at the corner points. First, of course, there are the four corner points themselves. Then there are the tangent vectors associated with the boundary curves. Finally, we introduce a *twist* vector at each corner, which control the rate of change of the tangent vectors of the interior isoparametric curves. We denote these twist vectors as $\mathbf{p}_{00}^{uw}$, $\mathbf{p}_{01}^{uw}$, $\mathbf{p}_{10}^{uw}$, and $\mathbf{p}_{11}^{uw}$. They are defined as

$$\mathbf{p}_{00}^{uw} = \frac{\partial^2 \mathbf{p}(u, w)}{\partial u \partial w}, \text{ evaluated at } u = 0 \text{ and } w = 0$$

$$\mathbf{p}_{01}^{uw} = \frac{\partial^2 \mathbf{p}(u, w)}{\partial u \partial w}, \text{ evaluated at } u = 0 \text{ and } w = 1$$

$$\mathbf{p}_{10}^{uw} = \frac{\partial^2 \mathbf{p}(u, w)}{\partial u \partial w}, \text{ evaluated at } u = 1 \text{ and } w = 0$$

$$\mathbf{p}_{11}^{uw} = \frac{\partial^2 \mathbf{p}(u, w)}{\partial u \partial w}, \text{ evaluated at } u = 1 \text{ and } w = 1$$

(16.13)

The calculus operation called for in Equation 16.13 is called *partial differentiation*, and is not very different from ordinary differentiation. In this case, we take the derivative of a function with two independent variables, treating each independent variable, in turn, as a constant while differentiating with respect to the other. Let us see how this works for $\mathbf{p}_{00}^{uw}$.

First, we compute $\partial \mathbf{p}(u, w)/\partial w$, treating u as a constant in Equation 16.9. This yields

$$\frac{\partial \mathbf{p}(u, w)}{\partial w} = 3\mathbf{a}_{33}u^3w^2 + 2\mathbf{a}_{32}u^3w + \mathbf{a}_{31}u^3$$
$$+ 3\mathbf{a}_{23}u^2w^2 + 2\mathbf{a}_{22}u^2w + \mathbf{a}_{21}u^2$$
$$+ 3\mathbf{a}_{13}uw^2 + 2\mathbf{a}_{12}uw + \mathbf{a}_{11}u$$
$$+ 3\mathbf{a}_{03}w^2 + 2\mathbf{a}_{02}w + \mathbf{a}_{01}$$

(16.14)

Next, we compute $\partial[\partial \mathbf{p}(u, w)/\partial w]/\partial u$, holding w constant in Equation (16.14).

$$\frac{\partial}{\partial u}\left(\frac{\partial \mathbf{p}(u, w)}{\partial w}\right) = \frac{\partial^2 \mathbf{p}(u, w)}{\partial u \partial w}$$

$$= 9\mathbf{a}_{33}u^2 w^2 + 6\mathbf{a}_{32}u^2 w + 3\mathbf{a}_{31}u^2 \qquad (16.15)$$

$$+ 6\mathbf{a}_{23}uw^2 + 4\mathbf{a}_{22}uw + 2\mathbf{a}_{21}u$$

$$+ 3\mathbf{a}_{13}w^2 + 2\mathbf{a}_{12}w + \mathbf{a}_{11}$$

Finally, we evaluate this last result at $u = 0$ and $w = 1$ to obtain

$$\mathbf{p}_{01}^{uw} = 3\mathbf{a}_{13} + 2\mathbf{a}_{12} + \mathbf{a}_{11} \qquad (16.16)$$

A similar procedure applies to the other twist vectors.

We are now ready to assemble all 16 boundary condition vectors into a matrix. The usual order of matrix elements is

$$\mathbf{B} = \begin{bmatrix} \mathbf{p}_{00} & \mathbf{p}_{01} & \mathbf{p}_{00}^{w} & \mathbf{p}_{01}^{w} \\ \mathbf{p}_{10} & \mathbf{p}_{11} & \mathbf{p}_{10}^{w} & \mathbf{p}_{11}^{w} \\ \mathbf{p}_{00}^{u} & \mathbf{p}_{01}^{u} & \mathbf{p}_{00}^{uw} & \mathbf{p}_{01}^{uw} \\ \mathbf{p}_{10}^{u} & \mathbf{p}_{11}^{u} & \mathbf{p}_{10}^{uw} & \mathbf{p}_{11}^{uw} \end{bmatrix} \qquad (16.17)$$

There are several things worth noting about the order of the matrix elements in Equation 16.17. The four corner points are in the upper left quadrant of the matrix. The four twist vectors are in the lower right quadrant, and the u and w tangent vectors are in the lower left and upper right quadrants, respectively.

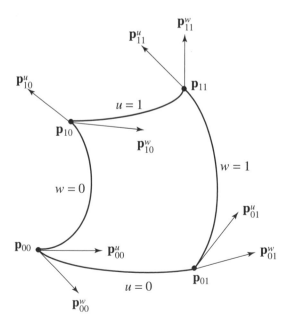

Figure 16.4 *Boundary conditions of a bicubic patch.*

There is even more information in this order. The elements comprising the first row are the geometric coefficients of the boundary curve corresponding to $u = 0$. The elements in the second row are the geometric coefficients of the boundary curve corresponding to $u = 1$. The first and second columns contain, in order, the geometric coefficients for the boundary curves corresponding to $w = 0$ and $w = 1$. Rows 3 and 4, and columns 3 and 4, define auxiliary curves, which describe how the interior isoparametric curve tangent vectors vary.

The complete matrix equation for a bicubic Hermite patch in terms of these geometric coefficients is

$$\mathbf{p}(u, w) = \mathbf{UMBM^T W^T} \tag{16.18}$$

where $\mathbf{M}$ is the same as that given for cubic Hermite curves in Equation 14.61.

The relationship between the algebraic and geometric coefficients is

$$\mathbf{A} = \mathbf{MBM^T} \tag{16.19}$$

and

$$\mathbf{B} = \mathbf{M^{-1}A(M^T)^{-1}} \tag{16.20}$$

There are other ways to define a bicubic surface. For example, a grid of 16 points is sufficient. These are usually the four corner points, two intermediate points on each boundary curve, and four interior points.

16.4 The Bézier Surface

A Bézier surface requires a grid of *control points* for its definition (Figure 16.5). These control points define a *characteristic polyhedron*, analogous to the characteristic polygon of the Bézier curve. Evaluating the following parametric function at a set of

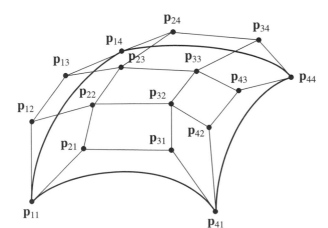

Figure 16.5 *Bicubic Bézier surface.*

u, w values generates a point on a Bézier surface:

$$\mathbf{p}(u, w) = \sum_{i=0}^{m} \sum_{j=0}^{n} \mathbf{p}_{ij} B_{i,m}(u) B_{j,n}(w) \tag{16.21}$$

where the control points $\mathbf{p}_{ij}$ are the vertices of the characteristic polyhedron that form an $(m + 1) \times (n + 1)$ rectangular grid of points, and the basis functions $B_{i,m}(u)$ and $B_{j,n}(w)$ are defined the same way as for Bézier curves (Equation 15.19 et seq.). Note that m and n define the degrees of the basis functions and that they are not necessarily equal. The advantage of this is that we can simplify the formulation somewhat if the surface behavior itself is simpler to describe in one of the parametric directions than in the other.

Expanding Equation 16.21 for a 4×4 grid of points produces a bicubic surface. The corresponding matrix equation is

$$\mathbf{p}(u, w) = \begin{bmatrix} (1-u)^3 \\ 3u(1-u)^2 \\ 3u^2(1-u) \\ u^3 \end{bmatrix}^{\mathrm{T}} \begin{bmatrix} \mathbf{p}_{11} & \mathbf{p}_{12} & \mathbf{p}_{13} & \mathbf{p}_{14} \\ \mathbf{p}_{21} & \mathbf{p}_{22} & \mathbf{p}_{23} & \mathbf{p}_{24} \\ \mathbf{p}_{31} & \mathbf{p}_{32} & \mathbf{p}_{33} & \mathbf{p}_{34} \\ \mathbf{p}_{41} & \mathbf{p}_{42} & \mathbf{p}_{43} & \mathbf{p}_{44} \end{bmatrix} \begin{bmatrix} (1-w)^3 \\ 3w(1-w)^2 \\ 3w^2(1-w) \\ w^3 \end{bmatrix} \tag{16.22}$$

Only the four corner points $\mathbf{p}_{11}$, $\mathbf{p}_{41}$, $\mathbf{p}_{14}$, and $\mathbf{p}_{44}$ actually lie on the surface. The other points control the boundary curves and interior of the surface, but do not lie on the surface.

16.5 Surface Normal

The vector product of the tangent vectors $\mathbf{p}^u(u, w)$ and $\mathbf{p}^w(u, w)$ at any point on a parametric surface produces the normal vector at that point (Figure 16.6). Thus

$$\mathbf{n}(u, w) = \mathbf{p}^u(u, w) \times \mathbf{p}^w(u, w) \tag{16.23}$$

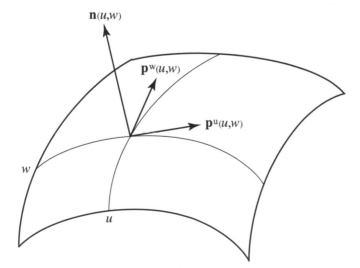

Figure 16.6 *Normal vector to a surface.*

The order in which the vector product is taken determines the direction of the normal vector. A consistent direction is necessary when the surface is part of a solid model, and when light reflecting off the surface is a consideration.

Exercises

16.1 Find the parametric equations for a plane surface patch whose four corner points are $\mathbf{p}_{00} = [4 \quad 3 \quad 1]$, $\mathbf{p}_{01} = [4 \quad 1 \quad 7]$, $\mathbf{p}_{10} = [4 \quad 5 \quad 3]$, and $\mathbf{p}_{11} = [4 \quad 5 \quad 7]$.

16.2 Find the four corner points of the plane surface patch given by $x = u - 2$, $y = 5$, and $z = 3w - 4$.

16.3 Find the parametric equations of the four boundary lines of the plane surface patch given in Exercise 16.2.

16.4 Find the general corner points $\mathbf{p}_{00}$, $\mathbf{p}_{01}$, $\mathbf{p}_{10}$, and $\mathbf{p}_{11}$ in terms of the algebraic coefficients.

16.5 Find each of the eight patch tangent vectors in terms of the algebraic coefficients.

CHAPTER 17

COMPUTER GRAPHICS DISPLAY GEOMETRY

The ability to create geometric models of objects and to display and interact with them on a computer-graphics monitor has brought about a revolution in science, industry, and entertainment, too. Computer graphics applications now quickly and accurately generate images for medical diagnostic devices, entertain us with challenging computer games, and let us create special effects for movies. Computer graphics gives doctors, scientists, and engineers a new way to look at the world, producing images of phenomena beyond their power to visualize just a few years ago. This chapter introduces some of the basic geometry and mathematics of computer graphics, including display coordinate systems, windows and viewports, line and polygon clipping, geometric-element display, and visibility.

17.1 Coordinate Systems

The most important characteristic of a coordinate system is the number of its dimensions. The simplest coordinate system is one dimensional, consisting of an unbounded straight line, a reference point O on it , called the *origin*, and a scale of measure, or *metric*. This is all we need to position, construct, and measure geometric objects, which in this one- dimensional system are limited to points and lines.

We locate a point $\mathbf{p}$ in this system by giving its distance x from the origin (Figure 17.1). This single number, a *coordinate*, describes and locates $\mathbf{p}$. A plus or minus sign on x tells us which direction to measure from the origin. Mathematicians sometimes refer to this one-dimensional coordinate system as the *real-number line*, because there is a one-to-one correspondence between points on this line and the set of real numbers.

Changing the location of the origin to O' does not change the geometric properties of objects defined in the system (Figure 17.2). For example, the distance between

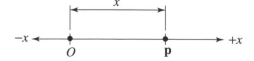

Figure 17.1 *One-dimensional coordinate system.*

286

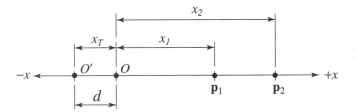

Figure 17.2 *Moving the origin.*

$\mathbf{p}_1$ and $\mathbf{p}_2$ is the same in either the O or O' system. If O and O' are separated by a distance d, then the coordinate of O' in the O system is $x_T = -d$. This means that in the O' system the coordinate of $\mathbf{p}_1$ is $x_1 + d$ and the coordinate of $\mathbf{p}_2$ is $x_2 + d$. The distance between $\mathbf{p}_1$ and $\mathbf{p}_2$ in the O system is $x_2 - x_1$, and in the O' system it is $(x_2 + d) - (x_1 + d)$; clearly, $x_2 - x_1 = (x_2 + d) - (x_1 + d)$. This process of changing the origin is called a *coordinate system transformation*. Coordinates in the two systems O and O' are related mathematically by a *transformation equation*: (review Chapter 3):

$$x' = x - x_T \tag{17.1}$$

where x_T is the location of O' in the O system. The concept of a transformation is an important one in computer graphics and geometric modeling.

The one-dimensional coordinate system has some interesting properties and gives rise to a surprisingly rich assortment of mathematical ideas. In a one-dimensional coordinate system, we can study relationships between points, points and lines, and lines. For example, does a given point lie between two other given points? Do two line segments overlap? What is the length of a line segment? Mathematicians and geometers do not limit one-dimensional coordinate systems to unbounded straight lines. A *half-line* (a *ray*, or semi-infinite line) or finite line segment can serve as a one-dimensional coordinate system. We can also construct coordinate systems and geometries on closed curves. There is more geometry here than you might have imagined.

Two-Dimensional Coordinate Systems

René Descartes (1596–1650), the French mathematician and philosopher, discovered that some algebraic equations could be interpreted geometrically by graphing them onto a two-dimensional coordinate system. His discovery followed earlier work by others who hypothesized a one-to-one correspondence between algebra and geometry. He came to the inescapable conclusion that not only could algebraic equations be interpreted as graphed curves, but that many geometric figures have simple algebraic equivalents. We now know that all geometric objects—curves, surfaces, solids, and more exotic things like *fractals* and *dusts*—can be described algebraically. Conversely, all algebraic functions have a geometric interpretation. Descartes had founded *analytic geometry*. The rectangular, two-dimensional system he used to graph algebraic functions bears his name: the *Cartesian coordinate system*.

The two-dimensional Cartesian coordinate system has well-defined properties, which are often taken for granted. To construct one, we begin with two (hypothetically)

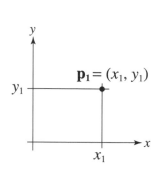

Figure 17.3 *Two-dimensional Cartesian coordinate system.*

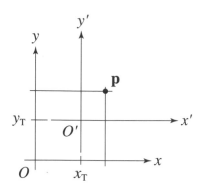

Figure 17.4 *Translating the origin of a two-dimensional coordinate system.*

unbounded straight lines intersecting at right angles to form the *principal axes x* and *y*, with positive and negative directions indicated (Figure 17.3). Their point of intersection O is the *origin*. We impose a grid of equally spaced lines parallel to and in the plane of the principal axes, which forms the basis of all measurements and analysis. Every point in the plane of this coordinate system is defined by a pair of numbers (x, y), its coordinates. The coordinates of $\mathbf{p}_1$ in the figure are (x_1, y_1). To locate $\mathbf{p}_1$ in the system, we construct a line parallel to the y axis through x_1 and parallel to the x axis through y_1. These two lines intersect at $\mathbf{p}_1$. (Note that a skewed-axes system, in which the axis are not perpendicular, is also possible.)

The location and orientation of a coordinate system are arbitrary, and there may be more than one. For example, we can express the coordinates of a point in terms of a second coordinate system, superimposed on the first, whose axes are translated x_T and y_T relative to the first system (Figure 17.4). The relationship between coordinates in the two systems is given by

$$x' = x - x_T$$
$$y' = y - y_T \tag{17.2}$$

We could also create a second coordinate system whose axes are rotated through an angle θ with respect to the first coordinate system (Figure 17.5). We could even apply a combination of translation and rotation transformations. Such transformations are discussed in Chapter 3 and elsewhere.

Figure 17.5 *Rotating a two-dimensional coordinate system.*

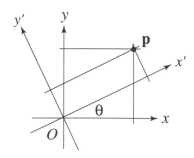

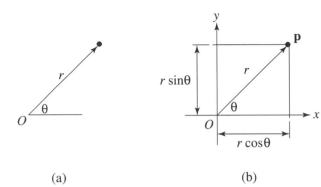

(a)   (b)

Figure 17.6 *Relationship between Cartesian and polar coordinate system.*

The polar coordinate system has limited use in computer graphics and geo-metric modeling, but its relationship to the Cartesian system suggests how we might rotate points. Figure 17.6a illustrates a polar coordinate system. Points in it are located by giving their distance r from the origin and the angular displacement θ of the line of r with respect to a reference line. Thus, the polar coordinates of a point are (r, θ).

We can convert between polar coordinates and Cartesian coordinates. Figure 17.6b shows the two systems superimposed so that their origins coincide and the polar reference line and x axis are collinear. The transformation equations from polar to Cartesian are derived using simple trigonometry. Thus,

$$x = r \cos \theta$$
$$y = r \sin \theta \tag{17.3}$$

Three-Dimensional Coordinate Systems

The three-dimensional Cartesian coordinate system is a simple extension of the two-dimensional Cartesian system (Figure 17.7). It consists of three mutually per-pendicular axes, called the *principal coordinate axes*, providing the three coordinates

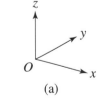

(a)

Figure 17.7 *Right- and left-handed three-dimensional coordinate systems.*

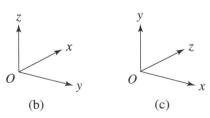

(b)   (c)

x, y, and z, for locating a point. The nature of three-dimensional Cartesian space allows two distinctly different relationships between the three principal axes. The relationship shown in Figure 17.7a is a right-hand system, and the relationship in Figure 17.7b is a left-hand system. It is called a right-hand system in Figure 17.7a because, if the fingers of your right hand curl in a direction that sends the positive x axis rotating about the z axis in the direction of the positive y axis, then your thumb points in the positive z direction. A similar relationship applies to a left-hand system. Many computer graphics systems use the left-hand system, where the x and y axes are in the plane of the display screen and the positive z axis points away from the viewer and into the screen (Figure 17.7c).

We can create a second three-dimensional system, related to the first, whose axes are parallel to those in the first system (Figure 17.8). If the origin O' of the second system is located at x_T, y_T, z_T in the first system, then the coordinates of any point x, y, z in the first system can be transformed into coordinates in the second system according to the following transformation equations:

$$x' = x - x_T$$
$$y' = y - y_T \tag{17.4}$$
$$z' = z - z_T$$

The coordinates of a point in a cylindrical system are r, θ, and z: a radial distance r from the origin, an angle θ between the line of r and some reference line, and a distance z from the plane defined by the reference line and line of r (Figure 17.9). If we embed the cylindrical system within the Cartesian system so that the reference line coincides with the x axis, if the line of r lies in the x, y plane, and if θ is measured counterclockwise about the z axis and in the x, y plane, then the Cartesian coordinates of a point whose coordinates are given in the cylindrical system are

$$x = r \cos \theta$$
$$y = r \sin \theta \tag{17.5}$$
$$z = z$$

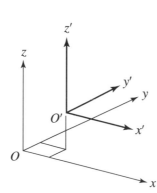

Figure 17.8 *Two three-dimensional coordinate systems.*

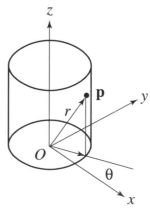

Figure 17.9 *Cylindrical coordinate system.*

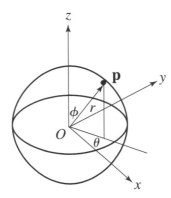

Figure 17.10 *Spherical coordinate system.*

The coordinates of a point in a spherical system are r, θ, and ϕ: a radial distance r from the origin, and two angles θ and ϕ measured as shown in Figure 17.10. Embedded in a Cartesian system, the two systems are related by the transformation equations

$$x = r \sin\phi \cos\theta$$
$$y = r \sin\phi \sin\theta \qquad\qquad (17.6)$$
$$z = r \cos\phi$$

Display Coordinate Systems

To create a computer-graphics display of an object, we must consider its geometric characteristics, such as size, shape, location, orientation, and its spatial relationship to other objects nearby. If we are to describe, measure, and analyze these characteristics, we must place the object within some frame of reference—in this case, a coordinate system. Computer-graphics displays require a variety of two- and three-dimensional coordinate systems, each identified by name and a unique set of coordinate axes. We often use the following five interrelated systems to create a picture on a computer monitor. In Figure 17.11 we can see each system in relation to the other four systems. They are:

1. The *global* or *world* coordinate system: This is the primary x, y, z (sometimes G_x, G_y, G_z) three-dimensional frame of reference in which we define and locate in space all of the geometric objects in a computer-graphics scene. We also locate the observer's position and viewing direction in the global coordinate system.

2. The *local* coordinate system: We use a local coordinate system L_x, L_y, L_z to define an object independent of the global coordinate system, without having to specify a location or orientation in the global system.

3. The *view* coordinate system: We locate objects relative to the observer in this V_x, V_y, V_z coordinate system. Doing this simplifies the mathematics of projecting the object's image onto the picture plane if we arrange it so that we look along one of the principal axes. We must change the location and orientation of the view coordinate system each time we change the view.

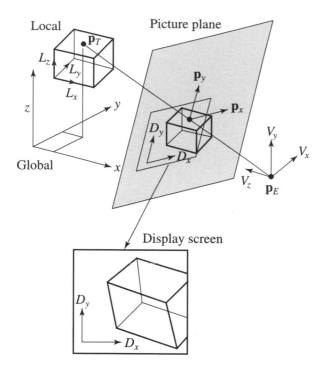

Figure 17.11 *Four related display coordinate systems.*

4. The *picture-plane* coordinate system: This is a two-dimensional system P_x, P_y for locating points on the picture plane. (We may also call this the *projection plane*.)

5. The *display-screen* coordinate system: This is a two-dimensional coordinate system for locating the image of an object on the surface of the computer-graphics monitor. It may consist of two parts, a *window* and a *viewport* (Section 17.2).

Let's see how these five systems are related to each other logically and mathematically, and how they are the basis of a three-dimensional computer-graphics system. The first thing we must do is create a mathematical description of the shape of an object located in three-dimensional space—its geometric model. For a cube, this may be nothing more than the coordinates of its eight corner points and some instructions on how to connect them with straight lines to form its edges (Figure 17.12). For a sphere, it might be its radius and the location of its center point. We could also define the object in a local coordinate system (Figure 17.13). We usually do this because we can place a local system in a position that simplifies defining the object. The local system, carrying the object with it, is then linked to the global system by a set of transformations, usually translation, rotation, and scaling (sizing).

Once the object is defined in the global system, we must decide from which point we want to view it. In other words, we must position the observer and the observer's line of sight. To do this, the observer's eye point $\mathbf{p}_E$ is given by specifying its coordinates in the global system (Figure 17.11). There are two ways to specify the line of sight: as a set of direction cosines in the global system or as a line joining $\mathbf{p}_E$ to some view point

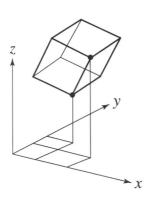

Figure 17.12 *Global coordinate system.*

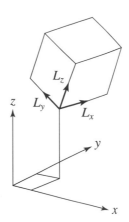

Figure 17.13 *Local coordinate system.*

$\mathbf{p}_T$ that lies on the line of sight. The final angular orientation of the view with respect to the observer is achieved by just an appropriate rotation about the line of sight.

Changing the view of the object depends on whether the object moves or the observer moves. To move the object, we transform all the points defining it to a new position. To move the observer requires only that we specify a new eye point location and a new orientation for the line of sight. It is the eye of the observer that moves to create a new computer-graphics scene in a static arrangement of objects, as in a computer-aided design program. Otherwise, objects move, as in computer games or dynamic simulations.

Once we have selected an eye point and viewpoint, we project the model onto a two-dimensional surface called a *picture plane*. We must do this each time we want to change the view. For a perspective projection, the picture changes if the location of the picture plane changes, even though $\mathbf{p}_E$ and the line of sight are unchanged. The picture plane is usually located between the object and the eye point and is perpendicular to the line of sight (as it is shown in Figure 17.11).

Of course, all images projected onto the picture plane are only mathematical images. They are data sets in computer memory that are, in turn, computed from the data set representing the geometric model of the object. We create the image of an object that we actually see on the computer-graphics monitor by first mathematically transferring the picture-plane data set of the object from the picture plane to the surface of the graphics monitor, and this requires a *window* or *viewport*.

17.2 Window and Viewport

Usually only part of the model's image projected on the picture plane is displayed on the computer monitor. This is done by specifying a rectangular boundary, called a *window*, that encloses that part of the picture-plane image that we want to display. The window boundaries W_L, W_R, W_T, and W_B are defined in the picture plane coordinate system (see Figure 17.14). The image in the window is then mapped onto

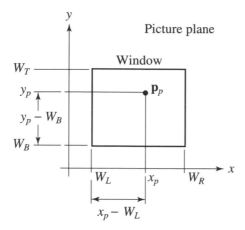

Figure 17.14 *Window.*

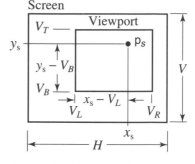

Figure 17.15 *Viewport.*

another bounded area, called the *viewport*, in the display-screen coordinate system (Figure 17.15). This is the image we actually see displayed on the computer monitor.

The monitor's screen coordinate system has horizontal and vertical limits, determined by the number of horizontal and vertical positions accessible to the electronics that generate the display. Usually two numbers specify a location on the screen. The first number indicates the horizontal displacement from some reference point, and the second number indicates vertical displacement. These numbers define a discrete, uniquely addressable picture element, a *pixel*. The complete rectangular array of pixels forms the physical basis of the screen coordinate system. Note that pixel addresses are always positive integers. The number of rows of pixels, V, defines the vertical dimension of a screen, and the number of pixels in each row, H, defines the horizontal dimension. H and V are not necessarily equal. There is a sequence of transformations from the infinite and continuous three-dimensional global coordinate system to the finite and discrete two-dimensional screen coordinate system.

To create the image $\mathbf{p}_S = x_S, y_S$ of a point $\mathbf{p}_P = x_P, y_P$ in the picture-plane coordinate system on the display screen, we must find a set of *transformation equations* relating x_P, y_P to x_S, y_S (Figures 17.14 and 17.15). To do this, we assume an $H \times V$ pixel screen, with a viewport defined and bounded by V_L, V_R, V_T, and V_B. The geometry of the window and viewport lets us establish a one-to-one correspondence between points in the picture-plane window and points in the screen viewport. We say that the equations describing this correspondence *map* points in the world system onto the screen.

Figures 17.14 and 17.15 suggest that we use some simple ratios of the geometric relationships of the points $\mathbf{p}_P$ and $\mathbf{p}_S$ to their locations within their respective regions. Thus,

$$\frac{x_S - V_L}{V_R - V_L} = \frac{x_P - W_L}{W_R - W_L}$$

$$\frac{y_S - V_B}{V_T - V_B} = \frac{y_P - W_B}{W_T - W_B}$$

$$(17.7)$$

Solving these equations for x_S and y_S produces

$$x_S = (x_R - W_L)\left(\frac{V_R - V_L}{W_R - W_L}\right) + V_L$$

$$y_S = (y_P - W_B)\left(\frac{V_T - V_B}{W_T - W_B}\right) + V_B \tag{17.8}$$

The terms $(V_R - V_L / W_R - W_L)$ and $(V_T - V_B / W_T - W_B)$ are scale factors, and we can denote them k_x and k_y, so that

$$k_x = \left(\frac{V_R - V_L}{W_R - W_L}\right) \quad \text{and} \quad k_y = \left(\frac{V_T - V_B}{W_T - W_B}\right) \tag{17.9}$$

There will be no horizontal or vertical distortion of the image if and only if $k_x = k_y$. This means, for example, that a square on the picture plane will appear as a square on the display screen if the scale factors are equal. In many displays the window and viewport coincide so that $k_x = k_y = 1.0$.

Some interesting mathematical considerations arise when we map points from a continuous three-dimensional space onto a discrete two-dimensional space. For example, computing x_S and y_S results in real, noninteger numbers for screen coordinates. This means that we must truncate them if the decimal part is less than 0.5, or round up if it is 0.5 or greater, which converts x_S and y_S to integers. (Other truncating rules may improve image quality, but they are not discussed here.) Computing and recovering x_P and y_P from the screen coordinates x_S and y_S is not a wise thing to do, because information was lost in the truncating operation as well as in the projection transformation (which creates a two-dimensional image of a three-dimensional model).

In an extensive graphics network with a variety of monitors and screen sizes, we use a technique called *normalization* (see Figure 17.16). We use an imaginary screen whose dimensions are $H = 1.0$ and $V = 1.0$. For any point $\mathbf{p}_P = x_P, y_P$ we compute x_N and y_N, the normalized coordinates:

$$x_N = (x_P - W_L)\left(\frac{N_R - N_L}{W_R - W_L}\right) + N_L$$

$$y_N = (y_P - W_B)\left(\frac{N_T - N_B}{W_T - W_B}\right) + N_B \tag{17.10}$$

where $0 \le N_L, N_R, N_T, N_B \le 1.0$, which means that $0 \le x_N, y_N \le 1.0$.

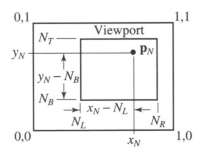

Figure 17.16 *Normalization.*

For a specific display screen in the network, we compute x_S and y_S by multiplying each normalized coordinate by horizontal and vertical scales H_S, V_S; thus,

$$x_S = H_S x_N \quad \text{and} \quad y_S = V_S y_N \tag{17.11}$$

If the screen size is 640 × 480 pixels, then

$$\begin{aligned} x_S &= 640 x_N \\ y_S &= 480 y_N \end{aligned} \tag{17.12}$$

17.3 Line Clipping

In a model or database with many line segments, only those within the computer-graphics display-screen window are displayed. This means that we need a test to decide if a line segment is entirely inside, partially inside, or entirely outside the window area. (Figure 17.17). A line segment is completely within the window area if, and only if, all the following inequalities are true:

$$W_L \leq x_0, x_1 \leq W_R \quad \text{and} \quad W_B \leq y_0, y_1 \leq W_T \tag{17.13}$$

Otherwise, the line is only partially inside the window or entirely outside it. Line 1 in the figure passes the test of Equation 17.13, which means that both endpoints are inside the window area.

If only one endpoint is contained by the window, as in the case of line 2, then we must compute the point of intersection between the line and the appropriate window boundary (Figure 17.18). In this case, $x_0 > W_R$ and $y_0 > W_B$, and we do not need to calculate intersections with W_L and W_B. Knowing that $y_0 > W_T$ does not add anything to our knowledge of the situation, because either of the two distinct line orientations is possible. This means that we must compute intersections with W_R and W_T. The u value closest to that of the contained point corresponds to the appropriate point of intersection.

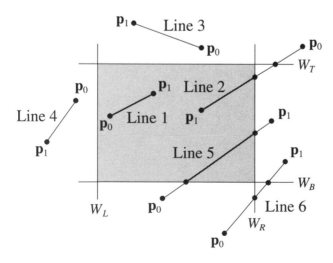

Figure 17.17 *Line containment and clipping.*

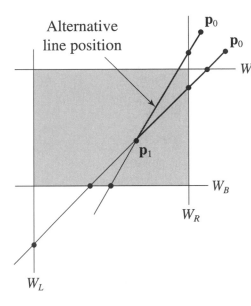

Figure 17.18 *Clipping.*

This point of intersection now becomes the new endpoint of the line, and this segment of the line is saved and displayed. In computer graphics, this process is called *clipping*, because that part of a line not within the window boundaries is "clipped" off. A line is completely outside the window if any one of the following sets of inequalities is true:

$$x_0, x_1 < W_L \quad x_0, x_1 < W_R$$
$$y_0, y_1 < W_B \quad y_0, y_1 < W_T$$

$$(17.14)$$

Other line and window arrangements are possible. If none of the preceding conditions is true, then we are faced with a condition analogous to one of the alternatives shown in Figure 17.19. One solution is to compute all possible intersection points, p_I,

Figure 17.19 *Other arrangements of lines and window boundaries.*

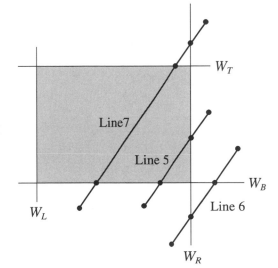

and then use the u_i, x_i, and y_i values of each of these points to determine the two points that appropriately bound that portion of the line contained within the window.

If we convert the window boundaries into parametric line segments (Figure 17.20), then the clipping is simplified, because all valid points of intersection on the window boundary lines must be within the unit interval on these lines. For example, the parametric equations for line B are

$$x = (W_R - W_L)u + W_L$$
$$y = W_B \tag{17.15}$$

The intersection of line B with some arbitrary line A is given by solving the two simultaneous equations

$$(x_1 - x_0)u_A + x_0 = (W_R - W_L)u_B + W_L$$
$$(y_1 - y_0)u_A + y_0 = W_B \tag{17.16}$$

for u_A and u_B, which yields

$$u_A = \frac{W_B - y_0}{y_1 - y_0}$$
$$u_B = \left(\frac{x_1 - x_0}{y_1 - y_0}\right)\left(\frac{W_B - y_0}{W_R - W_L}\right) + \left(\frac{x_0 - W_L}{W_R - W_L}\right) \tag{17.17}$$

To summarize line clipping: We perform two sets of tests on the endpoints of a line. If the first test shows that both $W_L \leq x_0$, $x_1 \leq W_R$ and $W_B \leq y_0$, $y_1 \leq W_T$ are true, then the line is contained in the window, and no intersection computations are required. If the first test shows that it is not true, then we perform a second set of tests: If any of the following four inequalities are true, that is if x_0, $x_1 < W_L$; x_0, $x_1 > W_R$; y_0, $y_1 < W_B$; or y_0, $y_1 > W_T$, then the line lies entirely outside the window. If neither of these two sets of tests is true, we must do the intersection computations. Obviously, this method applies equally well to polygons or to any figure defined by line segments.

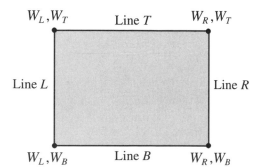

Figure 17.20 *Window boundaries as line segments.*

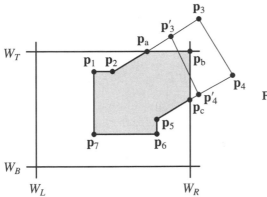

Figure 17.21 *Polygon clipping.*

17.4 Polygon Clipping

If a polygon is in the picture plane, it must be tested to see if it is contained within the display window. We apply the point-containment test to each of its vertex points (Figure 17.21). If the containment status changes between two consecutive vertex points, then the polygon edge bounded by those two points must intersect a window boundary. We clip the polygon to the window boundary by edges and edge segments that lie outside the window. A new sequence of vertices is established, defining the clipped polygon.

In the figure, $\mathbf{p}_2$ is inside the window and $\mathbf{p}_3$ is outside. Therefore, edge 2–3 intersects the window and must be clipped at $\mathbf{p}_a$, the point of intersection. Points $\mathbf{p}_3$ and $\mathbf{p}_4$ are both outside the window and so, too, is edge 3–4. However, if they had been at $\mathbf{p}'_3$ and $\mathbf{p}'_4$, the edge 3′–4′ would intersect window boundaries and would have to be clipped. This tells us that all edges joining sequential pairs of vertices outside the window must be tested using a line containment and clipping procedure.

For the polygon in the figure, then, $\mathbf{p}_1$ and $\mathbf{p}_2$ are inside the window, $\mathbf{p}_3$ and $\mathbf{p}_4$ are outside, and $\mathbf{p}_5$, $\mathbf{p}_6$, and $\mathbf{p}_7$ are inside. Therefore, there is an edge intersection between $\mathbf{p}_2$ and $\mathbf{p}_3$, as we have already seen, and also between $\mathbf{p}_4$ and $\mathbf{p}_5$. This means that we clip the polygon by deleting vertices 3 and 4 and by adding $\mathbf{p}_a$, $\mathbf{p}_b$, and $\mathbf{p}_c$ to the sequence of points defining it.

17.5 Displaying Geometric Elements

In geometric modeling and computer graphics displays, we use two kinds of points: *absolute points* and *relative points*. This distinction arises because of how we compute the point coordinates and how we plot and display the points. An absolute point is defined directly by its own unique coordinates. For example, a set of n absolute points is given by the following mathematical statement:

$$\text{Absolute points } \mathbf{p}_i = (x_i, y_i) \quad \text{for } i = 1, \ldots, n \qquad (17.18)$$

A relative point is defined with respect to the coordinates of the point preceding it. For example, a set of n relative points is given by the following mathematical expression:

$$\text{Relative points } \mathbf{p}_i = \mathbf{p}_{i-1} + \Delta_i \quad \text{for } i = 1, \ldots, n \tag{17.19}$$

A common point-display problem is that of defining a window that just encloses a given set of points. To do this, we investigate each point in the set in a search for the maximum and minimum x and y values, where $W_R = \max x$, $W_L = \min x$, $W_T = \max y$, and $W_B = \min y$, which then define the window boundaries.

Another interesting problem may arise when displaying three or more points. It may not always be possible to resolve every point in a set of points that are to be displayed. Let's consider, for example, the three points $\mathbf{p}_1, \mathbf{p}_2$, and $\mathbf{p}_3$ in Figure 17.22. For simplicity, we can arrange them on a horizontal line so that $y_1 = y_2 = y_3$. This means that only the x coordinates determine the separations. If the number of pixels, h, between $\mathbf{p}_1$ and $\mathbf{p}_3$ is less than the ratio of the separations of $\mathbf{p}_1$ and $\mathbf{p}_3$ to $\mathbf{p}_1$ and $\mathbf{p}_2$, then $\mathbf{p}_1$ and $\mathbf{p}_2$ will not be resolved. That is, $\mathbf{p}_1$ and $\mathbf{p}_2$ will not be displayed as two separate and distinct points. This relationship is expressed by the inequality

$$h < \frac{x_3 - x_1}{x_2 - x_1} \tag{17.20}$$

If this inequality is true, then $\mathbf{p}_1$ and $\mathbf{p}_2$ cannot be resolved.

We must also know if a given point in the picture plane is inside or outside the window region of a display (see the preceding discussion on line containment). If the coordinates of a point are x, y, then the point is inside the display window if and only if both of the following inequalities are true:

$$W_L \leq x \leq W_R \quad \text{and} \quad W_B \leq y \leq W_T \tag{17.21}$$

Lines

By connecting a set of points with straight lines we can form both open and closed figures for computer graphics displays and geometric models (Figure 17.23). There

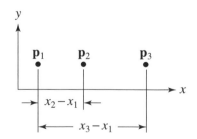

Figure 17.22 *Resolving points.*

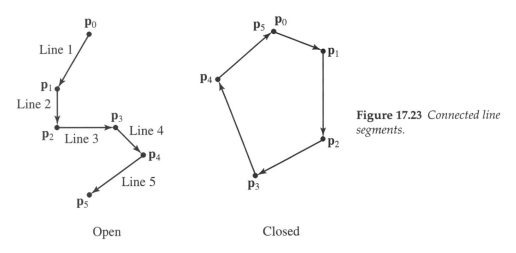

Figure 17.23 *Connected line segments.*

Open   Closed

is a brute-force approach that requires us to define each line separately. Thus, for n lines,

$$\text{Line } 1 = \mathbf{p}_0, \mathbf{p}_1$$
$$\text{Line } 2 = \mathbf{p}_1, \mathbf{p}_2$$
$$\text{Line } 3 = \mathbf{p}_2, \mathbf{p}_3$$
$$\vdots$$
$$\text{Line } n = \mathbf{p}_{n-1}, \mathbf{p}_n$$

where the equal sign here reads "is defined by." Using this approach, we must store the coordinates of $2n$ points in a database.

Another approach uses a single series of points to define a line string. A computer-graphics line-display algorithm might then interpret them as follows:

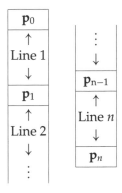

In this system, coordinates are not duplicated as they are in the previous database. Only $n + 1$ points are stored for the n lines.

Both of these methods use the absolute line representation, where the point coordinates are given directly. Using the relative line representation, we define relative offsets or displacements, so that each point in the sequence is derived relative to the previous point (Figure 17.24). This looks like:

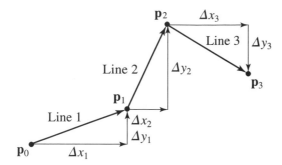

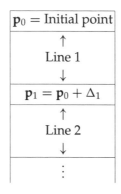
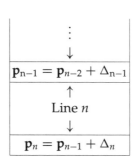

Figure 17.24 *Lines defined by a sequence of offsets.*

where $\Delta_i = (\Delta x_i, \Delta y_i)$. The database now contains the initial point $\mathbf{p}_0$ and the Δs. All the other points are computed sequentially from these data. This database is no more crowded than the previous one. It is easy to convert from relative lines to absolute lines and back. The relative line representation has a slight advantage in that, for translation transformations of the set of lines, only the initial point is translated, because the Δ offsets remain the same before and after a translation.

Polygons

The edges of a polygon are nothing more than a closed string of line segments. However, some situations may require that we not only display the edges of a polygon, but also fill its interior with color or shading. This means that we will want to turn on pixels on segments of raster scan lines that are inside the polygon. This is called *polygon filling* and it amounts to finding the intersections between the edges of the polygon and a series of horizontal lines.

The polygon in Figure 17.25 is intersected by horizontal lines A and B, which represent raster scan lines. We observe that a straight line always intersects a polygon an even number of times. In general, if n_i is the number of intersections between a line and a polygon, then $n_i/2$ is the number of segments of the line inside the polygon. If line A in the figure is a raster line, then its endpoint coordinates are $\mathbf{p}_0 = (0, y_A)$ and $\mathbf{p}_1 = (H, y_A)$, where H is the display width ($y = y_A$), a constant because the line is horizontal).

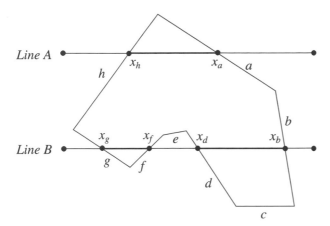

Figure 17.25 *Polygon filling.*

Its parametric equations are

$$x = x_H u$$
$$y = y_A$$

(17.22)

Each polygon edge is also expressed by a set of parametric equations.

Next, we search for and compute all intersections between line A and lines a through h of the polygon. We can speed up the search by considering only those polygon edges whose endpoint y coordinates are such that one is greater than y_A and the other is less than y_A. This eliminates edges b through g from further consideration. The parametric equations for edge a are

$$x = (x_1 - x_0)t + x_0$$
$$y = (y_1 - y_0)t + y_0$$

(17.23)

where $(x_1 - x_0)$ and $(y_1 - y_0)$ are understood to be the endpoint coordinates of edge a, and t denotes the parametric variable to distinguish it from u on line A. At a point of intersection of line A and edge a, we equate their y coordinates to obtain

$$(y_1 - y_0)t + y_0 = y_A$$

(17.24)

which we solve for t:

$$t = \frac{y_A - y_0}{y_1 - y_0}$$

(17.25)

Substituting this value of t into Equation 17.23 produces the x coordinate of the point of intersection:

$$x_a = \frac{(x_1 - x_0)(y_A - y_0)}{(y_1 - y_0) + x_0}$$

(17.26)

We repeat this method to find x_h, the x coordinate of the point of intersection of line A and edge b. Because there are only two points of intersection between line A

and the polygon, there is only one segment of the raster line represented by line A that is inside the polygon, and that is the segment between x_a and x_h.

There are four intersections between raster line B and the polygon: x_b, x_d, x_f, and x_g. We compute them (following the previous method) and pair them off in ascending order, which in this case is (x_g, x_f) and (x_d, x_b). These segments contribute to the polygon fill.

Curves

To display a curve, we must resolve a rather interesting sequence of problems. First, we must project the curve onto the picture plane; next, we must clip the projected curve, as necessary, within the window; finally, we must compute points along the curve from which to construct a displayable image.

We can simplify matters considerably by choosing as the picture plane one of the principal planes, say the x, y plane (Figure 17.26). Here a parametric cubic curve is defined by the four points $\mathbf{p}_1$, $\mathbf{p}_2$, $\mathbf{p}_3$, $\mathbf{p}_4$. We denote their projection onto the x, y plane as $\mathbf{p}_1'$, $\mathbf{p}_2'$, $\mathbf{p}_3'$, $\mathbf{p}_4'$, which define a plane curve. For an orthographic projection, we merely set equal to zero the z component of each of the four original points. Thus,

$$
\begin{aligned}
(\mathbf{p}_1)' &= [x_1 \quad y_1 \quad 0] \\
(\mathbf{p}_2)' &= [x_2 \quad y_2 \quad 0] \\
(\mathbf{p}_3)' &= [x_3 \quad y_3 \quad 0] \\
(\mathbf{p}_4)' &= [x_4 \quad y_4 \quad 0]
\end{aligned}
\tag{17.27}
$$

If the original curve is defined by its end points and tangent vectors, then for an orthographic projection (Section 18.1) onto the x, y plane it is also easy to show that

$$
\begin{aligned}
(\mathbf{p}_0)' &= [x_0 \quad y_0 \quad 0] \\
(\mathbf{p}_0^u)' &= [x_0^u \quad y_0^u \quad 0] \\
(\mathbf{p}_1)' &= [x_1 \quad y_1 \quad 0] \\
(\mathbf{p}_1^u)' &= [x_1^u \quad y_1^u \quad 0]
\end{aligned}
\tag{17.28}
$$

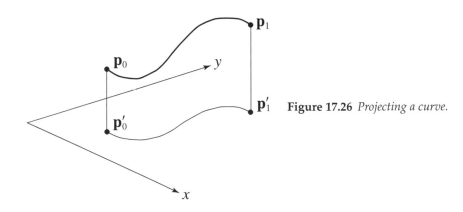

Figure 17.26 *Projecting a curve.*

Using the equations of the projected curve, we can determine if it is contained within the window. This means that we must search for valid intersections. Consider the curve and window shown in Figure 17.27. We compute its intersection with each of four lines: $x = W_L$, $x = W_R$, $y = W_B$, and $y = W_T$. These yield the following four cubic equations

$$a_x u^3 + b_x u^2 + c_x u + d_x = W_L$$
$$a_x u^3 + b_x u^2 + c_x u + d_x = W_R$$
$$a_y u^3 + b_y u^2 + c_y u + d_y = W_B$$
$$a_y u^3 + b_y u^2 + c_y u + d_y = W_T$$

(17.29)

The solutions to these equations yield six possible intersections for the example in Figure 17.27. All valid intersections must be in the unit interval of the curve, $u \in [0, 1]$, and within the appropriate window intervals. Clearly, not all of the intersection points meet this criteria.

The simplest way to display a curve is as a series of connected line segments. (This has certain disadvantages that will become apparent as we proceed and that can be overcome.) To do this, we compute a sequence of points along the curve. Then we connect these points, in turn, with line segments, forming an approximation of the curve that is displayed (Figure 17.28). The more points, the closer together they are, and the better the

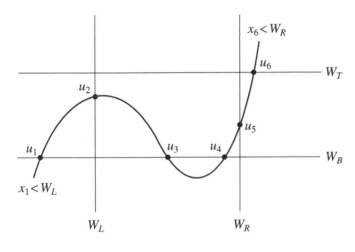

Figure 17.27 *Clipping a curve.*

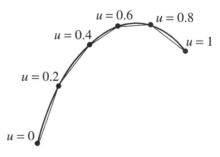

Figure 17.28 *Points on a curve with equal parametric intervals.*

approximation. There is a limit to this subdivision where we can no longer discern any improvement in the displayed image. This is a function of screen size and resolution and the degree of curvature of the curve. (These are subjects discussed in more advanced texts.)

If we divide the curve into equal increments, then where the curve is relatively flat, we may find too many points, or where it is more curved we will have too few points. One way to avoid this is by selecting points so that there is an equal change in slope between successive pairs of points. The angle θ shown in Figure 17.29 characterizes the change in slope between two adjacent points on the curve. The smaller the value we choose for θ, the more points are generated and the better the approximation to the true shape of the curve. (Here, again, the techniques for doing this are beyond the scope of this text.)

By one technique or another, we eventually end up with a set of u values, for each of which we must compute corresponding x, y coordinates. The problem of computing a point on a curve (or surface, for that matter) is reduced to calculating a polynomial.

One simple, straightforward way to calculate polynomials is *Horner's rule*. To apply this to the cubic polynomial $x = a_x u^3 + b_x u^2 + c_x u + d_x$, we write it so that only three multiplications and three additions are required to compute the solution for a given value of u

$$x = [(a_x u + b_x)u + c_x]u + d_x \tag{17.30}$$

We can extend this technique to the general polynomial of degree n:

$$x = a_n u^n + a_{n-1} u^{n-1} + \cdots + a_1 u + a_0 \tag{17.31}$$

An efficient procedure to use to compute x for values of u is as follows: we define the input as $a_0, a_1, \ldots, a_n$, and a value for the parametric variable u. The output

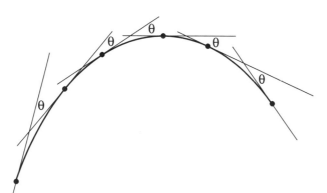

Figure 17.29 *Points on a curve at equal change in tangent angle.*

is, of course, x. We generalize Horner's rule so that

$$
\begin{aligned}
&\text{For } n = 1: \quad x(u) = a_1 u + a_0 \\
&\text{For } n = 2: \quad x(u) = (a_2 u + a_1)u + a_0 \\
&\text{For } n = 3: \quad x(u) = [(a_3 u + a_2)u + a_1]u + a_0 \\
&\text{For } n = 4: \quad x(u) = \{[(a_4 + a_3)u + a_2]u + a_1\}u + a_0
\end{aligned}
\tag{17.32}
$$

and so on.

For any n, we develop what is called a *straight-line program* of $2n$ steps to evaluate a general nth-degree polynomial. Clearly, n multiplications and n additions are necessary. Combining Horner's rule with a straight-line program yields, for $n = 1$, $n = 2$, and $n = 3$:

$$
\begin{aligned}
&\text{For } n = 1 \quad && t \leftarrow a_1 u \\
& && x \leftarrow t + a_0 \\[4pt]
&\text{For } n = 2 \quad && t \leftarrow a_2 u \\
& && t \leftarrow t + a_1 \\
& && t \leftarrow tu \\
& && x \leftarrow t + a_0 \\[4pt]
&\text{For } n = 3 \quad && t \leftarrow a_3 u \\
& && t \leftarrow t + a_2 \\
& && t \leftarrow t + tu \\
& && t \leftarrow t + a_1 \\
& && t \leftarrow tu \\
& && x \leftarrow t + a_0
\end{aligned}
\tag{17.33}
$$

This process is easily extended to higher-degree polynomials.

Surfaces

Many of the problems of displaying a surface are the same as those for a curve. To produce a simple wireframe image of a surface, we project, clip, and plot a set of representative curves lying on the surface. At the very minimum, this might be the four bounding curves at $u = 0$, $u = 1$, $w = 0$, and $w = 1$. As shown in Figure 17.30a, this is barely adequate. Including some intermediate curves, as in Figure 17.30, produces a more meaningful image of interior surface behavior.

The most realistic display of a surface requires that we introduce color, shading, and other lighting effects. This is called *rendering*. To do this we need the normal at enough points on the surface to adequately adjust each of its display pixels. The relationship of the surface normal to the line of sight, the incident light, and surface texture determine the quality and quantity of light emitted by the surface back to the viewer's eye, which in turn determines what we must do to each pixel. These techniques are thoroughly discussed in more advanced textbooks.

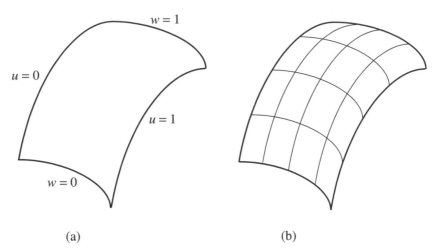

(a)   (b)

Figure 17.30 *Displaying a surface.*

17.6 Visibility

We will now explore problems in creating a realistic display of a polyhedron. Because each face of a polyhedron is a polygon, in order to project a polyhedron modeled in the world coordinate system onto a picture plane, we simply project each of its face polygons. Doing this and nothing more for a cube produces something like Figure 17.31a, where all edges are visible, as if the faces were transparent or as if the polyhedron were merely a wire framework construction or *wire frame* model. The wireframe image does give us a good idea of the polyhedron's overall shape, but a more realistic image discriminates between visible and hidden edges (Figure 17.30b).

Computing the *edge visibility* of a convex polyhedron is much simpler than for a concave polyhedron, and we do it by computing the visibility of each vertex. If a vertex of a convex polyhedron is hidden then all the edges radiating from it are also hidden. In Figure 17.31a, vertex 6 is hidden, therefore edges 6–2, 6–5, and 6–7 are hidden. The rule is easy to state, but many computations are necessary before we can make a

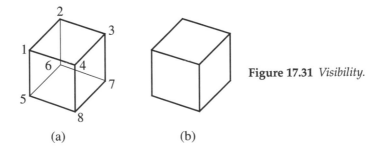

(a)   (b)

Figure 17.31 *Visibility.*

final determination of visibility. For example, here is an algorithm to compute vertex visibility:

1. Project the face polygons onto the picture plane.
2. Write the plane equation for each face of the polyhedron using the world system coordinates of any three vertex points for each face polygon.
3. Write the parametric equation (again in the world system) of the line between each vertex point V_i and the eye point p_E selected for the display. For each line, assume $u = 0$ at p_E and $u = 1$ at V_i. These are called the *vertex projection lines*.
4. Compute the point of intersection between each vertex projection line and all the face planes, excluding faces surrounding the vertex in question. If a vertex projection line intersects any face such that the point of intersection p_i lies between p_E and V_i, that is, if $0 < u_i < 1$, then vertex V_i is hidden. Otherwise V_i is visible. Not only must the vertex projection line intersect the face in the world system, but the point of intersection must be inside the projected polygon defining the boundary of the face in the picture plane, requiring a containment test in the picture-plane system. Thus, in Figure 17.31a it is the face surrounded by the vertices 4–3–7–8 that hides vertex 6.
5. Eliminate all edges radiating from each hidden vertex.

This method does not work for a concave polyhedron. For example, in Figure 17.32, vertex 9 is not visible, but edge 9–10 is partly visible. Compare this situation to convex polyhedra, where an edge is either entirely visible or entirely hidden. We can compute edge visibility for a concave polyhedron, but more computations are necessary.

Polyhedron Silhouette

A problem with applicability to the visible-edge problem is to find the set of edges defining the silhouette of a polyhedron. (The silhouette is the outermost boundary of a projected image.) For example, in Figure 17.33 the sequence of edges defining the cube's silhouette is (1–2), (2–3), (3–7), (7–8), (8–5), and (5–1). Here is a simple method to

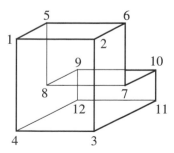

Figure 17.32 *Visible and hidden lines.*

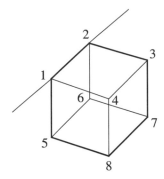

Figure 17.33 *Polyhedron silhouette.*

find the silhouette edges of a convex polyhedron:

1. Project the polyhedron onto the picture plane (say, the x, y plane).
2. Compute the implicit equation of each edge, $f_i(x, y) = 0$.
3. Test each edge against all vertices (except, of course, the two endpoints of the edge in question). If and only if $f_i(x, y)$ is the same sign for all vertices, then that edge is part of the silhouette.

Thus, by extending edge 1–2 in Figure 17.32 we see that all the vertices lie on the same side of it.

Directed Polyhedron Faces

On each face of a polyhedron in a computer-graphics scene we can construct a vector perpendicular to it (Figure 17.34). To do this, we use any three vertices in a counterclockwise sequence surrounding the face to define two vectors **a** and **b**. The vector product $\mathbf{a} \times \mathbf{b}$ produces an outward-pointing vector **c** that is perpendicular to the face. We then investigate the direction cosines of this vector to determine if the face

Figure 17.34 *Directed polyhedron faces.*

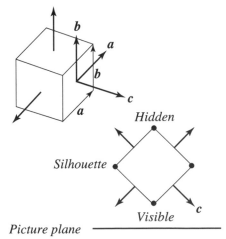

points toward or away from the viewer. An edge is hidden if both of its bounding faces point away from the viewer. An edge is visible if both bounding faces point toward the viewer. If one face points toward the viewer and the other face points away from the viewer, then the shared edge is on the silhouette.

Exercises

17.1 For the interval defined by a set of points x such that $x \in [-2, 5]$, determine if the following points are inside, outside, or on the boundary:
  a. 4   b. 7   c. −2   d. 0   e. −1
  f. 2.5   g. 5.1   h. −5   i. 3.6   j. −0.2

17.2 If line segments in a one-dimensional coordinate system are given as intervals, then describe the geometric relationship between each of the following five lines and $x_T \in [2, 7]$:
  a. $x_A \in [2, 4]$ d. $x_D \in [5.5, 9.5]$
  b. $x_B \in [0.5, 3.5]$ e. $x_E \in [2.5, 5.5]$
  c. $x_C \in [0, 4]$

17.3 For each of the points in Exercise 17.1, determine if it is inside or outside each of the following intervals:
  a. $x_A \in [-1, 3]$ d. $x_D \in [-5, 8]$
  b. $x_B \in [2, 5.5]$ e. $x_E \in [-4, 0.5]$
  c. $x_C \in [0, 4]$

17.4 Find the coordinates of each of the following ten points in the O' system whose origin is at $(2, -3)$ in the O system.
  a. (2, 3)   b. (0, 2)   c. (−2, −1)   d. (7, 3)   e. (6, −1)
  f. (−5, −3)   g. (−2, −6)   h. (3, −5)   i. (8, −3)   j. (−4, 2)

17.5 Find the inverse of Equation 17.3; that is, find the polar coordinates of a point x, y.

17.6 If the lower left corner of the window corresponds to the origin of the screen coordinate system, show how this affects Equation 17.8.

17.7 Which of the ten points in Exercise 17.4 are inside the window $W_L = 4$, $W_R = 7$, $W_B = -6$, and $W_T = 1$?

17.8 List the coordinates of the four corner points of the window described in Exercise 17.7.

17.9 Find the points of intersection of the window defined in Exercise 17.7 and the lines whose end points are, pairwise, from Exercise 17.4:
  a. a, b b. c, d c. e, f d. g, h e. i, j

17.10 Clip and redefine the endpoints of the following lines with respect to the window whose boundaries are $W_L = 1$, $W_R = 12$, $W_B = 1$, and $W_T = 9$:
  a. Line 1: $\mathbf{p}_0 = (-5, 4)$, $\mathbf{p}_1 = (-1, 10)$
  b. Line 2: $\mathbf{p}_0 = (8, 9)$, $\mathbf{p}_1 = (7, 4)$

c. Line 3: $\mathbf{p}_0 = (12, -1)$, $\mathbf{p}_1 = (12, 3)$
d. Line 4: $\mathbf{p}_0 = (4, 10)$, $\mathbf{p}_1 = (0, 2)$
e. Line 5: $\mathbf{p}_0 = (3, -1)$, $\mathbf{p}_1 = (6, 2)$
f. Line 6: $\mathbf{p}_0 = (4, 3)$, $\mathbf{p}_1 = (6, -1)$
g. Line 7: $\mathbf{p}_0 = (2, 0)$, $\mathbf{p}_1 = (2, 12)$
h. Line 8: $\mathbf{p}_0 = (7, 10)$, $\mathbf{p}_1 = (11, -2)$
i. Line 9: $\mathbf{p}_0 = (11, 7)$, $\mathbf{p}_1 = (15, 7)$
j. Line 10: $\mathbf{p}_0 = (15, -2)$, $\mathbf{p}_1 = (12, -5)$

17.11 Clip and redefine as necessary the endpoints of the following line segments with respect to a window whose boundaries are $W_L = 2$, $W_R = 14$, $W_T = 10$, and $W_B = -2$. Describe each line's relationship to this window.

a. $\mathbf{p}_0 = (-5, 4)$, $\mathbf{p}_1 = (1, 13)$ f. $\mathbf{p}_0 = (4, 3)$, $\mathbf{p}_1 = (8, -4)$
b. $\mathbf{p}_0 = (7, 4)$, $\mathbf{p}_1 = (5, 9)$ g. $\mathbf{p}_0 = (2, 1)$, $\mathbf{p}_1 = (2, 12)$
c. $\mathbf{p}_0 = (12, 6)$, $\mathbf{p}_1 = (12, 0)$ h. $\mathbf{p}_0 = (12, -6)$, $\mathbf{p}_1 = (6, 13)$
d. $\mathbf{p}_0 = (5, 12)$, $\mathbf{p}_1 = (-3, -4)$ i. $\mathbf{p}_0 = (11, 7)$, $\mathbf{p}_1 = (15, 7)$
e. $\mathbf{p}_0 = (2, -2)$, $\mathbf{p}_1 = (6, 2)$ j. $\mathbf{p}_0 = (13, -5)$, $\mathbf{p}_1 = (18, 4)$

17.12 Use the following sequence of point coordinates to construct each of the three line data formats discussed in this chapter: $\mathbf{p}_0 = (1, 3)$, $\mathbf{p}_1 = (4, 5)$, $\mathbf{p}_2 = (4, 8)$, $\mathbf{p}_3 = (9, 5)$, $\mathbf{p}_4 = (5, 1)$, $\mathbf{p}_5 = (7, -3)$.

CHAPTER 18

DISPLAY AND SCENE TRANSFORMATIONS

To generate the computer graphics display of an object we apply a sequence of transformations to the points defining the object's geometric model. This sequence is usually best described by a set of matrices establishing a particular view orientation followed by a projection transformation matrix. We change our view of an object by changing these matrices.

We distinguish between *scene transformations* and *display transformations*. Scene transformations are characteristically three dimensional and operate on model data to alter our point of view. Display transformations operate on the two-dimensional display data to change scale (zoom, for example) or to rotate the display around the line of sight. Figure 18.1 shows an example of this distinction. In (a) we see the displayed image of an L-shaped block. In (b) we see the image of the block rotated 30° about our line of sight. And in (c) we now see two new faces of the block, because our point of view has changed (a scene transformation).

The kinematic relationship between the observer and the scene is also something we must consider. Should the objects in the scene move, or should the viewer

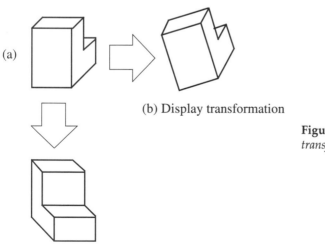

(a)

(b) Display transformation

(c) Scene transformation

Figure 18.1 *Display and scene transformations.*

move? Should the transformations be relative to a fixed global axis, or relative to the current display axis? Once we transform and display an object, the next view of that object is produced by assuming that either the object itself moves or the view point moves. This difference is expressed by the algebraic sign of each of the elements of the transformation matrix. These issues affect the design of the user interface for interactive geometric modeling systems, such as CAD/CAM, and to do this requires an understanding of the underlying mathematics.

This chapter examines orthographic and perspective transformations and explores several scene transformations: orbit, pan, and aim.

18.1 Orthographic Projection

Projections are transformations that produce two-dimensional representations of three-dimensional objects. They are perhaps the most important of all transformations. Without them, we could not construct a display image.

Consider the orthographic projection of any point in space **p** onto an arbitrary plane (Figure 18.2). We find the projected image, **p′**, by constructing a line through **p** perpendicular to the plane. This line intersects the plane at **p′**. Clearly, **p′** must satisfy

$$(\mathbf{p}' - \mathbf{p}) \times \hat{\mathbf{n}} = 0 \tag{18.1}$$

where $\hat{\mathbf{n}}$ is the unit normal vector to the plane.

If the plane is given by $Ax + By + Cz + d = 0$, then the following simple transformation applies

$$\mathbf{p}' = \mathbf{p} - d\hat{\mathbf{n}} \tag{18.2}$$

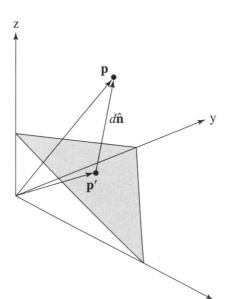

Figure 18.2 *Orthographic projection of a point.*

where

$$\hat{n}_x = \frac{A}{\sqrt{A^2 + B^2 + C^2}}$$

$$\hat{n}_y = \frac{B}{\sqrt{A^2 + B^2 + C^2}} \qquad (18.3)$$

$$\hat{n}_z = \frac{C}{\sqrt{A^2 + B^2 + C^2}}$$

and

$$d = \frac{Ax + By + Cz + D}{\sqrt{A^2 + B^2 + C^2}} \qquad (18.4)$$

where d is the directed distance from $\mathbf{p}$ to the plane along the direction $\hat{n}$ (i.e., the shortest distance). The algebraic sign associated with it indicates that it is in the same direction $(+)$ as $\hat{n}$ or in the opposite direction $(-)$.

We can also project curves this way, too. We merely project the points that define the curve; these then define the new (projected) curve. This is a very good approximation of the projected image for some curves, and exact for others (a subject for more advanced texts). Study Figure 18.3, which shows a method for projecting the $\mathbf{p}_i$ points defining a curve. First, rotate both the plane and each of the points defining the curve so that the plane is parallel to a selected principal plane. This requires finding the rotation transformation that will rotate the unit normal to the plane, $\hat{n}$, into coincidence with the x, y, or z axis. In this example, $\hat{n}$ rotates through the angle θ about the z axis, where $\theta = \tan^{-1}(\hat{n}_x/\hat{n}_z)$. This puts $\mathbf{p}$ into the x, z plane. Next, $\hat{n}$ rotates through the angle β about the x axis, where $\beta = \sin^{-1} n_y$. This puts $\hat{n}$ into coincidence with the z axis; so the plane is now perpendicular to the z axis and intersects it at $z = d$. We apply these same

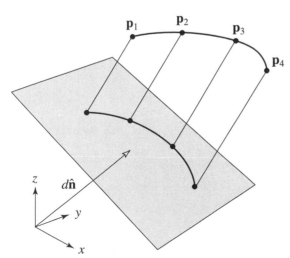

Figure 18.3 *Rotation and projection.*

Projection plane

two rotation transformations to the points. Next, we project the transformed points onto the plane in this new orientation, which simply results in setting the z coordinates equal to d. Finally, we reverse the rotation transformation, putting the plane back in its original position, along with the projected curve-defining points.

The first two rotations, θ and β, above, yield for each of the $\mathbf{p}_i$ points

$$\mathbf{p}_i' = \mathbf{R}_\beta \mathbf{R}_\theta \mathbf{p}_i \tag{18.5}$$

Projecting $\mathbf{p}_{\theta\beta}$ onto the $z = d$ plane yields

$$\mathbf{p}_i'' = [x_i' \quad y_i' \quad d] \tag{18.6}$$

Finally, reversing the rotations gives us

$$\mathbf{p}_i''' = \mathbf{R}_{-\beta} \mathbf{R}_{-\theta} \mathbf{p}_i'' \tag{18.7}$$

We are now ready for a more general approach to computing orthographic projections. The basic ingredients are: the points to be projected, $\mathbf{p}_i$, the projection plane defined in the global coordinate system by $\mathbf{p}_0 = x'\mathbf{u}_1 + y'\mathbf{u}_2$, and the direction of projection, $\mathbf{u}_3$ (Figure 18.4). The vector $\mathbf{p}_0$ defines the origin of a coordinate system on the projection plane (the picture plane coordinate system), with the unit vectors $\mathbf{u}_1$ and $\mathbf{u}_2$ defining the plane itself. If $\mathbf{p}'$ is the projection of $\mathbf{p}$ onto the plane, then

$$\mathbf{p}' = \mathbf{p} - z'\mathbf{u}_3 \tag{18.8}$$

where z' is to be determined. This is similar to Equation 18.2.

This lets us define the projected point $\mathbf{p}'$ in terms of the vectors defining the picture plane. Thus,

$$\mathbf{p}' = \mathbf{p}_0 + x'\mathbf{u}_1 + y'\mathbf{u}_2 \tag{18.9}$$

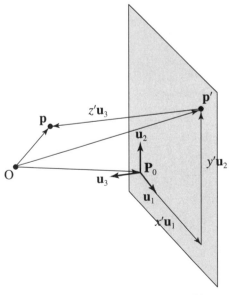

Figure 18.4 *Vector geometry of an orthographic projection.*

Picture plane

We combine Equations 18.8 and 18.9 to obtain

$$\mathbf{p} - z'\mathbf{u}_3 = \mathbf{p}_0 + x'\mathbf{u}_1 + y'\mathbf{u}_2 \tag{18.10}$$

or

$$x'\mathbf{u}_1 + y'\mathbf{u}_2 + z'\mathbf{u}_3 = \mathbf{p} - \mathbf{p}_0 \tag{18.11}$$

which we can solve for x', y', and z'. For example, to find x', we take the scalar product of all terms on both sides of the equation with $\mathbf{u}_2 \times \mathbf{u}_3$, producing

$$x'\mathbf{u}_1 \cdot (\mathbf{u}_2 \times \mathbf{u}_3) + y'\mathbf{u}_2 \cdot (\mathbf{u}_2 \times \mathbf{u}_3) + z'\mathbf{u}_3 \cdot (\mathbf{u}_2 \times \mathbf{u}_3) = (\mathbf{p} - \mathbf{p}_0) \cdot (\mathbf{u}_2 \times \mathbf{u}_3) \tag{18.12}$$

Since $\mathbf{u}_2 \cdot (\mathbf{u}_2 \times \mathbf{u}_3) = \mathbf{u}_3 \cdot (\mathbf{u}_2 \times \mathbf{u}_3) = 0$, we find that

$$x' = \frac{(\mathbf{p} - \mathbf{p}_0) \cdot (\mathbf{u}_2 \times \mathbf{u}_3)}{\mathbf{u}_1 \cdot (\mathbf{u}_2 \times \mathbf{u}_3)} \tag{18.13}$$

We apply a similar treatment to obtain y' and z':

$$y' = \frac{(\mathbf{p} - \mathbf{p}_0) \cdot (\mathbf{u}_3 \times \mathbf{u}_1)}{\mathbf{u}_2 \cdot (\mathbf{u}_3 \times \mathbf{u}_1)}$$
$$z' = \frac{(\mathbf{p} - \mathbf{p}_0) \cdot (\mathbf{u}_1 \times \mathbf{u}_2)}{\mathbf{u}_3 \cdot (\mathbf{u}_1 \times \mathbf{u}_2)} \tag{18.14}$$

Note that we have not yet specified a relationship between $\mathbf{u}_3$ and the plane defined by $\mathbf{u}_1$ and $\mathbf{u}_2$, nor have we specified any particular relationship between $\mathbf{u}_1$ and $\mathbf{u}_2$. Equations 18.13 and 18.14 are independent of the direction of these unit vectors. However, orthographic projection requires $\mathbf{u}_3$ to be perpendicular to the picture plane. Furthermore, it is obviously convenient to choose $\mathbf{u}_1$ and $\mathbf{u}_2$ so that they, too, are mutually perpendicular. This means that we can dramatically simplify these equations since, for example, $\mathbf{u}_1 \cdot (\mathbf{u}_2 \times \mathbf{u}_3) = 1$ and $\mathbf{u}_1 = \mathbf{u}_2 \times \mathbf{u}_3$, so that

$$x' = (\mathbf{p} - \mathbf{p}_0) \cdot \mathbf{u}_1$$
$$y' = (\mathbf{p} - \mathbf{p}_0) \cdot \mathbf{u}_2 \tag{18.15}$$
$$z' = (\mathbf{p} - \mathbf{p}_0) \cdot \mathbf{u}_3$$

Ordinarily z' helps to provide information on the visibility of $\mathbf{p}$ relative to other objects and the chosen picture plane. If we save z' as a data item, we can use it later to reconstruct the initial point, $\mathbf{p}$, if necessary. But, in orthographic projection we do not use z' directly in constructing the image on the picture plane.

Finally, note that $\mathbf{u}_1$, $\mathbf{u}_2$, and $\mathbf{u}_3$ form a right-hand coordinate system as defined above. Frequently, however, the picture-plane coordinate system is a left-hand system so that on a display screen the normal relationship between the x and y axes is preserved; the positive x axis points to the right, and the positive y axis points upward. This means that the positive z axis points generally along the line of sight toward the object being viewed. The farther a point is from the viewer, the higher the value of its z coordinate.

The world coordinate system is usually a right-hand one, so the transformation of object points to the picture-plane coordinate system must also transform a right-hand to a left-hand system, using a matrix that simply inverts the sign of the z coordinate:

$$\mathbf{T}_{R \to L} = \begin{bmatrix} 1 & 0 & 0 & 0 \\ 0 & 1 & 0 & 0 \\ 0 & 0 & -1 & 0 \\ 0 & 0 & 0 & 1 \end{bmatrix} \tag{18.16}$$

This matrix assumes that points are in homogeneous coordinates.

18.2 Perspective Projection

Perspective projection closely approximates the way an observer forms a visual image of an object. Objects in a scene are projected onto a picture plane from a central point, usually assumed to be the eye of the observer. The center of the picture plane is at the point of intersection between the plane and the normal to it from the observer, that is at 0 in Figure 18.5. This is called the *center of projection*. The figure shows two line segments defined by their end points and projected onto the picture plane.

Now let's consider some of the vector geometry of a perspective projection. Let λ denote the normal distance from the viewer to the picture plane (Figure 18.6). Using

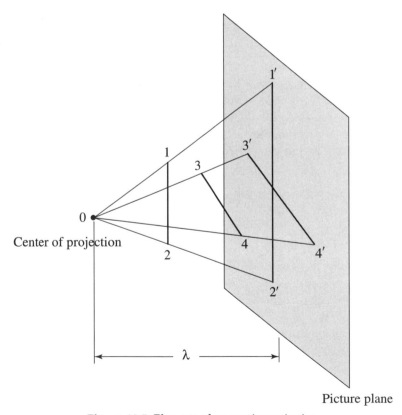

Figure 18.5 *Elements of perspective projection.*

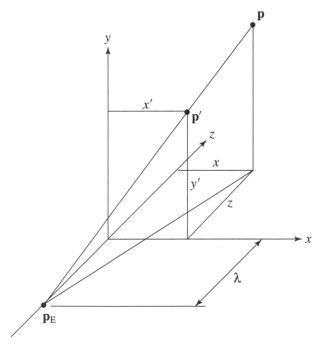

Figure 18.6 *Geometry of a perspective projection.*

some simple vector geometry, we see that $\mathbf{p}' = \mathbf{p}_E + k(\mathbf{p} - \mathbf{p}_E)$. We find that $k = \lambda/(z+\lambda)$, and the components of $\mathbf{p}'$ are

$$x' = \frac{\lambda x}{z + \lambda}$$

$$y' = \frac{\lambda y}{z + \lambda} \tag{18.17}$$

$$z' = 0$$

Here, homogeneous coordinates again prove to be very useful. Consider the following transformation:

$$\mathbf{p}' = \begin{bmatrix} 1 & 0 & 0 & 0 \\ 0 & 1 & 0 & 0 \\ 0 & 0 & 0 & 0 \\ 0 & 0 & r & 1 \end{bmatrix} \mathbf{p} \tag{18.18}$$

If $\mathbf{p} = [x \quad y \quad z \quad 1]$, then

$$\mathbf{p}' = [x \quad y \quad 0 \quad (rz + 1)] \tag{18.19}$$

or, in ordinary Cartesian coordinates

$$\mathbf{p}' = \left[\frac{x}{rz + 1} \quad \frac{y}{rz + 1} \quad 0 \right] \tag{18.20}$$

and if $r = 1/\lambda$, then

$$\mathbf{p}' = \begin{bmatrix} \dfrac{\lambda x}{z+\lambda} & \dfrac{\lambda y}{z+\lambda} & 0 \end{bmatrix} \tag{18.21}$$

These are the same values for x', y', and z' that we derived in Equation 18.17. The center of projection is located at $[0 \quad 0 \quad -\lambda]$, and the picture plane is, of course, the $z = 0$ plane.

Now we are ready to explore the vector geometry of a more general perspective projection (Figure 18.7). Here, $\mathbf{p}$ is the point to be projected, $\mathbf{p}_0 + a\mathbf{u}_1 + b\mathbf{u}_2$ defines the projection plane or picture plane, $\mathbf{p}'$ is the projection of point $\mathbf{p}$, and $\mathbf{p}_E$ is the eye point or center of projection. Once again, we let the vector $\mathbf{p}_0$ define the origin of the picture-plane coordinate system, with $\mathbf{u}_1$ and $\mathbf{u}_2$ mutually perpendicular unit vectors defining the picture plane and its coordinate system. The unit vector $\mathbf{u}_3$ is normal to the picture plane, so that $\mathbf{u}_1 \times \mathbf{u}_2 = \mathbf{u}_3$. We see that

$$\mathbf{p}' = \mathbf{p}_0 + a\mathbf{u}_1 + b\mathbf{u}_2 \tag{18.22}$$

and since $\mathbf{p}'$ is at the intersection of the projection line and a line joining $\mathbf{p}_E$ and $\mathbf{p}$, then

$$\mathbf{p}' = \mathbf{p}_E + c(\mathbf{p} - \mathbf{p}_E) \tag{18.23}$$

Combining Equations 18.22 and 18.23 yields

$$\mathbf{p}_0 + a\mathbf{u}_1 + b\mathbf{u}_2 = \mathbf{p}_E + c(\mathbf{p} - \mathbf{p}_E) \tag{18.24}$$

or

$$a\mathbf{u}_1 + b\mathbf{u}_2 + c(\mathbf{p}_E - \mathbf{p}) = \mathbf{p}_E - \mathbf{p}_0 \tag{18.25}$$

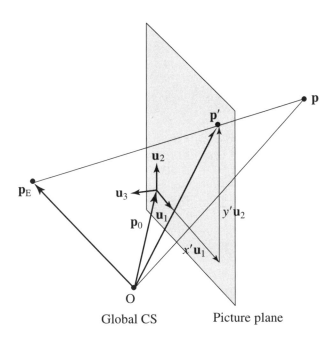

Figure 18.7 *General perspective projection.*

We solve for a and b by taking the appropriate scalar products of the equation. To solve for a, we take the scalar product with $\mathbf{u}_2 \times (\mathbf{p}_E - \mathbf{p})$, which is, of course, perpendicular to both $\mathbf{u}_2$ and $(\mathbf{p}_E - \mathbf{p})$; for b, we use $\mathbf{u}_1 \times (\mathbf{p}_E - \mathbf{p})$, which is perpendicular to $\mathbf{u}_1$ and $(\mathbf{p}_E - \mathbf{p})$. So we have for a

$$a = \frac{(\mathbf{p}_E - \mathbf{p}_0) \cdot [\mathbf{u}_2 \times (\mathbf{p}_E - \mathbf{p})]}{\mathbf{u}_1 \cdot [\mathbf{u}_2 \times (\mathbf{p}_E - \mathbf{p})]} \tag{18.26}$$

and for b

$$b = \frac{(\mathbf{p}_E - \mathbf{p}_0) \cdot [\mathbf{u}_1 \times (\mathbf{p}_E - \mathbf{p})]}{\mathbf{u}_2 \cdot [\mathbf{u}_1 \times (\mathbf{p}_E - \mathbf{p})]} \tag{18.27}$$

18.3 Orbit

Suppose we view an object from some arbitrary point in space, where the object remains in a fixed position and our viewing position moves from point to point. To describe this situation mathematically, we first define the initial eye point, $\mathbf{p}_E$. Next, we define the point toward which we are looking, $\mathbf{p}_V$, the view point. These two points define the line-of-sight vector, $\mathbf{p}_V - \mathbf{p}_E$, and the collinear $\mathbf{u}_3$ direction perpendicular to the picture plane (Figure 18.8).

The picture plane is perpendicular to the line of sight at $\mathbf{p}_0$ in the global coordinate system. And the unit vectors $\mathbf{u}_1$ and $\mathbf{u}_2$ lying in this plane form a left-handed coordinate system with $\mathbf{u}_3$. The viewing distance is d. Note that we do not have to specify

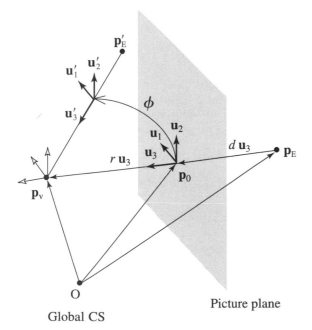

Figure 18.8 *Vector geometry of the orbit transformation.*

$\mathbf{p}_0$, because we can calculate it from the other known vectors as

$$\mathbf{p}_0 = \mathbf{p}_E + d\mathbf{u}_3 \tag{18.28}$$

Here the distance d is positive, although the coordinates of $\mathbf{p}_E$ in the picture plane coordinate system are $(0, 0, -d)$. So, first we specify $\mathbf{p}_E$, $\mathbf{p}_V$, and d.

Transformations to produce the initial view we might arrange as follows: Construct $\mathbf{u}_1$, $\mathbf{u}_2$, and $\mathbf{u}_3$ at the origin of the global coordinate system aligned with the positive x, y, and z axes. Then translate them to $\mathbf{p}_V$ using $\mathbf{T}_V$, where

$$\mathbf{T}_V = \begin{bmatrix} 1 & 0 & 0 & -x_V \\ 0 & 1 & 0 & -y_V \\ 0 & 0 & 1 & -z_V \\ 0 & 0 & 0 & 1 \end{bmatrix} \tag{18.29}$$

Next, we convert to a left-hand coordinate system using $\mathbf{T}_{R \to L}$, Equation 18.16. Then we compute the angles of rotation that will align $\mathbf{u}_3$ with $\mathbf{p}_V - \mathbf{p}_E$, and apply the required rotation transformations, say $\mathbf{R}_\alpha$ and $\mathbf{R}_\beta$. Usually, only two rotations are necessary. We translate this system to $\mathbf{p}_0$ using

$$\mathbf{T}_0 = \begin{bmatrix} 1 & 0 & 0 & 0 \\ 0 & 1 & 0 & 0 \\ 0 & 0 & 1 & r \\ 0 & 0 & 0 & 1 \end{bmatrix} \tag{18.30}$$

where $r = |\mathbf{p}_V - \mathbf{p}_0|$.

The perspective projection transformation becomes

$$\mathbf{T}_P = \begin{bmatrix} 1 & 0 & 0 & 0 \\ 0 & 1 & 0 & 0 \\ 0 & 0 & 1 & 0 \\ 0 & 0 & 1/d & 1 \end{bmatrix} \tag{18.31}$$

Finally, we can combine all of these transformations into a single expression giving the transformation of points in the global system into the picture plane:

$$\mathbf{p}' = \mathbf{T}_V \mathbf{T}_{R \to L} \mathbf{R}_\alpha \mathbf{R}_\beta \mathbf{T}_0 \mathbf{T}_P \mathbf{p} \tag{18.32}$$

As $\mathbf{p}_E$ orbits around point $\mathbf{p}_V$, we invoke orbit scene transformations to generate new images. The line of sight is always toward $\mathbf{p}_V$, although both d and r may vary. For example, if $\mathbf{p}_E$ orbits to the right relative to the picture-plane coordinate system, then the eye point, $\mathbf{p}_E$, must rotate about $\mathbf{p}_V$, say, through an angle ϕ in the plane defined by $\mathbf{u}_1$ and $\mathbf{u}_3$. This orbital motion can be described several ways, each leading to another set of transformation matrices appropriate for the new view.

The eye point may rotate around an axis in the direction of $\mathbf{u}_2$ and through $\mathbf{p}_V$ to produce $\mathbf{p}'_E$, a new eye point, repeating all the steps outlined above using $\mathbf{p}'_E$ to create the new view. The orbiting type of eye point movement can, of course, include orbiting "up" or "down" by rotating around an axis through $\mathbf{p}_V$ and normal to the plane of $\mathbf{u}_2$ and $\mathbf{u}_3$, as well as variations in r and d. We may also rotate the view around and axis collinear with $\mathbf{u}_3$, which does not change the projection. If we track the latest $\mathbf{p}'_E$ relative to the global coordinate system and retain the initial eye point, then we can compute a "home" transformation that will restore the initial view. Note that we can, of course, orbit to the far side of an object and obtain a back view of it.

18.4 Pan

Pan transformations result when we apply equal vector transformations of both the eye point, $\mathbf{p}_E$, and the view point, $\mathbf{p}_V$. The simplest of these are a pan to the right, left, up, or down. Figure 18.9 shows a pan to the right a distance c relative to the $\mathbf{u}_1$, $\mathbf{u}_2$, and $\mathbf{u}_3$ coordinate system. We calculate $\mathbf{p}'_E$ and $\mathbf{p}'_V$ from

$$\mathbf{p}'_E = \mathbf{p}_E + c\mathbf{u}_1$$
$$\mathbf{p}'_V = \mathbf{p}_V + c\mathbf{u}_1 \qquad (18.33)$$

We compute the other components just as easily, and thus complete the definition for the pan transformation. Notice that the $\mathbf{u}_1$, $\mathbf{u}_2$, $\mathbf{u}_3$ system is translated parallel to its initial orientation, which also means that the projection plane is similarly translated to the new viewing position.

Figure 18.10 shows the effect of a pan transformation on the projection of a simple object. A cross marks the location of the view point relative to the object in each scene. The eye point is located some arbitrary distance away from the picture plane, and

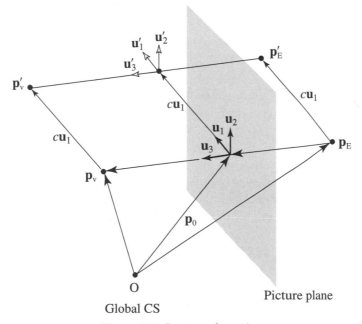

Figure 18.9 *Pan transformation.*

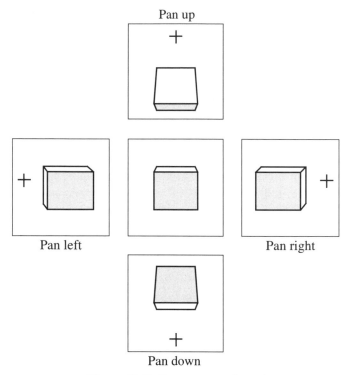

Pan up

Pan left   Pan right

Pan down

Figure 18.10 *Four pan transformations.*

on the normal through the view point. Notice that the pan transformation can never generate the view of the opposite of the object: A large enough pan motion will move the object out of the display in the opposite direction.

18.5 Aim

Aim transformations come about by moving the view point while maintaining the eye point in a fixed position. This is analogous to a person standing in one spot and looking to the right or left, up or down, forward or backward, or any rotation in between, to view the surroundings.

We relocate the view point by a sequence of rotations of the line of sight, $\mathbf{p}_V - \mathbf{p}_E$, while holding $\mathbf{p}_E$ fixed (Figure 18.11). After constructing the initial display, we define an aim movement by specifying an axis of rotation and an angle. The example in the figure shows a rotation ϕ around the $\mathbf{u}_2'$ axis. Thus, we determine new scene transformation components by first translating the $\mathbf{u}_1$, $\mathbf{u}_2$, $\mathbf{u}_3$ system to $\mathbf{p}_E$, and then rotating this system and $\mathbf{p}_V$ around $\mathbf{u}_2'$ through the angle ϕ. Following this rotation, we reverse the translation of the $\mathbf{u}_1$, $\mathbf{u}_2$, $\mathbf{u}_3$ system and all the vector elements are now in place for a new projection.

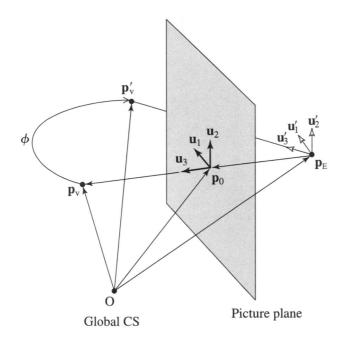

Global CS   Picture plane

Figure 18.11 *Aim transformation.*

BIBLIOGRAPHY

Akivis, M. A., and V. V. Goldberg. *An Introduction to Linear Algebra and Tensors*. New York: Dover, 1972. This is a translation of an outstanding Russian textbook. It is a more rigorous introduction than usual for this subject, because it leads directly into tensors.

Bachmann, F., H. Gehnke, K. Fladt, and W. Süss, eds. *Fundamentals of Mathematics, Volume III Analysis*. Cambridge: The MIT Press, 1988.

Bell, E. T. *Men of Mathematics*. New York: Simon and Schuster, 1965. This classic is more than a book about math. It also reveals the human side of this wonderful and ancient science through a series of unique biographies. Bell writes with wit, irony and sharp good sense. As history and Bell show us, most great mathematicians lived strange lives as soldiers, theologians, lawyers, diplomats, charlatans, mystics, drunkards, professors, and businessmen. Bell writes in a language and style easily accessible to students and teachers in the secondary level and up. Here be sure to read about William Rowan Hamilton, the discoverer of vectors, in the chapter aptly titled "An Irish Tragedy." Also, trace the history of functions, limit, and continuity through 2500 years of mathematical thinking and rethinking about these concepts in Bell's classic work.

Berlinski, David. *A Tour of the Calculus*. New York; Vintage Books, 1995. Berlinski offers a fresh view of real numbers, functions, and limits. This is really literature with equations added.

Bennett, M. K., *Affine and Projective Geometry*, Wiley-Interscience, 1995. This is a textbook for advanced undergraduate math students. It combines three different approaches to affine and projective geometry—algebraic, synthetic, and lattice theoretic.

Bézier, P. *The Mathematical Basis of the UNISURF CAD System*. London: Butterworths, 1986.

Burn, R. P. *Deductive Transformation Geometry*. Cambridge University Press, 1975.

Campbell, H. G. *An Introduction to Matrices, Vectors, and Linear Programming*. New York: Prentice Hall, 1977.

Coxeter, H. S. M. *Regular Polytopes*. New York: Dover, 1973. This is a another classic. Coxeter defines a polytope as a geometric figure bounded by lines, planes, and for the case of dimensions higher than three, hyperplanes. The text begins with the fundamental concepts of plane and solid geometry and then moves on to multi-dimensions. The coverage of polygons here is brief but crucial for a rigorous base from which to move on to constructions in higher dimensions.

Cromwell, Peter R. *Polyhedra*. Cambridge University Press, 1997. This text is a thought-provoking balance of the historical development of the mathematics of polyhedra and a modern rigorous treatment of the subject.

Crowe, Michael J. *A History of Vector Analysis*. New York: Dover, 1985. This is one of the first in-depth studies of the development of vectorial systems. Crowe traces the rise of the vector concept from the discovery of complex numbers through the systems of hypercomplex numbers created by Hamilton and Grassmann to the final acceptance in the early 20th century of the modern system of vector analysis.

Cullen, C. G. *Matrices and Linear Transformations*. 2d ed., Dover, 1990. de Casteljau, P. *Shape, Mathematics and CAD*. London: Kegan Page, 1986.

Dodge, C. W. *Euclidean Geometry and Transformations*. Reading: Addison-Wesley, 1972.

Farin, G. *Curves and Surfaces for Computer-Aided Geometric Design*. New York: Academic Press, 1993.

Faux, I. D., and M. J. Pratt. *Computational Geometry for Design and Manufacturing*. Chichester: Ellis Horwood Ltd., 1981.

Foley, J. D., A. Van Dam, and S. K. Feiner. *Fundamentals of Interactive Computer Graphics*. New York: Addison-Wesley, 1993.

Gans, D. *Transformations and Geometries*. New York: Appleton-Century-Crofts, 1969.

Ghyka, M. *The Geometry of Art and Life*. New York: Dover, 1977. This is the republication of a (now) fifty-year old essay which looks at geometry from a very broad perspective.

Golub, Gene H., and Charles F. Van Loan. *Matrix Computations*. Johns Hopkins U. Press, 1996. Covers computational aspects of matrices, applicable to numerical analysis and linear algebra.

Grünbaum, B., and G. C. Shephard. *Tilings and Patterns*. New York: W. H. Freeman, 1987. This is the first authoritative, comprehensive, and systematic treatment of the geometry of tilings. It focuses on the classification and enumeration of tilings, providing detailed surveys of coloring problems, tilings by polygons, and tilings by topologically unusual tiles.

Hoffman, Banesh. *About Vectors*. New York: Dover, 1975. This delightful, short and unconventional book is must reading for anyone with the slightest interest in vectors. Hoffmann presents a provocative discussion about defining a vector, and an easy introduction to tensors.

Holden, A. *Shapes, Space, and Symmetry*. Dover, 1991. Although the title is appropriate, the text is really about polyhedra and how shape, space, and symmetry are revealed in them.

Hsu, H. *Applied Vector Analysis*. New York: Harcourt, Brace, Jovanovich, 1991.

Infeld, Leopold. *Whom the Gods Love: The Story of Evariste Galois*. The National Council of Teachers, 1978. Evariste Galois (1811–1831) discovered the principles of group theory while still a teenager. Although he was killed in a duel when he was just twenty years old, the value of his contribution to mathematics exceeds that of most more fortunate and longer-lived mathematicians. The title is part of a Greek aphorism attributed to Menander (342–292 BCE), whose complete statement is "Whom the gods love dies young." Leopold Infeld, himself a mathematical physicist of no small reputation, has written a sensitive and fascinating story of Galois' tragic life as a mathematician and rebel during the revolution of 1830 in France. This biographical work is inspiring reading for both students and teachers of all ages.

Kappraff, J. *Connections: The Geometric Bridge between Art and Science*. New York: McGraw-Hill, 1991. Kappraff writes about the hidden harmony in the works of man and nature. He shows how both natural and artificial designs are all governed by geometrical laws, from the great pyramid of Cheops to patterns of plant growth. The role of symmetry is apparent throughout this work. It is written in a language accessible to students and teachers.

Kelly, P. J., and E. G. Strauss. *Elements of Analytic Geometry and Linear Transformations*. New York: Scott-Foresman, 1970.

Klein, F. *Elementary Mathematics from an Advanced Standpoint*. New York: Macmillan, 1939. Felix Klein (1849–1925) defined geometry to be "the study of those properties that are invariant when the elements (points) of a given set are subjected to the transformations of a given transformation group." His great work unified geometry and revealed a hierarchy of geometries, where each geometry is characterized by a particular group of transformations producing a set of invariant properties. This very readable and still timely textbook is the result of much of his work in this area.

Kramer, Edna E. *The Nature and Growth of Modern Mathematics*. Princeton University Press, 1982. The National Council of Teachers of Mathematics rightly identifies this fascinating book as one to be read and to browse, to use for reference and for review of various aspects of mathematics. For vectors and vector geometry read her chapters titled "Algebra from Hypatia to Hamilton" and "Twentieth-Century Vistas—Algebra."

Liebeck, P. *Vectors and Matrices*. New York: Pergamon Press, 1972.

Mortenson, M. E. *Geometric Modeling*. 2d ed. New York: John Wiley & Sons, Inc., 1997. Including the most recent developments, this text presents a comprehensive discussion of the core concepts of this subject. It describes and compares all the important mathematical methods for modeling curves, surfaces, and solids, and shows how to transform and assemble these elements into complex models.

Mortenson, M. E. *Geometric Transformations*. New York: Industrial Press, 1995. Written specifically for engineers and mathematicians working in computer graphics, geometric modeling,

CAD/CAM, virtual reality, computational geometry, robotics, kinematics, or scientific visualization, this text explores and develops the theory and application of transformations and gives the reader a full understanding of transformation theory, the role of invariants, the uses of various notation systems, and the relationships between transformations. Vectors are introduced early and used throughout this work.

Munkres, J. R. *Topology; A First Course*. Prentice Hall, 1974.

Newman, W. M., and R. F. Sproull. *Principles of Interactive Computer Graphics*. 2d ed. New York: McGraw-Hill, 1979.

Olver, P. J. *Equivalence, Invariants, and Symmetry*. Cambridge University Press, 1995. Here is a unique mix of methods used to study problems of equivalence and symmetry which are present in a variety of mathematical fields and physical applications.

Pedoe, D. *Geometry and the Visual Arts*. New York: Dover, 1983. This is a thought-provoking look at the importance of geometry in the development of aesthetics. Pedoe traces the effects of geometry on artistic achievement and discusses its fundamental importance to artists, scientists, architects, philosophers and others.

Pettofrezzo, A. J. *Matrices and Transformations*. New York: Dover, 1966.

Reinreich, G. *Geometrical Vectors*. University of Chicago Press, 1998. Reinreich presents a discussion of vectors that appeals to one's intuition and is highly geometrical, yet leads quite naturally into the study of tensors.

Rose, John S. *A Course on Group Theory*. New York: Dover, 1994.

Rosen, Joe. *Symmetry Discovered*. New York: Dover, 1998. This is a very accessible introduction to symmetry, successfully simplifying complex concepts.

Schey, Harry M. *Div, Grad, Curl, and All That: An Informal Text on Vector Calculus*. 3d ed., W. W. Norton, 1997. This book introduces the basics of vector calculus, with an emphasis on intuitive and geometrical understanding of the concepts.

Schneider, H. G. P. Barker, *Matrices and Linear Algebra*. New York: Dover, 1989.

Scott, W. R. *Group Theory*. New York: Dover, 1987.

Stevens, P. S. *Patterns in Nature*. New York: Little, Brown and Co., 1974. Stevens gives the reader an impressive synthesis of art and science, exploring the universal patterns in which nature expresses itself.

Stewart, I., and M. Golubitsky. *Fearful Symmetry*. Cambridge: Blackwell, 1992. This book offers a very original point of view of symmetry and symmetry breaking. It is full of thought-provoking illustrations and photos. It will make you take another look at nature, a deeper look.

Strang, Gilbert. *Calculus*. Wellesley: Wellesley-Cambridge Press, 1991. This is a thorough and rigorous textbook with a clear and lively presentation of basic classroom calculus. It is one of those rare works that is meant to be actually read and not just as a cookbook of mathematical recipes.

Weyl, H. *Symmetry*. Princeton University Press, 1980. This is a reissue of Weyl's classic work first published in 1952. Herman Weyl was a distinguished professor of mathematics at The Institute of Advanced Study in Princeton.

Wrede, Robert O. *Introduction to Vector and Tensor Analysis*. New York: Dover, 1972. This is perhaps the most advanced work on this list. Wrede is aiming at the advanced undergraduate student and beyond to graduate and professional readers. He stresses the relationships between algebra and geometry, and demonstrates that vector and tensor analysis provides a kind of bridge between elementary aspects of linear algebra, geometry, and analysis.

Yale, Paul B. *Geometry and Symmetry*. New York: Dover, 1988.

ANSWERS TO SELECTED EXERCISES

Chapter 1

1.1   a. $(3 + 2i) + (1 - 4i) = (4 - 2i)$ d. $(6 - 5i) \times i = (5 + 6i)$
       b. $(7 - i) + (1 + 3i) = (8 + 2i)$ e. $(a + bi)(a - bi) = a^2 + b^2$
       c. $4i + (-3 + 2i) = (-3 + 6i)$

1.3 $\mathbf{a} = [5 \quad 6]$ $\mathbf{d} = [-7 \quad 0]$
       $\mathbf{b} = [5 \quad -5]$ $\mathbf{e} = [-5 \quad 3]$
       $\mathbf{c} = [0 \quad 7]$

1.5 $\hat{\mathbf{a}} = [0.640 \quad 0.768]$ $\hat{\mathbf{d}} = [-1 \quad 0]$
       $\hat{\mathbf{b}} = [0.707 \quad -0.707]$ $\hat{\mathbf{e}} = [0.857 \quad 0.514]$
       $\hat{\mathbf{c}} = [0 \quad 1]$

1.7   The components of the unit vector are equal to the direction cosines of the vector.
       a. $\hat{\mathbf{a}} = [0.640 \quad 0.768]$ d. $\hat{\mathbf{d}} = [-1 \quad 0]$
       b. $\hat{\mathbf{b}} = [0.707 \quad -0.707]$ e. $\hat{\mathbf{e}} = [0.857 \quad 0.514]$
       c. $\hat{\mathbf{c}} = [0 \quad 1]$

1.9 $\hat{\mathbf{a}} = \hat{\mathbf{b}}$

1.11 a. $\mathbf{i} \cdot \mathbf{i} = 1$ d. $\mathbf{j} \cdot \mathbf{j} = 1$
       b. $\mathbf{i} \cdot \mathbf{j} = 0$ e. $\mathbf{j} \cdot \mathbf{k} = 0$
       c. $\mathbf{i} \cdot \mathbf{k} = 0$ f. $\mathbf{k} \cdot \mathbf{k} = 1$

1.13 a. $|\mathbf{a}| = 5, \quad \cos\alpha = 0.6, \quad \cos\beta = 0.8$
       b. $|\mathbf{b}| = 2, \quad \cos\alpha = 0, \quad \cos\beta = -1$
       c. $|\mathbf{c}| = \sqrt{34}, \quad \cos\alpha = \dfrac{-3}{\sqrt{34}}, \quad \cos\beta = \dfrac{-5}{\sqrt{34}}, \quad \cos\gamma = 0$
       d. $|\mathbf{d}| = \sqrt{26}, \quad \cos\alpha = \dfrac{1}{\sqrt{26}}, \quad \cos\beta = \dfrac{4}{\sqrt{26}}, \quad \cos\gamma = \dfrac{-3}{\sqrt{26}}$
       e. $|\mathbf{e}| = \sqrt{x^2 + y^2 + z^2}, \quad \cos\alpha = \dfrac{x}{|\mathbf{e}|}, \quad \cos\beta = \dfrac{y}{|\mathbf{e}|}, \quad \cos\gamma = \dfrac{z}{|\mathbf{e}|}$

1.15 a. $\mathbf{i} \times \mathbf{i} = 0$ f. $\mathbf{k} \times \mathbf{i} = \mathbf{j}$
    b. $\mathbf{j} \times \mathbf{j} = 0$ g. $\mathbf{j} \times \mathbf{i} = -\mathbf{k}$
    c. $\mathbf{k} \times \mathbf{k} = 0$ h. $\mathbf{k} \times \mathbf{j} = -\mathbf{i}$
    d. $\mathbf{i} \times \mathbf{j} = \mathbf{k}$ i. $\mathbf{i} \times \mathbf{k} = -\mathbf{j}$
    e. $\mathbf{j} \times \mathbf{k} = \mathbf{i}$

1.17 a. $\mathbf{a} \times \mathbf{a} = 0$ d. $\mathbf{b} \times \mathbf{c} = [-22 \quad -10 \quad 19]$
    b. $\mathbf{a} \times \mathbf{b} = [2 \quad -10 \quad -2]$ e. $\mathbf{c} \times \mathbf{a} = [-12 \quad 0 \quad -6]$
    c. $\mathbf{b} \times \mathbf{a} = [-2 \quad 10 \quad 2]$

1.19 a. $161.565°$ d. $58.249°$
    b. $40.601°$ e. $88.556°$
    c. $90°$

1.21 They are mutually perpendicular because $\mathbf{a} \cdot \mathbf{b} = 0$, $\mathbf{a} \cdot \mathbf{c} = 0$, and $\mathbf{b} \cdot \mathbf{c} = 0$.

1.23 The midpoint is $0.5(\mathbf{p}_0 + \mathbf{p}_1) = [0.5 \quad 5.5 \quad 2.5]$

1.25 a. $x = u, y = u, z = u$
    b. $x = -3 + 5u, y = 1 - u, z = 6 + u$
    c. $x = 1 + 4u, y = 1 - 4u, z = -4 + 13u$
    d. $x = 6 - 16u, y = 8 - 8u, z = 8 - 11u$
    e. $x = 0, y = 0, z = 1 - 2u$

1.27 The line segments are identical, with opposite directions of parameterization.

1.29 $\mathbf{p} = \mathbf{p}_0 + u\mathbf{s} + w\mathbf{t}$

where $\mathbf{p} = \begin{bmatrix} x \\ y \\ z \end{bmatrix}, \mathbf{p}_0 = \begin{bmatrix} 0 \\ 0 \\ 0 \end{bmatrix}, \mathbf{s} = \begin{bmatrix} s_x \\ 0 \\ s_z \end{bmatrix}, \mathbf{t} = \begin{bmatrix} t_x \\ 0 \\ t_z \end{bmatrix}$

and where s_x, s_z, t_x, t_z are arbitrary real number constants.

1.31 a. $\mathbf{p} - 2\mathbf{q} + \mathbf{r} = 0$
    b. $3\mathbf{p} + \mathbf{q} - \mathbf{r} = 0$
    c. $3\mathbf{p} - \mathbf{q} + 4\mathbf{r} = 0$

Chapter 2

2.1 a. $\mathbf{A} + \mathbf{B} = \begin{bmatrix} 8 & 8 & 5 \\ -2 & -7 & 9 \end{bmatrix}$

    b. $\mathbf{A} - \mathbf{B} = \begin{bmatrix} 6 & -2 & -7 \\ 6 & -3 & 3 \end{bmatrix}$

2.3 a. $\delta_{3,2} = 0$ d. $\delta_{7,10} = 0$
    b. $\delta_{1,4} = 0$ e. $\delta_{1,1} = 1$
    c. $\delta_{3,3} = 1$

2.5 $\begin{bmatrix} 0 & 0 \\ 0 & 0 \end{bmatrix}$

2.7 $-\mathbf{A} = [-3 \quad -7 \quad 2]$

2.9 $1.5\mathbf{A} = \begin{bmatrix} 1.5 & -6 \\ 4.5 & 0 \end{bmatrix}$

2.11 $\mathbf{AB} = \begin{bmatrix} 76 & 38 \\ 48 & 8 \\ 12 & 6 \end{bmatrix}$

2.13 $\mathbf{A}^T = \begin{bmatrix} 4 & 5 \\ 0 & 1 \\ 7 & 2 \end{bmatrix}$

2.15 $[(a_x t^2 + b_x t + c_x) \ (a_y t^2 + b_y t + c_y)]$

2.17 $\begin{bmatrix} 0 & 0 & 1 \\ 1 & 0 & 0 \\ 0 & -1 & 0 \end{bmatrix}^{-1} = \begin{bmatrix} 0 & 1 & 0 \\ 0 & 0 & -1 \\ 1 & 0 & 0 \end{bmatrix}$

2.18 $\mathbf{P}$ is 1×3.

2.19 $|\mathbf{A}^{-1}| = -1$.

2.21 a. $m_{11} = 4$ b. $m_{21} = 6$ c. $m_{31} = 2$ d. $m_{22} = 17$ e. $m_{12} = 8$
     f. $c_{11} = 4$ g. $c_{21} = -6$ h. $c_{31} = 2$ i. $c_{22} = 17$ j. $c_{12} = -8$

2.23 a. $\begin{bmatrix} 1 & 0 \\ 0 & 1 \end{bmatrix}^{-1} = \begin{bmatrix} 1 & 0 \\ 0 & 1 \end{bmatrix}$

     b. $\begin{bmatrix} 3 & -1 & 2 \\ 1 & 2 & 1 \\ -2 & 1 & 3 \end{bmatrix}^{-1} = \dfrac{1}{30} \begin{bmatrix} 5 & 5 & -5 \\ -5 & 13 & -1 \\ 5 & -1 & 7 \end{bmatrix}$

     c. $\begin{bmatrix} 1 & 0 & 0 \\ 2 & 1 & 3 \\ 1 & 1 & 2 \end{bmatrix}^{-1} = \begin{bmatrix} 1 & 0 & 0 \\ 1 & -2 & 3 \\ -1 & 1 & -1 \end{bmatrix}$

     d. Inverse does not exist, because $\begin{vmatrix} 3 & -1 & 2 \\ 1 & 2 & 1 \\ 3 & -1 & 2 \end{vmatrix} = 0$

2.25 a. The characteristic equation is $\lambda^2 - 4\lambda - 5 = 0$, and the eigenvalues are 5 and -1, corresponding to the eigenvectors $[k \quad 2k]^T$ and $[k \quad -k]^T$, respectively.
     b. The characteristic equation is $\lambda^3 - 6\lambda^2 + 11\lambda - 6 = 0$, and the eigenvalues are 2, 1, and 3, corresponding to the eigenvectors $[k \quad 0 \quad 0]^T, [0 \quad k \quad 0]^T,$ and $[0 \quad 0 \quad k]^T$, respectively.

Chapter 3

3.1 a. An expansion or contraction parallel to the y axis, followed by a reflection if a is negative.
  b. A combination of expansions and/or contractions parallel to the coordinate axes, followed by reflections in these axes if a and/or b are negative.
  c. A uniform expansion or contraction about the origin, followed by an inversion in the origin if a is negative.
  d. A rotation of the plane through $\pi/4$.

3.3 $\begin{bmatrix} 0 & 1 \\ 1 & 0 \end{bmatrix} \begin{bmatrix} 0 & -1 \\ -1 & 0 \end{bmatrix} = \begin{bmatrix} -1 & 0 \\ 0 & -1 \end{bmatrix}$
This is a rotation.

3.5 An inversion in the point of intersection of the two mutually perpendicular lines.

3.7 The y coordinate does not change because the lines $x_l = a$ and $x_m = b$ are perpendicular to the x axis, so we have (Figure 3.27)
$$x' = -x_0 + 2a$$
$$x'' = x_0 + 2(b - a)$$

3.9 $\mathbf{R}_{\theta\phi} = \begin{bmatrix} \cos\phi\cos\theta & -\sin\phi & \cos\phi\sin\theta \\ \sin\phi\cos\theta & \cos\phi & \sin\phi\sin\theta \\ -\sin\theta & 0 & \cos\theta \end{bmatrix}$ and

  $\mathbf{R}_{\phi\theta} = \begin{bmatrix} \cos\phi\cos\theta & -\sin\phi\cos\theta & \sin\theta \\ \sin\phi & \cos\phi & 0 \\ -\cos\phi\sin\theta & \sin\phi\sin\theta & \cos\theta \end{bmatrix}$ In general, $\mathbf{R}_{\theta\phi} \neq \mathbf{R}_{\phi\theta}$.

3.10 $\triangle AP'B$ is similar to $\triangle PP'P''$ (Figure 3.28). $PP' = 2AP'$ and, therefore, $PP'' = 2AB$.

3.11 See Figure 3.29.

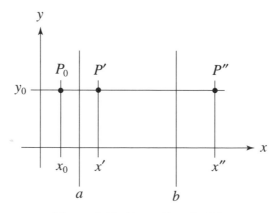

Figure 3.27 *Answer Exercise 3.7.*

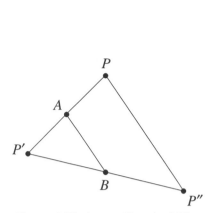

Figure 3.28 *Answer Exercise 3.10.*

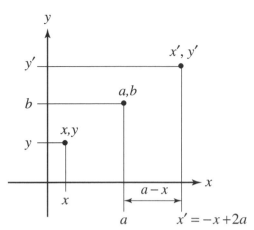

Figure 3.29 *Answer Exercise 3.11.*

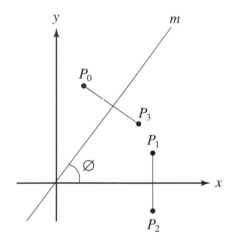

Figure 3.30 *Answer Exercise 3.12.*

3.12 See Figure 3.30.

a. Rotate m into coincidence with the x axis, taking P_0 with it. Thus,

$$\begin{bmatrix} x_1 \\ y_1 \end{bmatrix} = \begin{bmatrix} \cos\alpha & \sin\alpha \\ -\sin\alpha & \cos\alpha \end{bmatrix} \begin{bmatrix} x_0 \\ y_0 \end{bmatrix}$$

b. Reflect P_1 across the x axis.

$$\begin{bmatrix} x_2 \\ y_2 \end{bmatrix} = \begin{bmatrix} 1 & 0 \\ 0 & -1 \end{bmatrix} \begin{bmatrix} \cos\alpha & \sin\alpha \\ -\sin\alpha & \cos\alpha \end{bmatrix} \begin{bmatrix} x_0 \\ y_0 \end{bmatrix}$$

c. Rotate m back to its original position, taking P_2 with it.

$$\begin{bmatrix} x_3 \\ y_3 \end{bmatrix} = \begin{bmatrix} \cos\alpha & -\sin\alpha \\ \sin\alpha & \cos\alpha \end{bmatrix} \begin{bmatrix} 1 & 0 \\ 0 & -1 \end{bmatrix} \begin{bmatrix} \cos\alpha & \sin\alpha \\ -\sin\alpha & \cos\alpha \end{bmatrix} \begin{bmatrix} x_0 \\ y_0 \end{bmatrix}$$

$$\begin{bmatrix} x_3 \\ y_3 \end{bmatrix} = \begin{bmatrix} \cos 2\alpha & \sin 2\alpha \\ \sin 2\alpha & -\cos 2\alpha \end{bmatrix} \begin{bmatrix} x_0 \\ y_0 \end{bmatrix}$$

Chapter 4

4.1 The curve is a circle whose center is $(-1, 3)$. The lines pass through the center and are, therefore, lines of symmetry.

4.3 Yes.

4.5 A cube has nine planes of symmetry: three planes parallel to and midway between opposite faces, six planes containing opposite parallel edges.

4.7 It is symmetric about the origin and both axes, but not the lines $x = y$ and $x = -y$.

4.9 It is symmetric about the line $x = y$.

4.12 A 4-fold axis passes through each vertex and the center of the octahedron. A 3-fold axis passes through the center of each face and the center of the octahedron. A 2-fold axis passes through the center of each edge and the center of the octahedron.

Chapter 5

5.1 $\displaystyle\sum_{n=1}^{\infty} \frac{1}{n^2} = 1 + \frac{1}{4} + \frac{1}{9} + \frac{1}{16} + \frac{1}{25} + \cdots$

5.3 $\displaystyle\sum_{n=1}^{\infty} \frac{2n^2 - 9n + 13}{6(n-1)!} = 1 + \frac{1}{2} + \frac{1}{3} + \frac{1}{4} + \frac{1}{8} + \cdots$

5.5 $\displaystyle\sum_{n=2}^{\infty} \frac{1 + (-1)^n}{n^2 + 1} = \frac{2}{5} + 0 + \frac{2}{17} + 0 + \frac{2}{37} + \cdots$

5.7 $1 - \dfrac{1}{2} + \dfrac{1}{3} - \dfrac{1}{4} + \dfrac{1}{5} - \dfrac{1}{6} + \dfrac{1}{7} + \cdots + \dfrac{(-1)^{n-1}}{n}$

5.9 $x + x^2 + \dfrac{x^3}{1 \cdot 2} + \dfrac{x^4}{1 \cdot 2 \cdot 3} + \dfrac{x^5}{1 \cdot 2 \cdot 3 \cdot 4} + \dfrac{x^6}{1 \cdot 2 \cdot 3 \cdot 4 \cdot 5}$

$\qquad + \dfrac{x^7}{1 \cdot 2 \cdot 3 \cdot 4 \cdot 5 \cdot 6} + \cdots + \dfrac{nx^n}{n!}$

5.11 $\dfrac{x^2}{3} - \dfrac{x^3}{5} + \dfrac{x^4}{7} - \dfrac{x^5}{9} + \dfrac{x^6}{11} - \dfrac{x^7}{13} + \dfrac{x^8}{15} - \cdots + \dfrac{(-x)^{n+1}}{2n+1}$

5.13 $3 + \dfrac{4}{3} + \dfrac{5}{5} + \dfrac{6}{7} + \cdots + \dfrac{n+2}{2n-1}$

5.15 $1 + \dfrac{x}{\sqrt{2}} + \dfrac{x^2}{\sqrt{3}} + \dfrac{x^3}{\sqrt{4}} + \cdots + \dfrac{x^{n-1}}{\sqrt{n}}$

5.17 $\displaystyle\lim_{x \to \infty} \frac{5 - 2x^2}{3x + 5x^2} = -\frac{2}{5}$

Proof: $\displaystyle\lim_{x\to\infty} \frac{5-2x^2}{3x+5x^2} = \lim_{x\to\infty} \frac{\frac{5}{x^2}-2}{\frac{3}{x}+5}$

5.19 $\displaystyle\lim_{x\to 0} \frac{4x^2+3x+2}{x^3+2x-6} = -\frac{1}{3}$

5.21 $\displaystyle\lim_{x\to 2} \frac{x+3}{x^2-2} = \frac{5}{2}$

5.23 $\displaystyle\lim_{\theta\to 0} \frac{\sin^2\theta\cos^2\theta}{\theta^2} = 1$

5.25 For $\displaystyle\sum_{n=0}^{\infty} (-1)^n x^n$ the interval of convergence is $-1 < x < 1$.

5.27 For $\displaystyle\sum_{n=0}^{\infty} \frac{(-1)^n x^n}{(2n+1)^2\, 3^{n+1}}$ the interval of convergence is $-3 \le x \le 3$.

5.29 For $\displaystyle\sum_{n=0}^{\infty} (-1)^n n! x^n$ convergence is at $x = 0$.

5.31 $\displaystyle\sum_{n=0}^{\infty} \frac{x^n}{n!}$ converges for all values of x.

5.33 Given $f(x) = x^3 - 5x^2 - 4x + 20$, then
    a. $f(1) = 12$ c. $f(0) = 20$
    b. $f(5) = 0$ d. $f(a) = a^3 - 5a^2 - 4a + 20$

5.35 $f(t+1) = t^3 - 2t^2 - 11t + 12$

5.37 $f(x+\Delta x) - f(x) = -\dfrac{\Delta x}{x^2 + x\Delta x}$

5.39 $\displaystyle\lim_{x\to 3^+} \frac{x-3}{\sqrt{x^2-9}} = 0$

5.41 $\displaystyle\lim_{x\to 2^+} \frac{(x^4 - 4x^3 + 5x^2 - 4x + 4)^{\frac{1}{4}}}{(x^2 - 3x + 2)^{\frac{1}{2}}} = \sqrt[4]{5}$

5.43 $\displaystyle\lim_{\theta\to 0} \frac{\sin k\theta}{\theta} = k$

5.45 $\displaystyle\lim_{\theta\to 0} \frac{\tan\theta}{\theta} = 1$

5.47 $\displaystyle\lim_{\theta\to 0} \frac{\sin\theta}{\theta^2}$ has no limit.

5.49 $\displaystyle\lim_{x\to 0} \frac{\tan ax}{\sin bx} = \frac{a}{b}$

5.51 $\displaystyle\sum_{j=1}^{4} 1/j = 25/12$

5.53 $\displaystyle\sum_{j=0}^{6} 2^j = 127$

5.55 $\displaystyle\sum_{k=1}^{50} 2k = 2 + 4 + 6 + \cdots + 100 = 2550$

5.57 $\displaystyle\sum_{k=1}^{4} \frac{(-1)^{k+1}}{k} = 1 - \frac{1}{2} + \frac{1}{3} - \frac{1}{4} = \frac{7}{12}$

5.59 The function $\frac{x^2+5}{x^2-9}$ is discontinuous at $x = \pm 3$.

5.61 The function $\frac{x^2 - 2x}{x^3 - 2x^2 + 2x}$ is discontinuous at $x = 0$.

5.63 The function $\cot\phi$ is discontinuous at $n\pi$, where n is any integer.

5.65 $k = \sin 1$

5.67 No k

5.69 $k = 1$

Chapter 6

6.1   See Figure 6.21.

6.3   There are 16. The advanced student should refer to Coxeter and others.

6.5   Imagine that the region surrounding the vertex of a polyhedron is separated from the polyhedron itself. Cut along one of the edges to the vertex, and flatten the

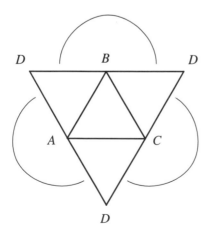

Figure 6.21 *Answer Exercise 6.1.*

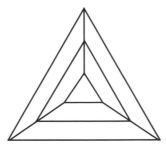

Figure 6.22 *Answer Exercise 6.9.*

faces onto a plane. Traverse this flat vertex from one cut edge to another. The only path constraints are that it must not be self-intersecting and that it begins and ends normal to a cut edge.

6.7 $K = 4\pi$ for all five regular polyhedra

6.9 $V = 9, E = 18, F = 9, \chi = V - E + F = 0$, and $K = 2\pi (9 - 18 + 9) = 0$. See Figure 6.22.

6.11 True

6.13 False. A curve that is a band around the torus, looping through its hole, does not.

6.15 True

6.17 False. $K = -4\pi$

6.19 True

6.21 True

6.23 True

6.25 False

Chapter 7

7.1 a. {2, 3, 4, 6, 8, 9, 10, 12, 14, 15, 16, 18, 20, 21, 22, 24}
     b. {6, 12, 18, 24}
     c. {2, 4, 8, 10, 14, 16, 20, 22}
     d. {3, 9, 15, 21}

7.3 {1, 3, 5, 6, 7, 9, 11, 12, 13, 15, 17, 18, 19, 21, 23, 24}

7.5 Draw appropriate Venn diagrams.

7.7 a. Inside   f. Inside
     b. Outside   g. Inside
     c. Outside   h. Inside
     d. Inside   i. Inside
     e. Inside   j. Boundary

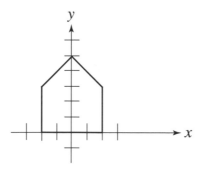

Figure 7.14 *Answer Exercise 7.15.*

7.9 a. Outside   f. Outside
     b. Outside   g. Outside
     c. Outside   h. Outside
     d. Inside   i. Outside
     e. Boundary   j. Boundary

7.11 a. Inside   f. Inside
     b. Outside   g. Outside
     c. Outside   h. Outside
     d. Outside   i. Outside
     e. Outside   j. Outside

7.13 $h(x, y) = -x^2 + y - 1$

7.15 The given set of six halfspaces is not the most efficient because $h_3(x, y)$ is not necessary. See Figure 7.14 above.

Chapter 8

8.1 a. 14.522   d. 10.00
     b. 6.557   e. 8.124
     c. 9.959

8.3 a. 14.522   d. 10.000
     b. 6.557   e. 8.124
     c. 9.959

8.5 a. (1.2, 0.4, 1.35)   d. (2.0, 0, 0)
     b. (2.5, 3.5, −1.5)   e. (6.5, 8.5, 1)
     c. (3.5, −0.65, 0.85)

8.7 (−3, −5.7 − 3)
     (10, −5.7, −3)
     (10, 9, −3)
     (−3, 9, −3)
     (−3, −5.7, 3)

(10, −5.7, 3)
(10, 9, 3)
(−3, 9, 3)

8.9 $\Delta_1 = (\Delta x_1 = 3, \ \Delta y_1 = 0)$
$\Delta_2 = (\Delta x_2 = 0, \ \Delta y_2 = 3)$
$\Delta_3 = (\Delta x_3 = -3, \ \Delta y_3 = 0)$

8.11 $\Delta_1 = (\Delta x_1 = 4, \ \Delta y_1 = 0)$
$\Delta_2 = (\Delta x_2 = 0, \ \Delta y_2 = 4)$
$\Delta_3 = (\Delta x_3 = -4, \ \Delta y_3 = 0)$

8.13 (−2, −4), (14, −4), (14, 8), (−2, 8)

8.15 $x' = (x + x_T)\cos\theta - (y + y_T)\sin\theta$
$y' = (x + x_T)\sin\theta + (y + y_T)\cos\theta$

8.17 The order in which two or more transformations are performed is important. Translating and then rotating a point does not, in general, produce the same result as rotating and then translating the point.

Chapter 9

9.1 From Equations 9.23, we can write
$$d_x^2 + d_y^2 + d_z^2 = \frac{(x_1 - x_0)^2 + (y_1 - y_0)^2 + (z_1 - z_0)^2}{L^2},$$
but $(x_1 - x_0)^2 + (y_1 - y_0)^2 + (z_1 - z_0)^2 = L^2$,
therefore, $d_x^2 + d_y^2 + d_z^2 = 1$.

9.3 a. $x = -2.8u + 3.7, \ y = -11.7u + 9.1, \ z = 2.4u + 0.2$
b. $x = 1.2u + 2.1, \ y = 7.1u - 6.4, \ z = -5.1u$
c. $x = -4.3u + 10.3, \ y = 6.1u + 4.2, \ z = 5.5u + 3.7$
d. $x = -5.3u + 5.3, \ y = 12u + -7.9, \ z = -.7u + 1.4$

9.5 a. (7, 3, 6), (7, 3, 3)
b. (−2.5, 7.25, −1.75), (−1, 8.5, −3.5), (0.5, 9.75, −5.25)
c. (1.5, 0.25, 3.25), (3, 0.5, 0.5), (4.5, 0.75, −2.25)

9.7 a. Line 1: $\begin{cases} x = 2u + 2 \\ y = 2u + 4 \\ z = -10u + 6 \end{cases}$ Line 2: $\begin{cases} x = 6u \\ y = 8u \\ z = -7u + 1 \end{cases}$ The two lines do not intersect.

b. Line 1: $\begin{cases} x = 2u + 2 \\ y = 2u + 4 \\ z = -10u + 6 \end{cases}$ Line 2: $\begin{cases} x = -1.5u + 4 \\ y = 1.5u + 3 \\ z = -1.5u + 5 \end{cases}$ The two lines do not intersect.

c. Line 1: $\begin{cases} x = 2u + 2 \\ y = 2u + 4 \\ z = -10u + 6 \end{cases}$ Line 2: $\begin{cases} x = -3u + 3 \\ y = -3u + 5 \\ z = 15u + 1 \end{cases}$ The two lines overlap.

d. Line 1: $\begin{cases} x = -11u + 10 \\ y = -9u + 8 \\ z = 0 \end{cases}$ Line 2: $\begin{cases} x = -9u + 13 \\ y = 5u + 2 \\ z = 0 \end{cases}$

Intersecting at $u_1 = 0.287$ and $u_2 = 0.683$

e. Line 1: $\begin{cases} x = -3u + 5 \\ y = 0 \\ z = 0 \end{cases}$ Line 2: $\begin{cases} x = 2u + 8 \\ y = 0 \\ z = 0 \end{cases}$

Both segments lie on the x axis. They do not overlap.

Chapter 10

10.1 Hint: Solve the following four simultaneous equations:
$Ax + By + Cz + D = 0$
$Aa + D = 0$
$Bb + D = 0$
$Cc + D = 0$

10.2 a. Same side as the reference point   d. Opposite side
      b. On the plane   e. Opposite side
      c. Same side

10.3 a. Same side   d. Opposite side
      b. On the plane   e. Opposite side
      c. On the plane

10.4 a. $u = +0.333$ d. The line lies in the plane.
      b. $u = -1.000$ e. $u = 0.5$
      c. $u = 0.666$

Chapter 11

11.1 a. Stellar   d. Concave
      b. Concave   e. Convex
      c. Convex

11.2 a. $(-2, 8)$, $(-2, 13)$, $(1, 13)$, $(2, 10)$
      b. $(4, 12)$, $(7, 15)$, $(9, 11)$
      c. $(-1, 2)$, $(2, 3)$, $(2, -1)$. A convex polygon is identical to its convex hull.
      d. $(4, 7)$, $(7, 7)$, $(7, 2)$, $(4, 2)$
      e. $(9, 5)$, $(12, 3)$, $(13, 8)$, $(9, 9)$

11.3 a. $120°$, $60°$ d. $60°$, $120°$
      b. $90°$, $90°$ e. $45°$, $135°$
      c. $72°$, $108°$

11.5 Perimeter $= 29.358$ cm

11.6 $x_{CG} = 6.6$, $y_{CG} = 6.8$

11.7 a. 20.040 cm, (0.2, 7.0)   d. 22.141 cm, (0.75, 0.50)
      b. 15.078 cm, (5.667, 6.333)   e. 18.000 cm, (10.25, −1.75)
      c. 17.211 cm, (11.5, 6.5)

11.9 A regular hexagon has 13 symmetry transformations: one inversion, six rotations, and six reflections.

11.11 Each vertex can be connected with a diagonal to $V - 3$ other vertices. However, the diagonal from vertex i to vertex j is the same as the diagonal from vertex j to vertex i. Therefore, the total number of distinct diagonals D for a polygon with V vertices is:

$$D = \frac{V(V-3)}{2}$$

Chapter 12

12.1 The answer to this question can take many forms. For example, three points or two intersecting lines determine a plane, whereas a four-sided polygon necessarily has four vertex points, any one of which may not lie in the plane of the other three. If that is the case, then the figure is a skewed polygon, and it does not define a legitimate face of a polyhedron.

12.3 An octahedron is produced.

12.5 A cube is produced.

12.7 a. Tetrahedron: $4 - 6 + 4 = 2$ d. Dodecahedron: $20 - 30 + 12 = 2$
      b. Cube: $8 - 12 + 6 = 2$ e. Icosahedron: $12 - 30 + 20 = 2$
      c. Octahedron: $6 - 12 + 8 = 2$

12.9 a. Tetrahedron: $180°$ d. Dodecahedron: $324°$
      b. Cube: $270°$ e. Icosahedron: $300°$
      c. Octahedron: $240°$

12.11 $450°$

12.12 a. Octahedron net (See Figure 12.18.)
      b. Dodecahedron net (See Figure 12.19.)

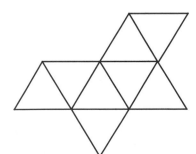

Figure 12.18 *Answer Exercise 12.12a.*

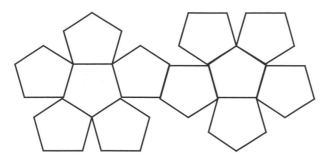

Figure 12.19 *Answer Exercise 12.12b.*

Figure 12.20 *Answer Exercise 12.12c.*

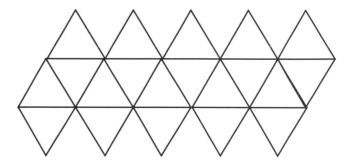

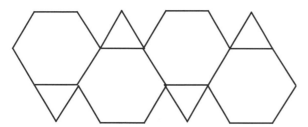

Figure 12.21 *Answer Exercise 12.13.*

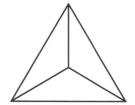

Figure 12.22 *Answer Exercise 12.16a.*

    c. Icosahedron net (See Figure 12.20.)

12.13 All edges are of equal length, with $V = 12$, $E = 18$, $F = 8$ (Figure 12.21).

12.15 Each four-fold axis passes through a vertex and the center of the octahedron. Each three-fold axis passes through the center of a face and the center of the octahedron. Each two-fold axis passes through the center of an edge and the center of the octahedron.

12.16 Tetrahedron (See Figure 12.22.)
    Cube (See Figure 12.23.)
    Octahedron (See Figure 12.24.)

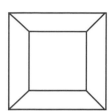

Figure 12.23 *Answer Exercise 12.16b.*

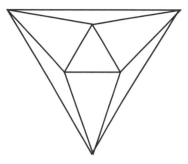

Figure 12.24 *Answer Exercise 12.16c.*

Chapter 13

13.1 $A \cup B = \{a, b, c, d, e, f, g, h\}$

13.3 $A - B = \{a, b, e\}$

13.5 a. $cA = \{d, e, f, g\}$
     b. $cB = \{a, b, c, f, g\}$
     c. $c(A \cup B) = \{f, g\}$

13.7 $h_1 = x$, $h_2 = -x + 2$, $h_3 = y$, $h_4 = -y + 5$

13.9 A complete quadtree of height k has $\sum_{i=0}^{k} 4^i$ nodes.

13.11 $C = A \cup B$
     $= (bA \cup iA) \cup (bB \cup iB)$
     $= (bA \cup bB) \cup (iA \cup bB) \cup (bA \cup iB) \cup (iA \cup iB)$

The associativity property of the union operator allows us to remove the parentheses without altering the meaning of this expression, so that

$$C = bA \cup bB \cup iA \cup b\underline{B} \cup b\underline{A} \cup iB \cup i\underline{A} \cup i\underline{B}.$$

This expression contains redundant elements (underlined). Removing them yields

$$C = bA \cup bB \cup iA \cup iB$$

which is identical to the second equation above. From this we can find the regularized boundary and interior subsets $b\hat{C}$ and $i\hat{C}$, respectively. We note that

$$i\hat{C} = iA \cup iB \cup [\text{Valid } i(bA \cap bB)]$$

and observe that some boundary points become interior points. If these points are not included, we will create a hole in $i\hat{C}$. (Note that it is redundant to $\cup(bA \cap iB) \cup (bB \cap iA)$ to the right side of the last equation above.)

From this we further observe that

$$b\hat{C} = \text{Valid } b(bA \cup bB)$$

where

$$\text{Valid } bA = bA \text{ not in } iB \text{ and part on } bB$$

or

$$\text{Valid } bA = bA - [(bA \cap iB) \cup \text{Valid } b(bA \cap bB)]$$

Similarly

$$\text{Valid } bB = bB - [(bB \cap iA) \cup \text{Valid } b(bA \cap bB)]$$

Again, note that there is an ambiguity in $(bA \cap bB)$, which we resolve by a test similar to that for the intersect operator. If we discard all of $(bA \cap bB)$, then $b\hat{C}$ is incomplete. Thus, the boundary of $\hat{C}$ is

$$b\hat{C} = bA \cup bB - [(bA \cap iB) \cup (bB \cap iA) \cup \text{Valid } b(bA \cap bB)]$$

13.12 First, we find the ordinary set theoretic difference (Figure 13.21a).
$$C = A - B$$
$$= (bA \cup iA) - (bB \cup iB)$$
$$= (bA - bB - iB) \cup (iA - bB - iB)$$
Then we look for the regularized difference (Figure 13.21b).

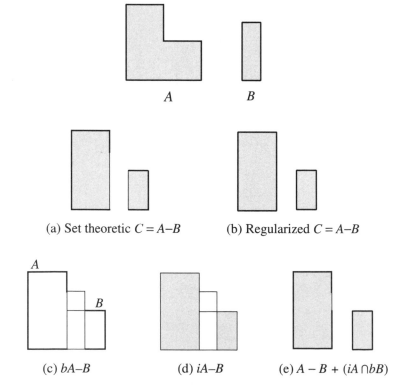

(a) Set theoretic $C = A{-}B$ (b) Regularized $C = A{-}B$

(c) $bA{-}B$ (d) $iA{-}B$ (e) $A - B + (iA \cap bB)$

Figure 13.21 *Answer Exercise 13.12.*

It is clear from the figure that $i\hat{C} = iA - bB - iB$, and, in the example, two disjoint sets result. Also, $\hat{C} \neq C$, because parts of $b\hat{C}$ are missing from C. Figures 13.21c and d show separately the boundary and interior of C. Adding $(iA \cap bB)$ to C as in Figure 21e still does not complete the boundary. A subset of $(bA \cap bB)$ is missing. We again must perform a test as we did for regularized intersection and union operators to find the valid subset. For the difference operator, $\text{Valid}(bA \cap bB)$ are those segments adjacent to only $i\hat{C}$ or $(iA - iB)$. Thus,

$$b\hat{C} = bC \cup (iA \cap bB) \cup \text{Valid}(bA \cap bB)$$
$$= (bA - bB - iB) \cup (iA \cap bB) \cup \text{Valid}(bA \cap bB)$$

so that $\hat{C}$ for $A - B$ is

$$\hat{C} = (bA - bB - iB) \cup (iA \cap bB) \cup \text{Valid}(bA \cap bB) \cup (iA - bB - iB)$$

Chapter 14

14.1   a. $\mathbf{a} = [2 \quad -10 \quad 16]$, $\mathbf{b} = [1 \quad 4 \quad -12]$, $\mathbf{c} = [0 \quad 2 \quad 2]$
     b. $\mathbf{a} = [-1 \quad -4 \quad 4]$, $\mathbf{b} = [3 \quad 2 \quad -10]$, $\mathbf{c} = [-1 \quad 0 \quad 4]$
     c. $\mathbf{a} = [-14 \quad 10 \quad -14]$, $\mathbf{b} = [23 \quad -17 \quad 13]$, $\mathbf{c} = [-3 \quad 7 \quad 1]$
     d. $\mathbf{a} = [10 \quad 6 \quad 6]$, $\mathbf{b} = [-15 \quad -17 \quad -13]$, $\mathbf{c} = [7 \quad 7 \quad 8]$
     e. $\mathbf{a} = [4 \quad 20 \quad -20]$, $\mathbf{b} = [-4 \quad -14 \quad 20]$, $\mathbf{c} = [0 \quad -1 \quad 2]$

14.3   Given that $\mathbf{UMP} = \mathbf{UA}$, and that $\mathbf{U} = [1 \times 3]$, $\mathbf{M} = [3 \times 3]$, $\mathbf{P} = [3 \times 3]$, $\mathbf{A} = [3 \times 3]$, then $[1 \times 3][3 \times 3][3 \times 3] = [1 \times 3][3 \times 3]$

14.5 $\mathbf{M}^{-1} = \begin{bmatrix} 0 & 0 & 1 \\ \frac{1}{4} & \frac{1}{2} & 1 \\ 1 & 1 & 1 \end{bmatrix}$

14.7   Since $\mathbf{B} = \begin{bmatrix} \mathbf{p}_0 \\ \mathbf{p}_1 \\ \mathbf{p}_0^u \\ \mathbf{p}_1^u \end{bmatrix} = \begin{bmatrix} x_0 & y_0 & z_0 \\ x_1 & y_1 & z_1 \\ x_0^u & y_0^u & z_0^u \\ x_1^u & y_1^u & z_1^u \end{bmatrix}$, and for curves in the x, y plane $z = 0$,

     then $\mathbf{B} = \begin{bmatrix} x_0 & y_0 & 0 \\ x_1 & y_1 & 0 \\ x_0^u & y_0^u & 0 \\ x_1^u & y_1^u & 0 \end{bmatrix}$

14.9   At $u = 0$: $G_1 = 1, G_2 = 0, G_3 = 0, G_4 = 0$

14.11   At $u = \frac{1}{3}$: $G_1 = 0, G_2 = 1, G_3 = 0, G_4 = 0$

14.13 $z_1 = 0, z_2 = 0, z_3 = 0, z_4 = 0$

14.15   At zero slope $dy/dx = 0$
     a. $8x = 0, x = 0, y = 0$
     b. $dy/dx = 1$; The function has a constant slope of 1, and therefore it is a straight line.

c. $3x^2 - 3 = 0, x = \pm 1, y = -1, 3$
d. $2x + 2 = 0, x = -1, y = 0$
e. $8x^3 + 3x^2 = 0, x = 0, 0, -3, y = 3, 3, 138$

14.17 a. $d^2y/dx^2 = 8$ d. $d^2y/dx^2 = 2$
   b. $d^2y/dx^2 = 0$ e. $d^2y/dx^2 = 24x^2 + 6x$
   c. $d^2y/dx^2 = 6x$

14.18 $\mathbf{B}_r = \begin{bmatrix} \mathbf{p}_1 & \mathbf{q}_0 & k_0\mathbf{p}_1^u & k_1\mathbf{q}_0^u \end{bmatrix}^T$

14.19 $\mathbf{B}_p$ and $\mathbf{B}_q$ each require that we specify 12 geometric coefficients (3 components for each of the 4 vectors) for a total of 24 coefficients. Only the two scalars k_0 and k_1 are required to define the curve $\mathbf{r}(u)$, since its end points and tangent vector direction are already determined by conditions at $\mathbf{p}_1$ and $\mathbf{q}_0$. Therefore, the composite set of three curves has 26 $(24 + 2)$ degrees of freedom.

Chapter 15

15.1 See Figure 15.10.
   a. It is a second-degree curve.
   b. $\mathbf{p}(0.5) = (9, 5.5)$

15.3 $B_{1,3}(u)$ is a maximum at $u = 1/3$.

15.5 $\mathbf{T} = \begin{bmatrix} \Delta u_i^3 & 0 & 0 & 0 \\ 3u_i\Delta u_i^2 & \Delta u_i^2 & 0 & 0 \\ 3u_i^2\Delta u_i & 2u_i\Delta u_i & \Delta u_i & 0 \\ u_i^3 & u_i^2 & u_i & 1 \end{bmatrix}$

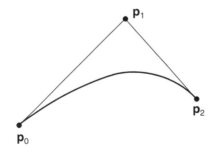

Figure 15.10 *Answer Exercise 15.1.*

Chapter 16

16.1 $x = 4, y = 4u + 1, z = 4w + 3$

16.2 $\mathbf{p}_{00} = \begin{bmatrix} -2 & 5 & -4 \end{bmatrix}, \mathbf{p}_{01} = \begin{bmatrix} -2 & 5 & -1 \end{bmatrix}, \mathbf{p}_{10} = \begin{bmatrix} -1 & 5 & -4 \end{bmatrix}, \mathbf{p}_{11} = \begin{bmatrix} -1 & 5 & -1 \end{bmatrix}$

16.3 For $u = 0: x = -2, y = 5, z = 3w - 4$
   For $u = 1: x = -1, y = 5, z = 3w - 4$
   For $w = 0: x = u - 2, y = 5, z = -4$
   For $w = 1: x = -2, y = 5, z = -1$

16.4 $\mathbf{p}_{00} = \mathbf{a}_{00}$

$\mathbf{p}_{01} = \mathbf{a}_{03} + \mathbf{a}_{02} + \mathbf{a}_{01} + \mathbf{a}_{00}$

$\mathbf{p}_{10} = \mathbf{a}_{30} + \mathbf{a}_{20} + \mathbf{a}_{10} + \mathbf{a}_{00}$

$\mathbf{p}_{11} = \mathbf{a}_{33} + \mathbf{a}_{32} + \mathbf{a}_{31} + \mathbf{a}_{30} + \mathbf{a}_{23} + \mathbf{a}_{22} + \mathbf{a}_{21} + \mathbf{a}_{20}$

$\qquad + \mathbf{a}_{13} + \mathbf{a}_{12} + \mathbf{a}_{11} + \mathbf{a}_{10} + \mathbf{a}_{03} + \mathbf{a}_{02} + \mathbf{a}_{01} + \mathbf{a}_{00}$

16.5 From Equation 16.9 we have

$$\left(\frac{d}{du}\right) p(u) = 3(\mathbf{a}_{33}u^2w^3 + \mathbf{a}_{32}u^2w^2 + \mathbf{a}_{31}u^2w + \mathbf{a}_{30}u^2)$$
$$+ 2(\mathbf{a}_{23}uw^3 + \mathbf{a}_{22}uw^2 + \mathbf{a}_{21}uw + \mathbf{a}_{20}u)$$
$$+ \mathbf{a}_{13}w^3 + \mathbf{a}_{12}w^2 + \mathbf{a}_{11}w + \mathbf{a}_{10}$$

so that:

$\mathbf{p}_{00}^{u} = \mathbf{a}_{10}$

$\mathbf{p}_{01}^{u} = \mathbf{a}_{13} + \mathbf{a}_{12} + \mathbf{a}_{11} + \mathbf{a}_{10}$

$\mathbf{p}_{10}^{u} = 3\mathbf{a}_{30} + 2\mathbf{a}_{20} + \mathbf{a}_{10}$

$\mathbf{p}_{11}^{u} = 3(\mathbf{a}_{33} + \mathbf{a}_{32} + \mathbf{a}_{31} + \mathbf{a}_{30}) + 2(\mathbf{a}_{23} + \mathbf{a}_{22} + \mathbf{a}_{21} + \mathbf{a}_{20}) + \mathbf{a}_{13} + \mathbf{a}_{12} + \mathbf{a}_{11} + \mathbf{a}_{10}$

For the four tangent vectors in the w direction, we have

$$\left(\frac{d}{dw}\right) p(u, w) = 3\mathbf{a}_{33}u^3w^2 + 2\mathbf{a}_{32}u^3w + \mathbf{a}_{31}u^3$$
$$+ 3\mathbf{a}_{23}u^2w^2 + 2\mathbf{a}_{22}u^2w + \mathbf{a}_{21}u^2$$
$$+ 3\mathbf{a}_{13}uw^2 + 2\mathbf{a}_{12}uw + \mathbf{a}_{11}u$$
$$+ 3\mathbf{a}_{03}w^2 + 2\mathbf{a}_{02}w + \mathbf{a}_{01}$$

so that

$\mathbf{p}_{00}^{w} = \mathbf{a}_{01}$

$\mathbf{p}_{01}^{w} = 3\mathbf{a}_{03} + 2\mathbf{a}_{02} + \mathbf{a}_{01}$

$\mathbf{p}_{10}^{w} = \mathbf{a}_{31} + \mathbf{a}_{21} + \mathbf{a}_{11} + \mathbf{a}_{01}$

$\mathbf{p}_{11}^{u} = 3(\mathbf{a}_{33} + \mathbf{a}_{23} + \mathbf{a}_{13} + \mathbf{a}_{03}) + 2(\mathbf{a}_{32} + \mathbf{a}_{22} + \mathbf{a}_{12} + \mathbf{a}_{02}) + \mathbf{a}_{31} + \mathbf{a}_{21} + \mathbf{a}_{11} + \mathbf{a}_{01}$

Chapter 17

17.1 a. Inside   b. Outside   c. Boundary   d. Inside   e. Inside

     f. Inside   g. Outside   h. Outside   i. Inside   j. Inside

17.2 a. x_A inside x_T b. x_B overlaps x_T c. x_C outside x_T

     d. x_D overlaps x_T e. x_E inside x_T

17.3 a. d, f, j inside; a, b, c, e, g, h, i outside

     b. a, f, g, i inside; b, c, d, e, h, j outside

     c. f, i inside; a, b, c, d, e, g, h, j outside

     d. All inside

     e. c, d, e, j inside; a, b, f, g, h, i outside

17.5 $r = \sqrt{x^2 + y^2}$, $\theta = \arctan\left(\dfrac{y}{x}\right)$

17.7 $c, e, g,$ and h

17.9   a. Line (a, b) lies outside the window.
       b. Line (c, d) intersects W_T at (2.5, 1).
       c. Line (e, f) intersects W_L at $(-4, -2.818)$.
       d. Line (g, h) lies inside the window.
       e. Line (i, j) intersects W_T at $(-1.6,\ 1)$ and W_R at $(7, -2.583)$.

17.11 a. Outside window
       b. Inside window
       c. Inside window
       d. Intersects W_T and W_L, Clipped to $\mathbf{p}_0 = (4, 10)$ and $\mathbf{p}_1 = (2, 6)$
       e. Inside window
       f. Intersects W_B
       g. Partially coincident with W_L
       h. Intersects W_B and W_T
       i. Intersects W_R
       j. Outside window